omedical Engineering

ief

him H. Nagel
iomedical Engineering
Stuttgart
36
art

bmt.uni-stuttgart.de

ional Federation for Medical and Biological Engineering, IFMBE, is a federation and transnational organizations representing internationally the interests of medical al engineering and sciences. The IFMBE is a non-profit organization fostering the semination and application of medical and biological engineering knowledge and the of technology for improved health and quality of life. Its activities include participation lation of public policy and the dissemination of information through publications and hin the field of medical, clinical, and biological engineering, IFMBE's aims are to esearch and the application of knowledge, and to disseminate information and promote n. The objectives of the IFMBE are scientific, technological, literary, and educational.

E is a WHO accredited NGO covering the full range of biomedical and clinical , healthcare, healthcare technology and management. It is representing through its 58 ieties some 120.000 professionals involved in the various issues of improved health and delivery.

ficers
Makoto Kikuchi, Vice-President: Herbert Voigt, Past-President: Joachim H. Nagel, Shankar M. Krishnan, Secretary-General: Ratko Magjarevic http://www.ifmbe.org

---

itions:

ds.): BIOMED, Biopacemaking, 2007, ISBN 978-3-540-72109-3

AF790107

Kensuke Sekihara and Nagarajan

Series Editor: Joachim H. Nagel

Series in Bi

Editor-in-Ch

Prof. Dr. Joac
Institute of B
University of
Seidenstrass
70174 Stuttg
Germany
E-mail: jn@

The Internat
of national
and biologi
creation, dis
managemen
in the form
forums. Wi
encourage
collaboratio

The IFMB
engineerin
member so
healthcare

IFMBE Of
President:
Treasurer:

Previous Ed
Spaan, J. (

Kensuke Sekihara · Srikatan S. Nagarajan

Adaptive Spatial Filters for Electromagnetic Brain Imaging

Springer

Kensuke Sekihara
Tokyo Metropolitan University
Dept. of Systems Design & Engineering
6-6 Asahigaoka
Hino
Tokyo
191-0065 Japan
ksekiha@cc.tmit.ac.jp

Srikatan S. Nagarajan
University of California
Biomagnetic Imaging Laboratory
Department of Radiology
513 Parnassus Avenue S362
San Francisco CA 94143
USA
sri@radiology.ucsf.edu

ISBN: 978-3-540-79369-4 e-ISBN: 978-3-540-79370-0

Series in Biomedical Engineering ISSN 1864-5763

Library of Congress Control Number: 2008925639

© 2008 Springer-Verlag Berlin Heidelberg

This work is subject to copyright. All rights are reserved, whether the whole or part of the material is concerned, specifically the rights of translation, reprinting, reuse of illustrations, recitation, broadcasting, reproduction on microfilm or in any other way, and storage in data banks. Duplication of this publication or parts thereof is permitted only under the provisions of the German Copyright Law of September 9, 1965, in its current version, and permission for use must always be obtained from Springer. Violations are liable to prosecution under the German Copyright Law.

The use of general descriptive names, registered names, trademarks, etc. in this publication does not imply, even in the absence of a specific statement, that such names are exempt from the relevant protective laws and regulations and therefore free for general use.

Cover design: deblik, Berlin

Printed on acid-free paper

9 8 7 6 5 4 3 2 1

springer.com

IFMBE

The International Federation for Medical and Biological Engineering (IFMBE) was established in 1959 to provide medical and biological engineering with a vehicle for international collaboration in research and practice of the profession. The Federation has a long history of encouraging and promoting international cooperation and collaboration in the use of science and engineering for improving health and quality of life.

The IFMBE is an organization with membership of national and transnational societies and an International Academy. At present there are 53 national members and 5 transnational members representing a total membership in excess of 120 000 professionals worldwide. An observer category is provided to groups or organizations considering formal affiliation. Personal membership is possible for individuals living in countries without a member society. The IFMBE International Academy includes individuals who have been recognized for their outstanding contributions to biomedical engineering.

Objectives

The objectives of the International Federation for Medical and Biological Engineering are scientific, technological, literary, and educational. Within the field of medical, clinical and biological engineering its aims are to encourage research and the application of knowledge, to disseminate information and promote collaboration.

In pursuit of these aims the Federation engages in the following activities: sponsorship of national and international meetings, publication of official journals, cooperation with other societies and organizations, appointment of commissions on special problems, awarding of prizes and distinctions, establishment of professional standards and ethics within the field as well as other activities which in the opinion of the General Assembly or the Administrative Council would further the cause of medical, clinical or biological engineering. It promotes the formation of regional, national, international or specialized societies, groups or boards, the coordination of bibliographic or informational services and the improvement of standards in terminology, equipment, methods and safety practices, and the delivery of health care.

The Federation works to promote improved communication and understanding in the world community of engineering, medicine and biology.

Activities

Publications of the IFMBE include: the journal Medical and Biological Engineering and Computing, the electronic magazine IFMBE News, and the Book Series on Biomedical Engineering. In cooperation with its international and regional conferences, IFMBE also publishes the IFMBE Proceedings Series. All publications of the IFMBE are published by Springer Verlag.

Every three years the IFMBE hosts a World Congress on Medical Physics and Biomedical Engineering in cooperation with the IOMP and the IUPESM. In addition, annual, milestone and regional conferences are organized in different regions of the world, such as Asia Pacific, Europe, the Nordic-Baltic and Mediterranean regions, Africa and Latin America.

The administrative council of the IFMBE meets once a year and is the steering body for the IFMBE. The council is subject to the rulings of the General Assembly, which meets every three years.

Information on the activities of the IFMBE are found on its web site at: http://www.ifmbe.org.

Contents

Chapter 1

Introduction

1.1 Functional brain mapping

Functional brain imaging allows us to look inside the human brain and study the neural mechanisms underlying human behavior. Functional brain imaging has revealed the neural correlates of several behaviors, such as our ability to make sense of and integrate large amounts of dynamic information in our environment, understand language, learn new skills and remember important facts. Mapping of normal and abnormal brain functions is also clinically important for patients with brain tumors or pharmacologically resistant intractable epilepsy, where surgery is often used in the management of patients. Generally, in such patients there is a trade-off between the margin of excision used to ensure complete removal, and the potential loss of function that may arise as a consequence of removing normal surrounding brain tissue.

The earliest studies of functional brain imaging were performed using positron emission tomography (PET), which is a nuclear-medicine-based imaging method that can track metabolic activity in the brain. During the past two decades, PET has largely been replaced by functional magnetic resonance imaging (fMRI) for studies of human brain mapping, and fMRI has nearly become synonymous with functional brain imaging in humans. fMRI is primarily based on measurement of blood oxygenation level dependent (BOLD) signals. Although extensive data relating to brain function has been obtained from fMRI and PET studies, it is important to note that fMRI and PET use metabolic or neurovascular activities; hence they do not directly measure neuronal activities. This fact potentially limits the time resolution and makes real-time imaging difficult in these brain-imaging methodologies.

Electrophysiological activity of neurons in the cerebral cortex generates both electric potentials on the scalp, as well as weak but measurable magnetic fields outside the scalp. Direct non-invasive measurements of these neuronal activities in the sub-millisecond time scale can be achieved with magnetoencephalography (MEG) [1][2] and electroencephalography (EEG) [3][4]. MEG measures tiny magnetic fields generated by the human brain and EEG measures electric potentials on the scalp due to brain activity. MEG is enabled by superconducting quantum interference devices (SQUIDs), used in conjunction with flux couplers and flux-locked loop circuitry, that have sensitivity on the order of tens of femto-Tesla (seven orders of magnitude smaller than the earth's magnetic field). Modern MEG systems are now capable of whole-head coverage with simultaneous measurements of more than 300 channels. EEG technology has matured to where modern systems are capable of measuring simultaneously from 512 electrodes on the scalp.

The past decade has shown rapid development of whole-head MEG/EEG sensor arrays that are capable of concurrent measurements of both EEG and MEG at a high spatial resolution. Furthermore, advances in signal processing algorithms now enable imaging of dynamic brain activity from MEG and EEG data – referred to as electromagnetic brain imaging[5]. Electromagnetic brain imaging is unique among functional imaging techniques for its ability to provide spatio-temporal brain activation profiles that reflect not only where the activity occurs in the brain but also when this activity occurs in relation to external and internal cognitive events, as well as to activity in other brain regions. The most important contributors to advances in electromagnetic brain imaging are algorithms that enable high-fidelity reconstruction of neuronal activities from MEG and EEG data.

1.2 Electromagnetic brain imaging

Neural activities in the human brain generate coherent synaptic and intracellular currents in cortical columns, which are thought to be major generators of MEG and EEG signals. Electromagnetic brain imaging is capable of non-invasively mapping these currents, and algorithms for electromagnetic brain imaging typically involve two major components – forward modeling and inverse modeling. Forward modeling derives the sensor outputs from a known distribution of these neural generators. Quasi-static approximations of Maxwell's equations fundamentally allow computation of the sensor outputs if the three-dimensional distribution of conductivity in a brain, referred to as the volume conductor, is known. Maxwell's equations can be solved analytically for special volume conductor geometries, such as a sphere. Forward modeling is embodied in the idea of the sensor lead field, which represents the sensitivity profile of a sensor array and describes a linear relationship between sources and measurements. The scope of this book does not allow a detailed exploration of forward modeling. Nevertheless, for completeness, we include several representative topics on forward modeling in the Appendix.

Inverse modeling refers to an algorithm or procedure that is used to reconstruct source distributions based on the sensor lead field and measurements. The inverse algorithms attempt to solve the bioelectromagnetic inverse problem, i.e., to estimate neural source spatio-temporal distributions from bioelectromagnetic measurements obtained outside the human head. Inverse algorithms can be broadly classified into two categories: Parameter-estimation and imaging methods. Parameter estimation methods assume that a small number of sources can adequately account for the observed sensor data. A typical example is the case in which the source spatial distribution is assumed to consist of a small number of point sources. In such a case, the locations, orientations, and strengths of these point sources form a set of unknown parameters that are estimated using a non-linear least-squares fit to the measured data. The model for such point sources is called the equivalent current dipole (ECD). When only a single source is assumed to exist, the method of estimating the source location, orientation, and strength is called the single-dipole search, which has been widely used in MEG/EEG source analysis. However, in parameter estimation methods, the estimation of the source parameters not only requires knowledge of the number of sources but also requires a $3Q$-dimensional nonlinear search where Q is the number of sources. Thus, as the number of sources increases, the search dimension becomes very high, and there are no numerical methods that can effectively solve such high-dimensional nonlinear optimization problems.

Imaging methods are an alternative approach. In contrast to the parameter estimation methods, the imaging methods do not require prior knowledge of the number of sources and can generally avoid the non-linear search in a high-dimensional parameter space. Imaging methods are further classified into two classes of algorithms: the tomographic reconstruction methods and spatial filters. The tomographic reconstruction methods involve voxel discretization over the reconstruction region, and assume a fixed source at each voxel. These methods estimate the amplitudes of the sources at voxels by the least-squares fit to the measured data. Because the number of voxels is much larger than the number of sensors, the tomographic reconstruction methods generally require some criterion other than the least-squares criterion to provide additional constraints on the source distribution. Various algorithms including the well-known minimum-norm method have been proposed, and many tomographic reconstruction methods can be reformulated as non-adaptive spatial filters. Typical tomographic reconstruction methods and their reformulation as non-adaptive spatial filters are discussed in Chapter 3.

1.3 Spatial filters

The other class of imaging algorithm is the spatial filter, which is a linear operator applied to the measured data, and is used to estimate the strength of activity at a particular spatial location. We refer to this spatial location as the filter-pointing location in this book. Since the filter-pointing location is a controllable parameter when computing the spatial-filter weight, the pointing location can be moved and scanned over the source space, simply by post-processing, to obtain the three-dimensional reconstruction of the source distribution. The spatial filter may be interpreted as a class of techniques that numerically control the sensor array sensitivity to form a virtual sensor whose sensitivity pattern is "preferably" localized around its pointing location. Quite often, the spatial filter is called the beamformer in the field of signal processing. If spatial filters only depend on the geometry of the measurements, they are referred to as non-adaptive spatial filters. In contrast to non-adaptive spatial filters, adaptive spatial filters depend not only on the measurement geometry but also on the measurement covariance matrix.

Adaptive spatial filters, also known as adaptive beamformers, were originally invented in the field of sensor-array signal processing in the 1960s. Seminal works by Capon[6] and numerous other investigators have led to major advances in these methods of array signal processing[7]. As a result, adaptive spatial filters continue to be used in the field of array signal processing. Stephan Robinson conducted some of the pioneering work in introducing adaptive spatial filters into the field of electromagnetic brain imaging in the early 1990s[8]. Since then, he and his colleague, Jiri Vrba, have made tremendous contributions to the application of adaptive spatial filters to electromagnetic brain imaging. Subsequently, many researchers, ourselves included, contributed to this research topic. Among these researchers, Barry Van Veen and Michel Spencer should particularly be noted for their efforts in developing the vector-type spatial filter, which is suitable for reconstructing 3D vector source distributions[9][10].

In recent years, adaptive spatial-filter brain imaging is increasingly being used both for basic human neuroscience studies and clinically for pre-operative functional brain imaging. When combined with magnetic resonance imaging (MRI) data, pre-operative functional localization with this brain imaging method can be integrated with neuro-navigational systems to provide intra-operative guidance to neurosurgeons. Such functional imaging has been shown to be important in pre-operative planning and complements intra-operative mapping by delineating retained areas of function non-invasively and in advance. As part of clinical routine, the somatosensory, auditory and motor cortices are mapped pre-operatively to aid in surgical navigation. An example of such pre-operative localization of the motor cortex is shown in Fig. 1.1.

Adaptive spatial filters are expected to finally realize the ultimate promise of electromagnetic brain imaging, which is the spatiotemporal localization of cortical networks involved in sensorimotor, language, memory, and higher cognitive functions. One example of these advances is a study on mapping functional connectivity which is described in Section 9.4. One example of an investigation toward this direction is shown in Fig. 1.2. To obtain the results in this figure, resting-state MEG was recorded from a patient with brain lesions. The hypothesis here is that such a patient has decreased connectivity around pathologic regions, and such decreased connectivity can be detected by mapping the mean imaginary coherence. In the results in Fig. 1.2, voxels showing decreased mean imaginary coherence are indicated by white squares, and such voxels are found near the brain tumor. Thus, mapping mean imaginary coherence can provide useful clinical information on brain lesions.

The aim of this book is to describe the technical advances of adaptive spatial filters in the context of electromagnetic brain imaging. This book integrates and synthesizes available information on adaptive spatial filters for electromagnetic brain imaging, and describes various aspects, including causes of performance degradation such as lead-field errors, brain background interference, and source

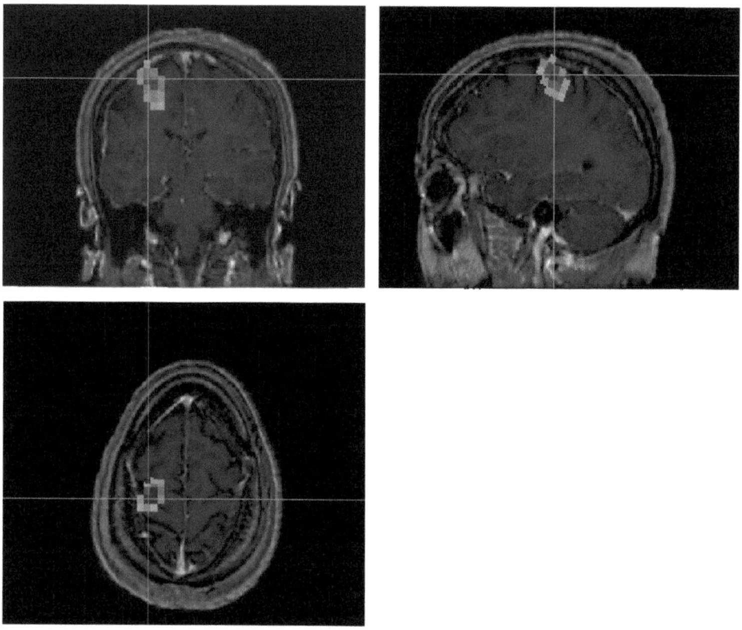

Figure 1.1: Results of adaptive spatial filter imaging of primary motor cortex. This imaging experiment was done for pre-operative surgical planning. MEG measurements were taken while the subject performed self-paced finger movements. Since movement-induced brain activity is not precisely time-locked, we used the imaging technique described in Section 8.4, and calculated covariance matrices from the beta-band (15-25Hz) data. The image above was obtained by applying the prewhitening adaptive spatial filter. The voxels indicated by gray squares show a significant decrease in beta-band activity between the task window (-300 ms – 0 ms) and the control window (1000 ms – 1300 ms), where the time origin is set to the onset of the finger movements.

correlations. The intended audience includes graduate students studying electromagnetic brain imaging, and researchers interested in the methodological aspects of electromagnetic brain imaging.

1.4 Book chapter organization

In Chapter 2, we give the basic definitions for describing the relationship between sensor outputs and sources in the brain, as well as several definitions needed to discuss spatial filters. Chapter 3 describes the basics of conventional least-squares-based tomographic reconstruction algorithms, which can be considered as non-adaptive spatial filters. In Chapter 4, we give the various formulations for adaptive

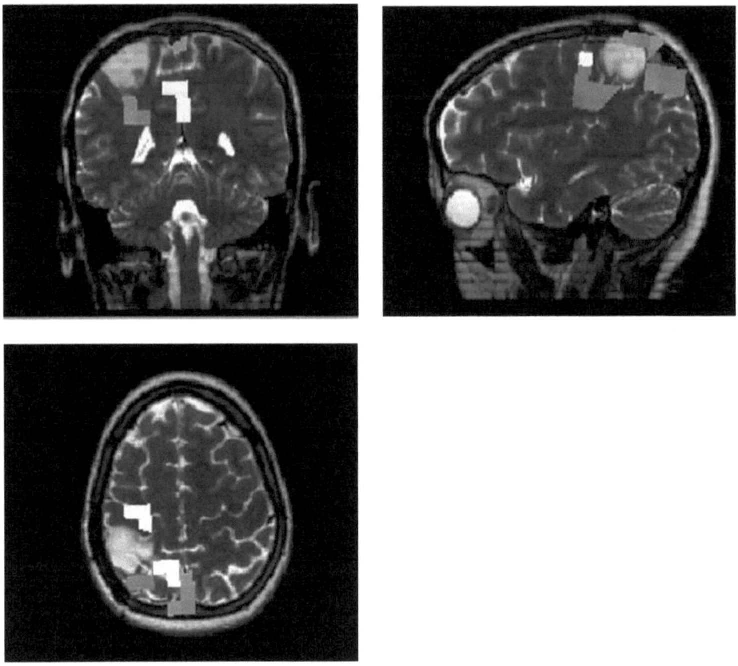

Figure 1.2: Results of imaging the mean imaginary coherence measured from a patient with brain lesions. MEG recordings during the resting state were measured and neural activity was estimated using adaptive spatial-filter imaging. The mean imaginary coherence in the frequency band from 1 to 20 Hz between brain voxels was calculated as an index of functional connectivity in the alpha-band. The voxels indicated by the white squares show a statistically-significant decrease in the mean imaginary coherence. Note that such voxels are located near the tumor.

spatial filters, and show the relationships among these formulations.

Two important properties for any inverse algorithm are the localization bias and the spatial resolution. Chapter 5 discusses these properties for various adaptive and non-adaptive spatial filters. Chapter 6 presents arguments on the signal-to-noise ratio (SNR) of the spatial filter output. Chapter 6 also discusses the degradation of the output SNR caused by errors in the forward-field calculations, and presents methods that are robust to these errors. Chapters 7 and 8 discuss the effects of external interference on the performance of adaptive spatial filters. In Chapter 9, we present an analysis on how adaptive spatial filters are affected by source correlations. Chapter 9 also presents arguments on how adaptive spatial filters can be used for imaging correlated activity across different brain regions. Chapter 10 discusses the effects of using sample covariance matrices on the performance of adaptive spatial filters.

Brain imaging studies often require statistical evaluation of the reconstructed results. In Chapter 11, we present non-parametric approaches to evaluating the statistical significance of adaptive spatial-filter images. Chapter 12 presents several algorithms that are closely related to adaptive spatial filters, and discusses their relationship to adaptive spatial filters. Chapter 13 contains the Appendix. The first six sections in the Appendix provide several supplementary discussions related to the mathematical arguments used in the book. The final section in the Appendix presents several selected topics on forward modeling for bioelectromagnetic measurements. It also presents the closed-form formulae for a spherical volume conductor used to calculate the lead field for an MEG or EEG sensor array. In addition, it briefly mentions how to extend the computation of the lead field based on the spherical conductor model to one based on more realistic volume conductor models.

1.5 Acknowledgements

We have been fortunate to have many wonderful colleagues and collaborators who have contributed to the work presented in this book. We are deeply indebted and thankful for their contributions. We have had the luxury of incredibly dedicated and competent research associates such as Susanne Honma, Mary Mantle and Anne Findlay. Graduate students who helped us in this work include Sarang Dalal, Johanna Zumer and Dosik Hwang. Invaluable contributions have come from postdoctoral fellows, Maneesh Sahani, Kenneth Hild, Adrian Guggisberg, and David Wipf. We particularly thank Kenneth Hild for his dedicated proofreading. His detailed comments greatly improved the quality of this book. We would also like to thank Daniel Palomo for his tremendous efforts on editing the whole manuscript.

Furthermore, several of our collaborators have greatly helped us by enhancing our own understanding of this field. They include: Hagai Attias, Barry Van Veen, Tulaya Limpiti, Stephan Robinson, and Jiri Vrba. We would also like to thank the following collaborators for their valuable support over many years: John Houde, David Poeppel, Alec Marantz, Tim Roberts, Hideaki Koizumi, Yasushi Miyashita, and Isao Hashimoto. One of the authors (KS) thanks colleagues at the KRENIKON project with whom he worked in the early 90s at Basic Development of Siemens Medical Engineering. These colleagues include: Arnulf Oppelt, Bernhard Scholz, Herbert Bruder, Rainer Graumann, and Klaus Abraham-Fuchs. KS especially thanks Dr. Arnulf Oppelt who drew his attention to Stephan Robinson's pioneering work. It was his suggestion that led to the course of investigations now detailed in this book.

Chapter 2

Sensor array outputs and spatial filters

In this chapter, we first define various quantities necessary for arguments in this book, and then discuss several properties of the data covariance matrix and those of low-rank signals. We introduce the basic formulations of spatial filters, and explain how the spatial filter is extended in order to reconstruct vector sources. The basic notion of resolution kernel analysis is also presented.

2.1 Neuromagnetic signals as sensor-array outputs

2.1.1 Definitions

Let us define the magnetic field measured by the mth sensor at time t as $b_m(t)$, and the column vector

$$\boldsymbol{b}(t) = \begin{bmatrix} b_1(t) \\ b_2(t) \\ \vdots \\ b_M(t) \end{bmatrix}, \tag{2.1}$$

as a set of measured data where M is the total number of sensors. This column vector $\boldsymbol{b}(t)$ is called the measurement vector, the data vector, or the array output.

A spatial location is represented by a three-dimensional vector $\boldsymbol{r}$: $\boldsymbol{r} = (x, y, z)$. A source vector at $\boldsymbol{r}$ and time t is defined as a three-dimensional column vector $\boldsymbol{s}(\boldsymbol{r}, t)$, i.e.,

$$\boldsymbol{s}(\boldsymbol{r}, t) = \begin{bmatrix} s_x(\boldsymbol{r}, t) \\ s_y(\boldsymbol{r}, t) \\ s_z(\boldsymbol{r}, t) \end{bmatrix}, \tag{2.2}$$

where $s_x(\boldsymbol{r},t)$, $s_y(\boldsymbol{r},t)$, and $s_z(\boldsymbol{r},t)$ are the x, y, and z components of the source vector $\boldsymbol{s}(\boldsymbol{r},t)$. The physical nature of the source vector for bioelectromagnetic measurements is the electro-motive force generated by neuronal activities in the brain. Additional discussion regarding the source nature is presented in Section 13.7 of the Appendix. The magnitude of the source vector is denoted as a scalar $s(\boldsymbol{r},t)$, and the orientation of the source is defined as a three-dimensional unit vector $\boldsymbol{\eta}(\boldsymbol{r}) = [\eta_x(\boldsymbol{r}), \eta_y(\boldsymbol{r}), \eta_z(\boldsymbol{r})]^T$, where the superscript T indicates the matrix transpose. Then, we have the relationship,

$$\boldsymbol{s}(\boldsymbol{r},t) = s(\boldsymbol{r},t)\boldsymbol{\eta}(\boldsymbol{r}) = s(\boldsymbol{r},t)\begin{bmatrix} \eta_x(\boldsymbol{r}) \\ \eta_y(\boldsymbol{r}) \\ \eta_z(\boldsymbol{r}) \end{bmatrix}. \tag{2.3}$$

We assume that the source orientation $\boldsymbol{\eta}(\boldsymbol{r})$ is time-independent throughout this book.

2.1.2 Sensor lead field

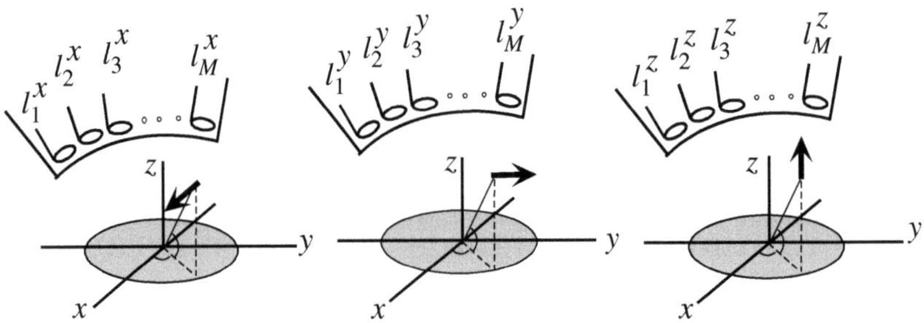

Figure 2.1: Sensor outputs when a unit-magnitude source directed in the x, y, and z directions exists at $\boldsymbol{r}$.

As depicted in Fig. 2.1, we assume that a single unit-magnitude source exists at $\boldsymbol{r}$. We denote the outputs of the mth sensor as $l_m^x(\boldsymbol{r})$, $l_m^y(\boldsymbol{r})$, and $l_m^z(\boldsymbol{r})$ when the unit-magnitude source is directed in the x, y, and z directions, respectively. Then, the three-dimensional row vector $\boldsymbol{l}_m(\boldsymbol{r})$,

$$\boldsymbol{l}_m(\boldsymbol{r}) = [l_m^x(\boldsymbol{r}), l_m^y(\boldsymbol{r}), l_m^z(\boldsymbol{r})] \tag{2.4}$$

indicates the sensitivity of the mth sensor to a source located at $\boldsymbol{r}$. The column vectors $\boldsymbol{l}_x(\boldsymbol{r})$, $\boldsymbol{l}_y(\boldsymbol{r})$, and $\boldsymbol{l}_z(\boldsymbol{r})$ are defined as

$$\begin{aligned} \boldsymbol{l}_x(\boldsymbol{r}) &= [l_1^x(\boldsymbol{r}), l_2^x(\boldsymbol{r}), \cdots, l_M^x(\boldsymbol{r})]^T, \\ \boldsymbol{l}_y(\boldsymbol{r}) &= [l_1^y(\boldsymbol{r}), l_2^y(\boldsymbol{r}), \cdots, l_M^y(\boldsymbol{r})]^T, \\ \boldsymbol{l}_z(\boldsymbol{r}) &= [l_1^z(\boldsymbol{r}), l_2^z(\boldsymbol{r}), \cdots, l_M^z(\boldsymbol{r})]^T. \end{aligned}$$

These vectors express the sensor array sensitivity for a source located at $\boldsymbol{r}$ and directed in the x, y, and z directions. Then, using these column vectors, the sensitivity of the whole sensor array for a source at $\boldsymbol{r}$ is expressed using the $M \times 3$ matrix

$$\boldsymbol{L}(\boldsymbol{r}) = [\boldsymbol{l}_x(\boldsymbol{r}), \boldsymbol{l}_y(\boldsymbol{r}), \boldsymbol{l}_z(\boldsymbol{r})]. \tag{2.5}$$

This matrix $\boldsymbol{L}(\boldsymbol{r})$ is called the lead-field matrix. We also define the lead-field vector, $\boldsymbol{l}(\boldsymbol{r})$, that expresses the sensitivity of the sensor array in a particular direction $\boldsymbol{\eta}(\boldsymbol{r})$, such that

$$\boldsymbol{l}(\boldsymbol{r}) = \boldsymbol{L}(\boldsymbol{r})\boldsymbol{\eta}(\boldsymbol{r}). \tag{2.6}$$

The problem of estimating the sensor lead field is referred to as the forward problem. The estimation of the sensor lead field is described in Section 13.7 of the Appendix.

Using the lead-field matrix in Eq. (2.5), the relationship between the measurement vector, $\boldsymbol{b}(t)$, and the three-dimensional source vector, $\boldsymbol{s}(\boldsymbol{r},t)$, is expressed as

$$\boldsymbol{b}(t) = \int_\Omega \boldsymbol{L}(\boldsymbol{r})\boldsymbol{s}(\boldsymbol{r},t)\, d^3r. \tag{2.7}$$

Here, d^3r indicates the volume element, and the integral is performed over a volume where sources can exist. This volume is called the source space, which is denoted Ω in this paper. Using the lead-field vector defined in Eq. (2.6), $\boldsymbol{l}(\boldsymbol{r})$, the above equation can be rewritten as

$$\boldsymbol{b}(t) = \int_\Omega \boldsymbol{L}(\boldsymbol{r})\boldsymbol{\eta}(\boldsymbol{r})s(\boldsymbol{r},t)\, d^3r = \int_\Omega \boldsymbol{l}(\boldsymbol{r})s(\boldsymbol{r},t)\, d^3r. \tag{2.8}$$

Note that, because the source orientation is included in $\boldsymbol{l}(\boldsymbol{r})$, the scalar $s(\boldsymbol{r},t)$ is used on the right-hand side of the equation above.

We define, for later use, an $M \times M$ matrix called the gram matrix; its (m,n) element is equal to the spatial similarity between the mth and the nth sensor lead fields. The gram matrix is denoted $\boldsymbol{G}$, and its (m,n) element is expressed as

$$G_{m,n} = \int_\Omega \boldsymbol{l}_m(\boldsymbol{r})\boldsymbol{l}_n^T(\boldsymbol{r})\, d^3r, \tag{2.9}$$

where $\boldsymbol{l}_m(\boldsymbol{r})$ is defined in Eq. (2.4). Thus, $\boldsymbol{G}$ is expressed as

$$\boldsymbol{G} = \int_\Omega \boldsymbol{L}(\boldsymbol{r})\boldsymbol{L}^T(\boldsymbol{r})\, d^3r. \tag{2.10}$$

The gram matrix plays an important role when formulating non-adaptive spatial filters in Chapter 3.

2.1.3 Linear independence of lead-field vectors

We briefly argue that the lead-field vectors are linearly independent. The fact that lead-field vectors at different locations or different orientations are linearly independent is implicitly assumed throughout this book. For example, in Section 2.4.1, we discuss that the lead-field vectors, $\boldsymbol{l}(\boldsymbol{r}_1), \ldots, \boldsymbol{l}(\boldsymbol{r}_Q)$, span the Q-dimensional signal subspace, where $\boldsymbol{r}_1, \ldots, \boldsymbol{r}_Q$ are the locations of Q sources. The fundamental assumption here is that the vectors $\boldsymbol{l}(\boldsymbol{r}_1), \ldots, \boldsymbol{l}(\boldsymbol{r}_Q)$ are linearly independent; i.e., none of them can be created by a linear combination of the other lead-field vectors.

This linear independence of the lead-field vectors is a direct result of the superposition law. The lead-field vector, $\boldsymbol{l}(\boldsymbol{r}_0)$, represents a magnetic field created by a unit-magnitude point source located at $\boldsymbol{r}_0$. That is, using Eq. (2.8), we can write

$$\boldsymbol{l}(\boldsymbol{r}_0) = \int_\Omega \boldsymbol{L}(\boldsymbol{r})\boldsymbol{\eta}_0\delta(\boldsymbol{r}-\boldsymbol{r}_0)\,d^3r, \tag{2.11}$$

where $\boldsymbol{\eta}_0$ represents the source-orientation vector for the source at $\boldsymbol{r}_0$, and $\delta(\boldsymbol{r})$ is the delta function that takes a three-dimensional vector as its argument. Let us assume that this lead field vector $\boldsymbol{l}(\boldsymbol{r}_0)$ can be expressed as a weighted sum of other lead field vectors $\boldsymbol{l}(\boldsymbol{r}_j)$ where $j = 1, \ldots, J_0$ and $J_0 < M$, such that

$$\boldsymbol{l}(\boldsymbol{r}_0) = \sum_{j=1}^{J_0} A_j \boldsymbol{l}(\boldsymbol{r}_j). \tag{2.12}$$

Here A_j is a scalar constant and $\boldsymbol{l}(\boldsymbol{r}_j)$ is the lead-field vector created by a point source at $\boldsymbol{r}_j$, such that

$$\boldsymbol{l}(\boldsymbol{r}_j) = \int_\Omega \boldsymbol{L}(\boldsymbol{r})\boldsymbol{\eta}_j\delta(\boldsymbol{r}-\boldsymbol{r}_j)\,d^3r, \tag{2.13}$$

where $\boldsymbol{\eta}_j$ represents the orientation vector of the source at $\boldsymbol{r}_j$. We exclude "electromagnetically silent" sources from the argument here, i.e., we assume that the lead-fled vectors, $\boldsymbol{l}(\boldsymbol{r}_1), \ldots, \boldsymbol{l}(\boldsymbol{r}_{J_0})$, have a non-zero norm. Using Eqs. (2.11), (2.12) and (2.13), we obtain

$$\begin{aligned}\int_\Omega \boldsymbol{L}(\boldsymbol{r})\boldsymbol{\eta}_0\delta(\boldsymbol{r}-\boldsymbol{r}_0)\,d^3r &= \sum_{j=1}^{J_0} A_j \int_\Omega \boldsymbol{L}(\boldsymbol{r})\boldsymbol{\eta}_j\delta(\boldsymbol{r}-\boldsymbol{r}_j)\,d^3r \\ &= \int_\Omega \boldsymbol{L}(\boldsymbol{r}) \left[\sum_{j=1}^{J_0} A_j\boldsymbol{\eta}_j\delta(\boldsymbol{r}-\boldsymbol{r}_j)\right] d^3r. \end{aligned}\tag{2.14}$$

As a result, the relationship,

$$\boldsymbol{\eta}_0\delta(\boldsymbol{r}-\boldsymbol{r}_0) = \sum_{j=1}^{J_0} A_j\boldsymbol{\eta}_j\delta(\boldsymbol{r}-\boldsymbol{r}_j) \tag{2.15}$$

should hold. However, Eq. (2.15) never holds, because the point source at $\boldsymbol{r}_0$ is never equal to a weighted sum of the point sources at the other locations, i.e., the delta function cannot be expressed as a sum of the other delta functions. Therefore, Eq. (2.12) is not valid and the lead-field vectors for different locations are linearly independent.

The same discussion holds for the linear independence among the lead-field vectors with linearly independent orientations. That is, we define three linearly independent orientations as $\boldsymbol{\eta}_0$, $\boldsymbol{\eta}_1$, and $\boldsymbol{\eta}_2$, and, to indicate the orientation dependence explicitly, we denote the associated lead-field vectors at the same location (that is denoted $\boldsymbol{r}_0$) as $\boldsymbol{l}(\boldsymbol{r}_0,\boldsymbol{\eta}_0)$, $\boldsymbol{l}(\boldsymbol{r}_0,\boldsymbol{\eta}_1)$, and $\boldsymbol{l}(\boldsymbol{r}_0,\boldsymbol{\eta}_2)$. These lead-field vectors are expressed as

$$\boldsymbol{l}(\boldsymbol{r}_0,\boldsymbol{\eta}_0)= \int_\Omega \boldsymbol{L}(\boldsymbol{r})\boldsymbol{\eta}_0\delta(\boldsymbol{r}-\boldsymbol{r}_0)\,d^3r = \boldsymbol{L}(\boldsymbol{r}_0)\boldsymbol{\eta}_0, \tag{2.16}$$

$$\boldsymbol{l}(\boldsymbol{r}_0,\boldsymbol{\eta}_1)= \int_\Omega \boldsymbol{L}(\boldsymbol{r})\boldsymbol{\eta}_1\delta(\boldsymbol{r}-\boldsymbol{r}_0)\,d^3r = \boldsymbol{L}(\boldsymbol{r}_0)\boldsymbol{\eta}_1, \tag{2.17}$$

$$\text{and}\quad \boldsymbol{l}(\boldsymbol{r}_0,\boldsymbol{\eta}_2)= \int_\Omega \boldsymbol{L}(\boldsymbol{r})\boldsymbol{\eta}_2\delta(\boldsymbol{r}-\boldsymbol{r}_0)\,d^3r = \boldsymbol{L}(\boldsymbol{r}_0)\boldsymbol{\eta}_2. \tag{2.18}$$

According to the argument above, if the following relationship holds,

$$\boldsymbol{l}(\boldsymbol{r}_0,\boldsymbol{\eta}_0) = A_1\boldsymbol{l}(\boldsymbol{r}_0,\boldsymbol{\eta}_1) + A_2\boldsymbol{l}(\boldsymbol{r}_0,\boldsymbol{\eta}_2), \tag{2.19}$$

where A_1 and A_2 are some scalar constants, we have

$$\boldsymbol{L}(\boldsymbol{r}_0)\boldsymbol{\eta}_0 = A_1\boldsymbol{L}(\boldsymbol{r}_0)\boldsymbol{\eta}_1 + A_2\boldsymbol{L}(\boldsymbol{r}_0)\boldsymbol{\eta}_2 = \boldsymbol{L}(\boldsymbol{r}_0)(A_1\boldsymbol{\eta}_1 + A_2\boldsymbol{\eta}_2). \tag{2.20}$$

Since the above equation holds for an arbitrary location, i.e., for an arbitrary $\boldsymbol{r}_0$, the relationship

$$\boldsymbol{\eta}_0 = A_1\boldsymbol{\eta}_1 + A_2\boldsymbol{\eta}_2 \tag{2.21}$$

should hold. However, the relationship above never holds, because of the linear independence among the three orientation vectors. Therefore, the lead-field vectors, $\boldsymbol{l}(\boldsymbol{r}_0,\boldsymbol{\eta}_0)$, $\boldsymbol{l}(\boldsymbol{r}_0,\boldsymbol{\eta}_1)$, and $\boldsymbol{l}(\boldsymbol{r}_0,\boldsymbol{\eta}_2)$, are linearly independent if $\boldsymbol{\eta}_0$, $\boldsymbol{\eta}_1$, and $\boldsymbol{\eta}_2$ are linearly independent.

2.2 Bioelectromagnetic inverse problem

Returning to the equations in Eq. (2.7) and (2.8), we restate these equations with noise. We assume that additive noise exists and denote $n_m(t)$ as the noise amplitude added to the mth sensor output. We define the noise vector $\boldsymbol{n}(t)$ such that

$$\boldsymbol{n}(t) = \begin{bmatrix} n_1(t) \\ n_2(t) \\ \vdots \\ n_M(t) \end{bmatrix}. \tag{2.22}$$

When such additive noise is taken into account, Eq. (2.7) and (2.8) respectively become

$$\boldsymbol{b}(t) = \int_\Omega \boldsymbol{L}(\boldsymbol{r})\boldsymbol{s}(\boldsymbol{r},t)\,d^3r + \boldsymbol{n}(t), \tag{2.23}$$

and

$$\boldsymbol{b}(t) = \int_{\Omega} \boldsymbol{l}(\boldsymbol{r}) s(\boldsymbol{r}, t)\, d^3 r + \boldsymbol{n}(t). \tag{2.24}$$

Equations (2.23) and (2.24) are the fundamental equations for measurements performed with a sensor array.

The bioelectromagnetic inverse problem is the problem of estimating the source-vector spatial distribution, $\boldsymbol{s}(\boldsymbol{r}, t)$, from the measurements, $\boldsymbol{b}(t)$. Here, we can assume that we know the sensor lead field $\boldsymbol{L}(\boldsymbol{r})$, although our knowledge of the sensor lead field is to some degree imperfect because it should be estimated by using an analytical model or numerical computations, as described in Section 13.7. The spatial distribution of the source orientation, $\boldsymbol{\eta}(\boldsymbol{r})$, may be a known quantity, if accurate subject anatomical information (such as high-precision subject MRI) can be obtained with accurate co-registration to the sensor coordinate space. (In this case, the inverse problem is one of estimating the source magnitude, $s(\boldsymbol{r}, t)$, instead of the source vector, $\boldsymbol{s}(\boldsymbol{r}, t)$.) However, it is generally difficult to obtain such accurate anatomical information for each subject and the current co-registration methods have only limited accuracy. Therefore, in this book, we treat $\boldsymbol{\eta}(\boldsymbol{r})$ as an unknown quantity unless otherwise noted. When estimating $\boldsymbol{s}(\boldsymbol{r}, t)$ from $\boldsymbol{b}(t)$, $\boldsymbol{s}(\boldsymbol{r}, t)$ is continuous in space, while $\boldsymbol{b}(t)$ is discrete in space. Thus, estimation of $\boldsymbol{s}(\boldsymbol{r}, t)$ is intrinsically an ill-posed problem, which cannot be solved unless some constraints are imposed on $\boldsymbol{s}(\boldsymbol{r}, t)$.

We therefore introduce the discrete source model, and assume that the bioelectromagnetic signal is generated from a total of Q discrete sources. We retain this discrete source model throughout the book. If the number of sources Q is smaller than the number of sensors M, the signal is referred to as a low-rank signal. We have derived an interesting property of low-rank signals, and a discussion regarding these low-rank signals appears in Section 2.4. We denote the locations of the Q sources as $\boldsymbol{r}_1, \boldsymbol{r}_2, \ldots, \boldsymbol{r}_Q$. Then, the source spatial distribution is expressed as

$$\boldsymbol{s}(\boldsymbol{r}, t) = \sum_{q=1}^{Q} \boldsymbol{s}(\boldsymbol{r}_q, t)\delta(\boldsymbol{r} - \boldsymbol{r}_q). \tag{2.25}$$

Substituting the above equation into Eq. (2.23) and (2.24), we obtain the discrete versions of the measurement equations, such as

$$\boldsymbol{b}(t) = \sum_{q=1}^{Q} \boldsymbol{L}(\boldsymbol{r}_q)\boldsymbol{s}(\boldsymbol{r}_q, t) + \boldsymbol{n}(t), \tag{2.26}$$

and

$$\boldsymbol{b}(t) = \sum_{q=1}^{Q} \boldsymbol{l}(\boldsymbol{r}_q) s(\boldsymbol{r}_q, t) + \boldsymbol{n}(t). \tag{2.27}$$

Let us define a $Q \times 1$ column vector $\boldsymbol{\nu}(t)$ such that

$$\boldsymbol{\nu}(t) = \begin{bmatrix} s(\boldsymbol{r}_1, t) \\ s(\boldsymbol{r}_2, t) \\ \vdots \\ s(\boldsymbol{r}_Q, t) \end{bmatrix}. \tag{2.28}$$

We define a composite lead field matrix $\boldsymbol{L}_C$ such that

$$\boldsymbol{L}_C = [\boldsymbol{L}(\boldsymbol{r}_1), \boldsymbol{L}(\boldsymbol{r}_2), \ldots, \boldsymbol{L}(\boldsymbol{r}_Q)], \tag{2.29}$$

where $\boldsymbol{L}(\boldsymbol{r}_q)$ is the lead field matrix at the qth source location, $\boldsymbol{r}_q$. Using the orientation of the qth source, $\boldsymbol{\eta}(\boldsymbol{r}_q)$, we also define a $3Q \times Q$ matrix $\boldsymbol{\Psi}$ such that

$$\boldsymbol{\Psi} = \begin{bmatrix} \boldsymbol{\eta}(\boldsymbol{r}_1) & 0 & \cdots & 0 \\ 0 & \boldsymbol{\eta}(\boldsymbol{r}_2) & \cdot & \vdots \\ \vdots & \cdot & \ddots & 0 \\ 0 & \cdots & 0 & \boldsymbol{\eta}(\boldsymbol{r}_Q) \end{bmatrix}. \tag{2.30}$$

Then, we can derive a relationship equivalent to Eq. (2.26), such that

$$\boldsymbol{b}(t) = \boldsymbol{L}_C \boldsymbol{\Psi} \boldsymbol{\nu}(t) + \boldsymbol{n}(t). \tag{2.31}$$

Let us define $\boldsymbol{L}_D$ as a composite lead-field matrix whose columns are the lead-field vectors, which is a function of both source location and orientation:

$$\boldsymbol{L}_D = \boldsymbol{L}_C \boldsymbol{\Psi} = [\boldsymbol{L}(\boldsymbol{r}_1)\boldsymbol{\eta}(\boldsymbol{r}_1), \ldots, \boldsymbol{L}(\boldsymbol{r}_Q)\boldsymbol{\eta}(\boldsymbol{r}_Q)] = [\boldsymbol{l}(\boldsymbol{r}_1), \ldots, \boldsymbol{l}(\boldsymbol{r}_Q)]. \tag{2.32}$$

We can obtain a relationship corresponding to Eq. (2.27), such that

$$\boldsymbol{b}(t) = \boldsymbol{L}_D \boldsymbol{\nu}(t) + \boldsymbol{n}(t). \tag{2.33}$$

Equations (2.31) and (2.33) are also the discrete versions of the measurement equations.

2.3 Expressions of data covariance matrices

2.3.1 Data and source covariance relationship

We define the covariance matrix of measurements $\boldsymbol{R}$ such that

$$\boldsymbol{R} = \langle \boldsymbol{b}(t)\boldsymbol{b}^T(t) \rangle = \begin{bmatrix} \langle b_1(t)^2 \rangle & \langle b_1(t)b_2(t) \rangle & \ldots & \langle b_1(t)b_M(t) \rangle \\ \langle b_2(t)b_1(t) \rangle & \langle b_2(t)^2 \rangle & \ldots & \langle b_2(t)b_M(t) \rangle \\ \vdots & \vdots & \ddots & \vdots \\ \langle b_M(t)b_1(t) \rangle & \langle b_M(t)b_2(t) \rangle & \ldots & \langle b_M(t)^2 \rangle \end{bmatrix}, \tag{2.34}$$

where $\langle\cdot\rangle$ indicates the expectation operator. This matrix $\boldsymbol{R}$ is also called the data covariance matrix. The matrix $\boldsymbol{R}$ is not exactly equal to the covariance matrix but equal to the second-order moment matrix. However, $\boldsymbol{R}$ is customarily called the covariance matrix because the relationship $\langle\boldsymbol{b}(t)\rangle \approx 0$ holds in many applications. In practice, the expectation is computed using a time average over a certain time window. Therefore, in this book, $\langle\cdot\rangle$ also indicates the time average. The data covariance matrix, $\boldsymbol{R}$, plays a key role in the adaptive-spatial-filter source imaging.

We next define the covariance matrix of the source activity $\boldsymbol{R}_\nu$ such that

$$\boldsymbol{R}_\nu = \langle\boldsymbol{\nu}(t)\boldsymbol{\nu}^T(t)\rangle = \begin{bmatrix} \langle s(\boldsymbol{r}_1,t)^2\rangle & \langle s(\boldsymbol{r}_1,t)s(\boldsymbol{r}_2,t)\rangle & \dots & \langle s(\boldsymbol{r}_1,t)s(\boldsymbol{r}_Q,t)\rangle \\ \langle s(\boldsymbol{r}_2,t)s(\boldsymbol{r}_1,t)\rangle & \langle s(\boldsymbol{r}_2,t)^2\rangle & \dots & \langle s(\boldsymbol{r}_2,t)s(\boldsymbol{r}_Q,t)\rangle \\ \vdots & \vdots & \ddots & \vdots \\ \langle s(\boldsymbol{r}_Q,t)s(\boldsymbol{r}_1,t)\rangle & \langle s(\boldsymbol{r}_Q,t)s(\boldsymbol{r}_2,t)\rangle & \dots & \langle s(\boldsymbol{r}_Q,t)^2\rangle \end{bmatrix}. \tag{2.35}$$

Using Eq. (2.31) and assuming that $\boldsymbol{\nu}(t)$ and $\boldsymbol{n}(t)$ are uncorrelated, we can derive a relationship between the data covariance matrix, $\boldsymbol{R}$, and the source covariance matrix, $\boldsymbol{R}_\nu$, such that

$$\boldsymbol{R} = \langle\boldsymbol{b}(t)\boldsymbol{b}^T(t)\rangle = [\boldsymbol{L}_C\boldsymbol{\Psi}]\boldsymbol{R}_\nu[\boldsymbol{\Psi}^T\boldsymbol{L}_C^T] + \boldsymbol{R}_n. \tag{2.36}$$

Here, $\boldsymbol{R}_n$ is the noise covariance matrix, which is obtained as

$$\boldsymbol{R}_n = \langle\boldsymbol{n}(t)\boldsymbol{n}^T(t)\rangle. \tag{2.37}$$

Using Eq. (2.33), we can derive the relationship equivalent to Eq. (2.36), as

$$\boldsymbol{R} = \boldsymbol{L}_D\boldsymbol{R}_\nu\boldsymbol{L}_D^T + \boldsymbol{R}_n, \tag{2.38}$$

where $\boldsymbol{L}_D$ is defined in Eq. (2.32).

If noise is generated by the sensor hardware, it can be approximated as zero-mean white Gaussian noise uncorrelated between different sensor channels. In that case, we have the relationship

$$\boldsymbol{R}_n = \sigma_0^2\boldsymbol{I}, \tag{2.39}$$

where σ_0^2 is the variance of the noise and $\boldsymbol{I}$ is the identity matrix[1]. In deriving the relationship in Eq. (2.39), we also assume that the noise variance is equal for all sensor channels. Throughout this book, $\boldsymbol{n}(t)$ indicates sensor noise which has the property expressed in Eq. (2.39). In bioelectromagnetic measurements, however, in addition to the sensor noise, interference signals may arise from outside the measurement hardware. These interferences with external origins often overlap with the signal of interest and they may distort the final source imaging results. The influence of these external interferences will be discussed later in Chapters 7 and 8. Using Eq. (2.39), Eqs. (2.36) and (2.38) are, respectively, converted to

$$\boldsymbol{R} = [\boldsymbol{L}_C\boldsymbol{\Psi}]\boldsymbol{R}_\nu[\boldsymbol{\Psi}^T\boldsymbol{L}_C^T] + \sigma_0^2\boldsymbol{I}, \tag{2.40}$$

[1] In this case, $\boldsymbol{I}$ is the $M \times M$ identity matrix. The size of the identity matrix is not explicitly stated in this book unless otherwise needed.

and

$$\boldsymbol{R} = \boldsymbol{L}_D \boldsymbol{R}_\nu \boldsymbol{L}_D^T + \sigma_0^2 \boldsymbol{I}. \tag{2.41}$$

Equations (2.40) and (2.41) are the basic equations for discussing the main properties of low-rank signals in Section 2.4.

So far, we explicitly describe pairs of mathematically-equivalent equations: those using $\boldsymbol{L}_C$ (e.g., Eq. (2.40)) and others using $\boldsymbol{L}_D$ (e.g., Eq. (2.41)). Although these two groups of equations are equivalent, we usually prefer to use the equations with $\boldsymbol{L}_D$ in the arguments in this book, because these equations facilitate simpler mathematical argumentation.

2.3.2 Formulation for uncorrelated sources

We next assume that the source time courses are uncorrelated. Substituting $\langle s(\boldsymbol{r}_p, t) s(\boldsymbol{r}_q, t) \rangle = 0$ for $p \neq q$ and $\sigma_q^2 = \langle s(\boldsymbol{r}_q, t)^2 \rangle$, where σ_q^2 represents the power of the qth source, into Eq. (2.35), $\boldsymbol{R}_\nu$ is expressed as

$$\boldsymbol{R}_\nu = \begin{bmatrix} \sigma_1^2 & 0 & \dots & 0 \\ 0 & \sigma_2^2 & \dots & 0 \\ \vdots & \vdots & \ddots & \vdots \\ 0 & 0 & \dots & \sigma_Q^2 \end{bmatrix}. \tag{2.42}$$

Substituting the equation above into Eq. (2.41), the measurement covariance matrix, $\boldsymbol{R}$, is expressed as

$$\boldsymbol{R} = \sum_{q=1}^{Q} \sigma_q^2 \boldsymbol{l}(\boldsymbol{r}_q) \boldsymbol{l}^T(\boldsymbol{r}_q) + \sigma_0^2 \boldsymbol{I}. \tag{2.43}$$

The power of the measurements is given by:

$$\langle \|\boldsymbol{b}(t)\|^2 \rangle = \mathrm{tr}\{\boldsymbol{R}\} = \sum_{q=1}^{Q} \sigma_j^2 \|\boldsymbol{l}(\boldsymbol{r}_q)\|^2 + M\sigma_0^2, \tag{2.44}$$

where $\mathrm{tr}\{\cdot\}$ indicates the trace operation applied to the matrix between the curly braces. From Eq. (2.44), we can see that the ratio between the signal power and the noise power,

$$\frac{\sigma_q^2 \|\boldsymbol{l}(\boldsymbol{r}_q)\|^2}{M\sigma_0^2}, \tag{2.45}$$

can be a measure of the strength of the qth source. In many signal-processing literatures (for example, [11]), α_q defined as

$$\alpha_q = \frac{\sigma_q^2 \|\boldsymbol{l}(\boldsymbol{r}_q)\|^2}{\sigma_0^2}, \tag{2.46}$$

is often called the input signal-to-noise ratio (SNR) (of the qth source), and this α_q is used as the measure of the strength of the qth source. Note that when the

qth source is so weak that the average power of the magnetic field generated from this source is equal to the average noise power, i.e., when the relationship

$$\frac{1}{M}\sigma_q^2\|\boldsymbol{l}(\boldsymbol{r}_q)\|^2 = \sigma_0^2$$

holds, α_q equals the number of sensors M. We assume in this book that sources to be reconstructed have values of α_q greater than M.

2.4 Low-rank signal modeling

A low-rank signal is a signal generated from Q discrete sources in which Q is smaller than the number of sensors M[12]. The low-rank signal model plays an important role in discussions of adaptive spatial filters. The assumption that $M > Q$ is fundamental, and we maintain it throughout this book. Here, we present a major property of low-rank signals. Some related arguments are also found in Section 12.2.

2.4.1 Definition of noise and signal subspaces

The important property of the low-rank signal is that the signal and the noise subspaces exist; these subspaces play a key role in discussions of adaptive spatial filters. Our arguments begin with Eq. (2.27), namely,

$$\boldsymbol{b}(t) = \sum_{q=1}^{Q}\boldsymbol{l}(\boldsymbol{r}_q)s(\boldsymbol{r}_q,t) + \boldsymbol{n}(t) = \boldsymbol{b}_s(t) + \boldsymbol{n}(t),$$

where $\boldsymbol{b}_s(t)$, called the signal vector, is the signal part in the whole measurement $\boldsymbol{b}(t)$. Using $\boldsymbol{L}_D$ in Eq. (2.32), and denoting $s(\boldsymbol{r}_q,t)$ as s_q, the signal vector $\boldsymbol{b}_s$ is expressed as

$$\begin{aligned}\boldsymbol{b}_s &= \sum_{j=1}^{Q}\boldsymbol{l}(\boldsymbol{r}_j)s_j = s_1\boldsymbol{l}(\boldsymbol{r}_1) + \cdots + s_Q\boldsymbol{l}(\boldsymbol{r}_Q)\\ &= [\boldsymbol{l}(\boldsymbol{r}_1),\boldsymbol{l}(\boldsymbol{r}_2),\ldots,\boldsymbol{l}(\boldsymbol{r}_Q)]\begin{bmatrix} s_1\\ s_2\\ \vdots\\ s_Q\end{bmatrix} = \boldsymbol{L}_D\boldsymbol{\nu},\end{aligned} \tag{2.47}$$

where $\boldsymbol{\nu} = [s_1,\ldots,s_Q]^T$, and the time notation is omitted for simplicity. Equation (2.47) indicates that the signal vector $\boldsymbol{b}_s$ lies within a subspace spanned by the linearly independent vectors $\boldsymbol{l}(\boldsymbol{r}_1),\ldots,\boldsymbol{l}(\boldsymbol{r}_Q)$, which is the column span of $\boldsymbol{L}_D$. (The linear independence of these lead field vectors is argued in Section 2.1.3.)

This column span is called the signal subspace and is denoted $\mathcal{E}_S$, i.e.,

$$\mathcal{E}_S = span\{\boldsymbol{l}(\boldsymbol{r}_1),\ldots,\boldsymbol{l}(\boldsymbol{r}_Q)\} \tag{2.48}$$

Thus, the relationship

$$\boldsymbol{b}_s \in \mathcal{E}_S \tag{2.49}$$

holds. The left-hand null space of $\boldsymbol{L}_D$ is called the noise subspace and is denoted $\mathcal{E}_N$, i.e.,

$$\mathcal{E}_N = \{\boldsymbol{x} | \boldsymbol{x}^T \boldsymbol{L}_D = 0\}. \tag{2.50}$$

The dimension of the signal subspace is Q and that of the noise subspace is $M - Q$. We can prove that $\mathcal{E}_S$ and $\mathcal{E}_N$ are the orthogonal complements, and that the relationship

$$\mathcal{E}_S \cup \mathcal{E}_N = \Re^M \tag{2.51}$$

holds where $\Re^M$ is the space of the M-dimensional real vectors[13].

2.4.2 Property of the data covariance matrix

We next discuss an important property of the measurement covariance matrix, a property which is closely related to the noise and the signal subspaces. Unless some source activities are perfectly correlated with each other, the rank of $\boldsymbol{R}_\nu$ is equal to the number of sources Q. Therefore, according to Eq. (2.41), the $M \times M$ matrix $\boldsymbol{L}_D \boldsymbol{R}_\nu \boldsymbol{L}_D^T$ $(= [\boldsymbol{L}_C \boldsymbol{\Psi}] \boldsymbol{R}_\nu [\boldsymbol{\Psi}^T \boldsymbol{L}_C^T])$ is a positive semi-definite matrix of rank Q and has the form

$$\boldsymbol{L}_D \boldsymbol{R}_\nu \boldsymbol{L}_D^T = \\ [\boldsymbol{e}_1, \ldots, \boldsymbol{e}_M] \begin{bmatrix} \lambda_1' & 0 & \cdots & \cdot & \cdots & 0 \\ 0 & \ddots & \cdot & \cdot & \cdot & 0 \\ \vdots & \cdot & \lambda_Q' & \cdot & \cdot & \vdots \\ \cdot & \cdot & \cdot & 0 & \cdot & \cdot \\ \vdots & \cdot & \cdot & \cdot & \ddots & 0 \\ 0 & \cdot & \cdots & \cdot & 0 & 0 \end{bmatrix} [\boldsymbol{e}_1, \ldots, \boldsymbol{e}_M]^T, \tag{2.52}$$

where λ_j' and $\boldsymbol{e}_j$ are the jth eigenvalue and its corresponding eigenvector of the matrix $\boldsymbol{L}_D \boldsymbol{R}_\nu \boldsymbol{L}_D^T$. Here, these eigenvalues are numbered in decreasing order. Then, the eigendecomposition of $\boldsymbol{R}$ is expressed as

$$\boldsymbol{R} = \boldsymbol{L}_D \boldsymbol{R}_\nu \boldsymbol{L}_D^T + \sigma_0^2 \boldsymbol{I} \\ = [\boldsymbol{e}_1, \ldots, \boldsymbol{e}_M] \begin{bmatrix} \lambda_1 & 0 & \cdots & \cdot & \cdots & 0 \\ 0 & \ddots & \cdot & \cdot & \cdot & 0 \\ \vdots & \cdot & \lambda_Q & \cdot & \cdot & \vdots \\ \cdot & \cdot & \cdot & \sigma_0^2 & \cdot & \cdot \\ \vdots & \cdot & \cdot & \cdot & \ddots & 0 \\ 0 & \cdot & \cdots & \cdot & 0 & \sigma_0^2 \end{bmatrix} [\boldsymbol{e}_1, \ldots, \boldsymbol{e}_M]^T, \tag{2.53}$$

where λ_j is the jth eigenvalue of $\boldsymbol{R}$ and $\lambda_j = \lambda'_j + \sigma_0^2$. The equation above indicates that $\boldsymbol{R}$ has Q eigenvalues greater than σ_0^2, and $M - Q$ eigenvalues equal to σ_0^2. The eigenvalues greater than σ_0^2 are called the signal-level eigenvalues and those equal to σ_0^2 are called the noise-level eigenvalues. Equation (2.53) indicates that $\boldsymbol{e}_j$ is also the jth eigenvector of the matrix $\boldsymbol{R}$, and because $\boldsymbol{R}$ is a real symmetric matrix, the eigenvectors form an orthonormal basis.

Let us define the matrix $\boldsymbol{E}_S$ as $\boldsymbol{E}_S = [\boldsymbol{e}_1, \ldots, \boldsymbol{e}_Q]$ and $\boldsymbol{E}_N$ as $\boldsymbol{E}_N = [\boldsymbol{e}_{Q+1}, \ldots, \boldsymbol{e}_M]$. We then show that the column span of $\boldsymbol{E}_S$ is equal to the signal subspace, and the span of $\boldsymbol{E}_N$ is equal to the noise subspace. To show this, we first rewrite Eqs. (2.41) as

$$(\boldsymbol{R} - \sigma_0^2 \boldsymbol{I}) = \boldsymbol{L}_D \boldsymbol{R}_\nu \boldsymbol{L}_D^T, \tag{2.54}$$

and then multiply each side of this equation by one of the noise-level eigenvectors, i.e.,

$$(\boldsymbol{R} - \sigma_0^2 \boldsymbol{I})\boldsymbol{e}_j = \boldsymbol{L}_D \boldsymbol{R}_\nu \boldsymbol{L}_D^T \boldsymbol{e}_j, \tag{2.55}$$

where $j = Q+1, \ldots, M$. Considering the eigendecomposition in Eq. (2.52), and the fact that the eigenvectors are orthogonal, the left-hand side of the above equation is found to be zero, and we have

$$\boldsymbol{L}_D \boldsymbol{R}_\nu \boldsymbol{L}_D^T \boldsymbol{e}_j = 0, \quad \text{for} \quad j = Q+1, \ldots, M. \tag{2.56}$$

Since $\boldsymbol{L}_D$ is a full column-rank matrix, and we assume that $\boldsymbol{R}_\nu$ is a full-rank matrix, the above equation results in

$$\boldsymbol{L}_D^T \boldsymbol{e}_j = [\boldsymbol{l}(\boldsymbol{r}_1), \boldsymbol{l}(\boldsymbol{r}_2), \ldots, \boldsymbol{l}(\boldsymbol{r}_Q)]^T \boldsymbol{e}_j = 0 \quad \text{for} \quad j = Q+1, \ldots, M. \tag{2.57}$$

This implies that the lead-field vectors at the true source locations are orthogonal to the eigenvectors $\boldsymbol{e}_{Q+1}, \ldots, \boldsymbol{e}_M$, i.e.,

$$\boldsymbol{l}^T(\boldsymbol{r}_q)\boldsymbol{E}_N = 0 \quad \text{for} \quad q = 1, \ldots, Q. \tag{2.58}$$

This equation indicates that the columns of $\boldsymbol{E}_N$ are the basis vectors of the noise subspace. The fact that the column span of $\boldsymbol{E}_S$ is the complementary subspace of the column span of $\boldsymbol{E}_N$ indicates that the columns of $\boldsymbol{E}_S$ are the basis vectors of the signal subspace. That is, we have the relationships

$$\mathcal{E}_S = span\{\boldsymbol{e}_1, \ldots, \boldsymbol{e}_Q\}, \tag{2.59}$$

$$\text{and} \quad \mathcal{E}_N = span\{\boldsymbol{e}_{Q+1}, \ldots, \boldsymbol{e}_M\}. \tag{2.60}$$

When $\boldsymbol{R}$ is the sample covariance matrix, the column span of $\boldsymbol{E}_S$ is the maximum-likelihood estimate of the signal subspace of $\boldsymbol{R}$, and the span of $\boldsymbol{E}_N$ is that of the noise subspace. (The proof for this is given in Section 13.1 of the Appendix.) This orthogonality relationship in Eq. (2.58) is the basis of the eigenspace-projection adaptive spatial filter described in Section 6.8, as well as the basis of the well-known MUSIC algorithm[14][15][16], which is described in

Section 12.2. In the low-rank signal model, the measurement covariance matrix $\boldsymbol{R}$ can be decomposed into noise and signal subspace components, such that

$$\boldsymbol{R} = \boldsymbol{E}_S\boldsymbol{\Lambda}_S\boldsymbol{E}_S^T + \boldsymbol{E}_N\boldsymbol{\Lambda}_N\boldsymbol{E}_N^T, \tag{2.61}$$

where $\boldsymbol{\Lambda}_S$ and $\boldsymbol{\Lambda}_N$ are diagonal matrices defined as

$$\boldsymbol{\Lambda}_S = \mathrm{diag}[\lambda_1, \ldots, \lambda_Q] \quad \text{and} \quad \boldsymbol{\Lambda}_N = \mathrm{diag}[\lambda_{Q+1}, \ldots, \lambda_M], \tag{2.62}$$

where $\mathrm{diag}[\cdots]$ indicates a diagonal matrix whose diagonal elements are equal to the entries in the square brackets. Using the orthogonality between $\boldsymbol{E}_S$ and $\boldsymbol{E}_N$, we can derive

$$\boldsymbol{R}^{-1} = \boldsymbol{E}_S\boldsymbol{\Lambda}_S^{-1}\boldsymbol{E}_S^T + \boldsymbol{E}_N\boldsymbol{\Lambda}_N^{-1}\boldsymbol{E}_N^T, \tag{2.63}$$

and

$$\boldsymbol{R}^{-2} = \boldsymbol{E}_S\boldsymbol{\Lambda}_S^{-2}\boldsymbol{E}_S^T + \boldsymbol{E}_N\boldsymbol{\Lambda}_N^{-2}\boldsymbol{E}_N^T. \tag{2.64}$$

These relationships are used in the arguments of Section 6.8.

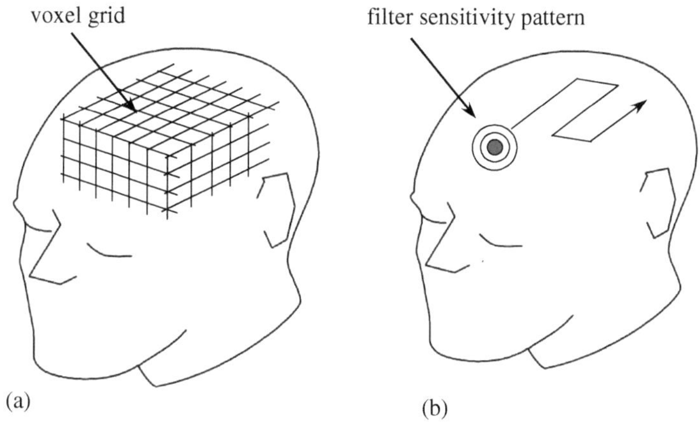

Figure 2.2: Conceptual view of tomographic-imaging methods and spatial-filters. (a) Tomographic source reconstruction method, which introduces voxel discretization and assumes a source at each voxel. The orientations and magnitudes of the sources are estimated using the least-squares fit to the measured data with a method-specific constraint. (b) Spatial-filter imaging. The spatial filter artificially controls the sensor array sensitivity to form a virtual sensor whose sensitivity pattern is expected to be localized at a specific location in the source space. This focused sensitivity pattern is scanned, in a completely post-processed manner, over the source space to obtain the three-dimensional reconstruction of the source distribution.

2.5 Spatial filters

2.5.1 Source reconstruction using a spatial filter

The spatial filter is a technique that numerically controls the sensor array sensitivity to form a virtual sensor whose sensitivity pattern is expected to be localized at the filter pointing location. The principle of spatial filter source imaging is conceptually illustrated, in contrast to tomographic imaging, in Fig. 2.2. Using the sensor array output, $\boldsymbol{b}(t)$, the spatial filter reconstructs the source magnitude, $s(\boldsymbol{r},t)$, by using

$$\widehat{s}(\boldsymbol{r},t) = \boldsymbol{w}^T(\boldsymbol{r})\boldsymbol{b}(t), \tag{2.65}$$

where $\widehat{s}(\boldsymbol{r},t)$ is the estimated or reconstructed source magnitude at location $\boldsymbol{r}$ and time t. (In this book, we use "^" for estimated quantities in order to distinguish them from their true values.) We may call the reconstructed source magnitude, $\widehat{s}(\boldsymbol{r},t)$, the spatial filter output. In Eq. (2.65), $\boldsymbol{w}(\boldsymbol{r})$ is an $M \times 1$ column vector called the weight vector. The weight vector characterizes the properties of the spatial filter. There are two types of spatial filters. One is the non-adaptive spatial filter in which the weight vector depends solely on the lead field of the sensor array. The other is the adaptive spatial filter in which the weight depends on the measured data as well as the lead field of the sensor array. In this book, in Chapter 3, we first describe several representative non-adaptive spatial filters used for bioelectromagnetic brain imaging. We then introduce adaptive spatial filters in Chapter 4, and discuss their properties in the following chapters.

Using Eq. (2.65), the power of the spatial-filter output is given by:

$$\langle \widehat{s}(\boldsymbol{r},t)^2 \rangle = \boldsymbol{w}^T(\boldsymbol{r})\langle \boldsymbol{b}(t)\boldsymbol{b}^T(t)\rangle \boldsymbol{w}(\boldsymbol{r}) = \boldsymbol{w}^T(\boldsymbol{r})\boldsymbol{R}\boldsymbol{w}(\boldsymbol{r}). \tag{2.66}$$

Assuming that the noise, $\boldsymbol{n}(t)$, is zero-mean white Gaussian noise with variance σ_0^2 and is uncorrelated between different sensor channels, the noise power of the spatial filter output is given by:

$$\boldsymbol{w}^T(\boldsymbol{r})\langle \boldsymbol{n}(t)\boldsymbol{n}^T(t)\rangle \boldsymbol{w}(\boldsymbol{r}) = \sigma_0^2 \boldsymbol{w}^T(\boldsymbol{r})\boldsymbol{I}\boldsymbol{w}(\boldsymbol{r}) = \sigma_0^2 \|\boldsymbol{w}(\boldsymbol{r})\|^2. \tag{2.67}$$

That is, the input noise power σ_0^2 is multiplied by $\|\boldsymbol{w}(\boldsymbol{r})\|^2$ in the spatial filter output. Therefore $\|\boldsymbol{w}(\boldsymbol{r})\|^2$ is called the noise-power gain or the white-noise gain of the spatial filter. The signal-to-noise ratio (SNR) of the spatial filter output can be defined as[2]

$$\mathcal{Z} = \frac{\boldsymbol{w}^T(\boldsymbol{r})\boldsymbol{R}\boldsymbol{w}(\boldsymbol{r})}{\sigma_0^2 \|\boldsymbol{w}(\boldsymbol{r})\|^2}. \tag{2.68}$$

[2]The numerator of Eq. (2.68) contains the noise contribution, so the $\mathcal{Z}$ value is not exactly equal to the SNR. The relationship between the $\mathcal{Z}$ value and the theoretical SNR is discussed in Section 6.1

2.5.2 Scalar and vector spatial filters

Scalar spatial filter

Because the bioelectromagnetic source is a three-dimensional vector quantity, we need to know the source orientation $\boldsymbol{\eta}$ when deriving the weight vector $\boldsymbol{w}(\boldsymbol{r})$. The source orientation may be predetermined if an accurate three-dimensional anatomical information of the subject is available. Generally, however, the source orientation $\boldsymbol{\eta}(\boldsymbol{r})$ should be estimated from the data. When $\boldsymbol{\eta}(\boldsymbol{r})$ is unknown, one strategy for obtaining a reasonable estimate of $\boldsymbol{\eta}(\boldsymbol{r})$ is to first derive the weight $\boldsymbol{w}(\boldsymbol{r},\boldsymbol{\eta})$ that depends on both the location $\boldsymbol{r}$ and orientation $\boldsymbol{\eta}(\boldsymbol{r})$. This weight $\boldsymbol{w}(\boldsymbol{r},\boldsymbol{\eta})$ provides the source magnitude at location $\boldsymbol{r}$ projected in the direction $\boldsymbol{\eta}$. We denote the output of this weight $\widehat{s}(\boldsymbol{r},\boldsymbol{\eta},t)$, i.e.,

$$\widehat{s}(\boldsymbol{r},\boldsymbol{\eta},t) = \boldsymbol{w}^T(\boldsymbol{r},\boldsymbol{\eta})\boldsymbol{b}(t). \tag{2.69}$$

The source orientation at $\boldsymbol{r}$ is estimated by maximizing the power of $\widehat{s}(\boldsymbol{r},\boldsymbol{\eta},t)$ with respect to $\boldsymbol{\eta}$. That is, the optimum source orientation $\boldsymbol{\eta}_{opt}(\boldsymbol{r})$ is obtained from

$$\boldsymbol{\eta}_{opt}(\boldsymbol{r}) = \underset{\boldsymbol{\eta}(\boldsymbol{r})}{\arg\max}\, \langle \widehat{s}(\boldsymbol{r},\boldsymbol{\eta},t)^2 \rangle = \underset{\boldsymbol{\eta}(\boldsymbol{r})}{\arg\max}\, \boldsymbol{w}^T(\boldsymbol{r},\boldsymbol{\eta})\boldsymbol{R}\boldsymbol{w}(\boldsymbol{r},\boldsymbol{\eta}). \tag{2.70}$$

Then, the weight vector $\boldsymbol{w}(\boldsymbol{r})$ is redefined as the weight associated with this optimum orientation $\boldsymbol{\eta}_{opt}(\boldsymbol{r})$, i.e.,

$$\boldsymbol{w}(\boldsymbol{r}) = \boldsymbol{w}(\boldsymbol{r},\boldsymbol{\eta}_{opt}(\boldsymbol{r})). \tag{2.71}$$

We use this $\boldsymbol{w}(\boldsymbol{r})$ to reconstruct the source intensity at location $\boldsymbol{r}$ such that

$$\widehat{s}(\boldsymbol{r},t) = \boldsymbol{w}^T(\boldsymbol{r})\boldsymbol{b}(t). \tag{2.72}$$

The spatial filter that reconstructs source spatial distributions in this manner is called the scalar spatial filter. The output power of the scalar spatial filter is sometimes denoted $\widehat{P}_s(\boldsymbol{r})$ in this book, i.e.,

$$\widehat{P}_s(\boldsymbol{r}) = \boldsymbol{w}^T(\boldsymbol{r})\boldsymbol{R}\boldsymbol{w}(\boldsymbol{r}),$$

where the weight vector $\boldsymbol{w}(\boldsymbol{r})$ is obtained using Eq. (2.71).

Vector spatial filter

Another type of spatial filter used in bioelectromagnetic source imaging is the vector spatial filter, in which a set of three weight vectors $\boldsymbol{w}_x(\boldsymbol{r})$, $\boldsymbol{w}_y(\boldsymbol{r})$, and $\boldsymbol{w}_z(\boldsymbol{r})$, respectively, detect the x, y, and z components of the source vector $\boldsymbol{s}(\boldsymbol{r},t)$. That is, using the weight matrix $\boldsymbol{W}(\boldsymbol{r})$, defined as

$$\boldsymbol{W}(\boldsymbol{r}) = [\boldsymbol{w}_x(\boldsymbol{r}),\boldsymbol{w}_y(\boldsymbol{r}),\boldsymbol{w}_z(\boldsymbol{r})], \tag{2.73}$$

the source vector can be reconstructed by using

$$\widehat{\boldsymbol{s}}(\boldsymbol{r},t) = [\widehat{s}_x(\boldsymbol{r},t), \widehat{s}_y(\boldsymbol{r},t), \widehat{s}_z(\boldsymbol{r},t)]^T = [\boldsymbol{w}_x(\boldsymbol{r}), \boldsymbol{w}_y(\boldsymbol{r}), \boldsymbol{w}_z(\boldsymbol{r})]^T \boldsymbol{b}(t) = \boldsymbol{W}^T(\boldsymbol{r})\boldsymbol{b}(t). \tag{2.74}$$

The source power matrix $\widehat{\boldsymbol{\Sigma}}_s(\boldsymbol{r})$ is defined as

$$\widehat{\boldsymbol{\Sigma}}_s(\boldsymbol{r}) = \langle \widehat{\boldsymbol{s}}(\boldsymbol{r},t)\widehat{\boldsymbol{s}}^T(\boldsymbol{r},t)\rangle = \boldsymbol{W}^T(\boldsymbol{r})\boldsymbol{R}\boldsymbol{W}(\boldsymbol{r}). \tag{2.75}$$

In the vector formulation, there are two ways to compute the source power estimate $\langle \widehat{s}(\boldsymbol{r},t)^2\rangle$. Probably, the most natural way is to calculate the trace of $\widehat{\boldsymbol{\Sigma}}_s(\boldsymbol{r})$, such that

$$\langle \widehat{s}(\boldsymbol{r},t)^2\rangle = \mathrm{tr}\{\widehat{\boldsymbol{\Sigma}}_s(\boldsymbol{r})\} = \langle \widehat{s}_x(\boldsymbol{r},t)^2\rangle + \langle \widehat{s}_y(\boldsymbol{r},t)^2\rangle + \langle \widehat{s}_z(\boldsymbol{r},t)^2\rangle. \tag{2.76}$$

This power estimate from the vector formulation is called the power estimate of the first kind and is denoted $\widehat{P}_V^{(\mathrm{I})}(\boldsymbol{r})$, i.e.,

$$\widehat{P}_V^{(\mathrm{I})}(\boldsymbol{r}) = \mathrm{tr}\{\widehat{\boldsymbol{\Sigma}}_s(\boldsymbol{r})\}. \tag{2.77}$$

An alternative way is to compute the projection of the vector output onto the direction that gives the maximum power output. The power estimate computed in this manner is called the power estimate of the second kind, denoted as $\widehat{P}_V^{(\mathrm{II})}(\boldsymbol{r})$. To compute $\widehat{P}_V^{(\mathrm{II})}(\boldsymbol{r})$, the optimal direction $\bar{\boldsymbol{\eta}}_{opt}$ is first obtained using

$$\bar{\boldsymbol{\eta}}_{opt}(\boldsymbol{r}) = \underset{\boldsymbol{\eta}(\boldsymbol{r})}{\arg\max}\, \langle (\boldsymbol{\eta}^T \widehat{\boldsymbol{s}}(\boldsymbol{r},t))^2\rangle = \underset{\boldsymbol{\eta}(\boldsymbol{r})}{\arg\max} \left[\boldsymbol{\eta}^T \widehat{\boldsymbol{\Sigma}}_s(\boldsymbol{r})\boldsymbol{\eta}\right], \tag{2.78}$$

and the power estimate of the second kind is given by:

$$\widehat{P}_V^{(\mathrm{II})}(\boldsymbol{r}) = \bar{\boldsymbol{\eta}}_{opt}^T(\boldsymbol{r})\widehat{\boldsymbol{\Sigma}}_s(\boldsymbol{r})\bar{\boldsymbol{\eta}}_{opt}(\boldsymbol{r}) = \langle (\bar{\boldsymbol{\eta}}_{opt}^T(\boldsymbol{r})\widehat{\boldsymbol{s}}(\boldsymbol{r},t))^2\rangle. \tag{2.79}$$

These two power estimates are generally not equal for given vector spatial filters.

Equivalence between scalar and vector spatial filters

The scalar and the vector formulations can be considered equivalent if the relationships

$$\bar{\boldsymbol{\eta}}_{opt} = \boldsymbol{\eta}_{opt} \tag{2.80}$$

and

$$\widehat{P}_s(\boldsymbol{r}) = \widehat{P}_V^{(\mathrm{II})}(\boldsymbol{r}) \tag{2.81}$$

hold. Some types of spatial filters have this equivalence. For example, the minimum-norm filter in Chapter 3 and the unit-gain constraint minimum-variance filter in Chapter 4 are shown to have this equivalence. Nonetheless, the two types of formulations are not equivalent for most adaptive and non-adaptive spatial filters. On the other hand, our numerical experiments show that practical differences between the two formulations are generally small.

2.5.3 Resolution kernel, point-spread function, and beam response

Combining Eqs. (2.8) and (2.65), we derive (omitting explicit time notation t)

$$\widehat{s}(\boldsymbol{r}) = \int_{\Omega} \boldsymbol{w}^T(\boldsymbol{r})\boldsymbol{l}(\boldsymbol{r}')s(\boldsymbol{r}')\, d^3r' = \int_{\Omega} \mathbb{R}(\boldsymbol{r},\boldsymbol{r}')s(\boldsymbol{r}')\, d^3r'. \tag{2.82}$$

Here, we define $\mathbb{R}(\boldsymbol{r},\boldsymbol{r}')$ to be

$$\mathbb{R}(\boldsymbol{r},\boldsymbol{r}') = \boldsymbol{w}^T(\boldsymbol{r})\boldsymbol{l}(\boldsymbol{r}'). \tag{2.83}$$

This $\mathbb{R}(\boldsymbol{r},\boldsymbol{r}')$ is called the resolution kernel[17][18], and expresses the relationship between the original and the reconstructed source distributions. There are two ways to interpret the resolution kernel. One interpretation is to consider $\mathbb{R}(\boldsymbol{r},\boldsymbol{r}')$ a point-spread function, which is very useful for evaluating the location bias and the spatial resolution of various spatial filters. Assuming that a single point source exists at $\boldsymbol{r}_1$ and substituting $s(\boldsymbol{r}') = \delta(\boldsymbol{r}' - \boldsymbol{r}_1)$ into Eq. (2.82), we derive $\widehat{s}(\boldsymbol{r}) = \mathbb{R}(\boldsymbol{r},\boldsymbol{r}_1)$. Namely, if we plot the resolution kernel with respect to $\boldsymbol{r}$ (with a fixed $\boldsymbol{r}_1$), this plot, defined as

$$\mathcal{F}(\boldsymbol{r}) = \mathbb{R}(\boldsymbol{r},\boldsymbol{r}_1), \tag{2.84}$$

expresses the reconstruction of the point source located at $\boldsymbol{r}_1$, and this $\mathcal{F}(\boldsymbol{r})$ is called the point-spread function. In Chapter 5, we use the point-spread function to analyze the properties of the source bias and the spatial resolution of various types of spatial filters.

The other way of interpreting the resolution kernel is to consider $\mathbb{R}(\boldsymbol{r},\boldsymbol{r}')$ as the beam response. In this case, the resolution kernel is plotted with respect to $\boldsymbol{r}'$ (with a fixed $\boldsymbol{r}$), i.e.,

$$\mathcal{H}(\boldsymbol{r}') = \mathbb{R}(\boldsymbol{r},\boldsymbol{r}'). \tag{2.85}$$

The beam response, $\mathcal{H}(\boldsymbol{r}')$, indicates the sensitivity of the spatial filter that is pointing to $\boldsymbol{r}$ to a source located at $\boldsymbol{r}'$. In other words, the beam response represents the gain for the unwanted leakage signal from sources located at places other than the filter pointing location. In Section 5.4, we discuss the beam response further, and present some numerical examples.

Chapter 3

Tomographic reconstruction and nonadaptive spatial filters

This chapter introduces representative nonadaptive spatial filters used in electromagnetic source imaging. We first describe the minimum-norm least-squares method, which is one of most popular methods in electromagnetic source imaging. We then reformulate this minimum-norm method as a non-adaptive spatial filter, and introduce two variants of the minimum-norm spatial filter: the weight-normalized minimum-norm filter and standardized low-resolution electromagnetic tomography (sLORETA). With the reformulation as a spatial filter, we can compare these least-squares-based methods with the adaptive spatial filters using a common basis, such as the point-spread function or output SNR. Such comparisons are presented in Chapters 5 and 6.

3.1 Minimum-norm method

3.1.1 Tomographic reconstruction formulation

As shown in Fig. 2.2 (a), tomographic reconstruction methods introduce voxel discretization over the source space. By introducing the voxels, the source reconstruction problem can be solved based on the linear least-squares inverse. Let us define the number of voxels as N, and denote the locations of the voxels as $\boldsymbol{r}_1, \boldsymbol{r}_2, \ldots, \boldsymbol{r}_N$. Then, the discrete form of Eq. (2.7) is expressed as

$$\boldsymbol{b}(t) = \sum_{n=1}^{N} \boldsymbol{L}(\boldsymbol{r}_n)\boldsymbol{s}(\boldsymbol{r}_n, t). \tag{3.1}$$

We introduce a composite lead-field matrix over all voxel locations as

$$\boldsymbol{L}_V = [\boldsymbol{L}(\boldsymbol{r}_1), \boldsymbol{L}(\boldsymbol{r}_2), \ldots, \boldsymbol{L}(\boldsymbol{r}_N)]. \tag{3.2}$$

We define a $3N \times 1$ column vector containing the source vectors at all voxel locations, $\boldsymbol{\nu}_{vox}(t)$, such that

$$\boldsymbol{\nu}_{vox}(t) = \begin{bmatrix} \boldsymbol{s}(\boldsymbol{r}_1, t) \\ \boldsymbol{s}(\boldsymbol{r}_2, t) \\ \vdots \\ \boldsymbol{s}(\boldsymbol{r}_N, t) \end{bmatrix}. \tag{3.3}$$

Readers should not confuse the above $\boldsymbol{\nu}_{vox}(t)$ with $\boldsymbol{\nu}(t)$ in Eq. (2.28). The elements of $\boldsymbol{\nu}(t)$ are the source magnitudes at the source locations, whereas these of $\boldsymbol{\nu}_{vox}(t)$ are the source magnitudes at the voxel locations.

Equation (3.1) is then rewritten as

$$\boldsymbol{b}(t) = [\boldsymbol{L}(\boldsymbol{r}_1), \boldsymbol{L}(\boldsymbol{r}_2), \ldots, \boldsymbol{L}(\boldsymbol{r}_N)] \begin{bmatrix} \boldsymbol{s}(\boldsymbol{r}_1, t) \\ \boldsymbol{s}(\boldsymbol{r}_2, t) \\ \vdots \\ \boldsymbol{s}(\boldsymbol{r}_N, t) \end{bmatrix} = \boldsymbol{L}_V \boldsymbol{\nu}_{vox}(t). \tag{3.4}$$

Here, since the composite lead-field matrix $\boldsymbol{L}_V$ is a known quantity, the only unknown quantity is the $3N \times 1$ column vector, $\boldsymbol{\nu}_{vox}(t)$, and this can be estimated as the solution of the linear-least-squares inverse of Eq. (3.4). That is, defining the estimate of $\boldsymbol{\nu}_{vox}(t)$ as $\widehat{\boldsymbol{\nu}}_{vox}(t)$:

$$\widehat{\boldsymbol{\nu}}_{vox}(t) = \begin{bmatrix} \widehat{\boldsymbol{s}}(\boldsymbol{r}_1, t) \\ \widehat{\boldsymbol{s}}(\boldsymbol{r}_2, t) \\ \vdots \\ \widehat{\boldsymbol{s}}(\boldsymbol{r}_N, t) \end{bmatrix}, \tag{3.5}$$

and the least-squares cost function as

$$\mathcal{F} = \|\boldsymbol{b}(t) - \boldsymbol{L}_V \widehat{\boldsymbol{\nu}}_{vox}(t)\|^2, \tag{3.6}$$

the least-squares solution that minimizes the above cost function $\mathcal{F}$ is given by

$$\widehat{\boldsymbol{\nu}}_{vox}(t) = \boldsymbol{L}_V^{+} \boldsymbol{b}(t), \tag{3.7}$$

where $\boldsymbol{L}_V^{+}$ indicates the generalized inverse of $\boldsymbol{L}_V$. Since $\boldsymbol{L}_V$ is an $M \times 3N$ matrix and generally $M < 3N$ holds, the system of linear equations in Eq. (3.4) is underdetermined. In such a case, the generalized inverse $\boldsymbol{L}_V^{+}$ is expressed as $\boldsymbol{L}_V^{+} = \boldsymbol{L}_V^T [\boldsymbol{L}_V \boldsymbol{L}_V^T]^{-1}$, and $\widehat{\boldsymbol{\nu}}_{vox}(t)$ is given by:

$$\widehat{\boldsymbol{\nu}}_{vox}(t) = \boldsymbol{L}_V^T [\boldsymbol{L}_V \boldsymbol{L}_V^T]^{-1} \boldsymbol{b}(t). \tag{3.8}$$

In the above equation, $\boldsymbol{L}_V \boldsymbol{L}_V^T$ is equal to the gram matrix defined in Eq. (2.10) if we ignore the errors resulting from the voxel discretization. That is,

$$\boldsymbol{G} = \int_\Omega \boldsymbol{L}(\boldsymbol{r})\boldsymbol{L}^T(\boldsymbol{r})d\boldsymbol{r} \approx \sum_{n=1}^{N} \boldsymbol{L}(\boldsymbol{r}_n)\boldsymbol{L}^T(\boldsymbol{r}_n) = \boldsymbol{L}_V \boldsymbol{L}_V^T. \tag{3.9}$$

Therefore, Equation (3.8) can be expressed as

$$\widehat{\boldsymbol{\nu}}_{vox}(t) = \boldsymbol{L}_V^T \boldsymbol{G}^{-1} \boldsymbol{b}(t). \tag{3.10}$$

The source reconstruction method represented by Eqs. (3.8) and (3.10) is called the minimum-norm reconstruction method, which has been introduced into the field of bioelectromagnetic imaging by Hämäläinen and Ilmoniemi[19] and is one of the most popular methods.

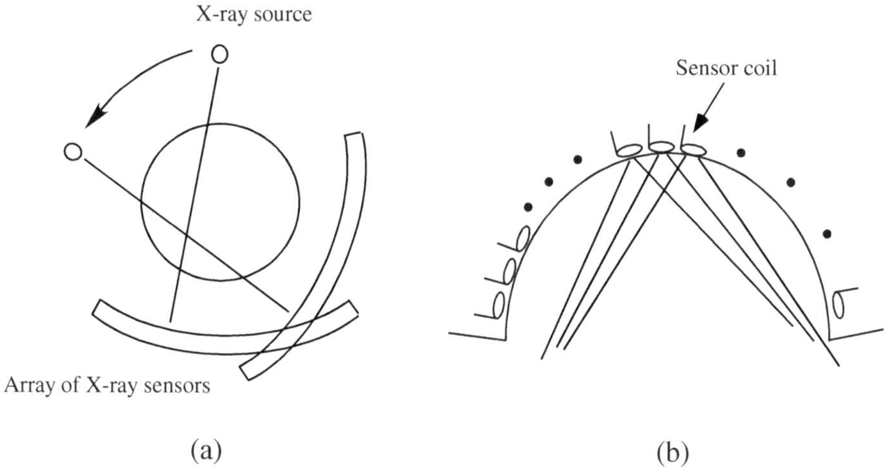

Figure 3.1: Schematic views of the sensitivity profiles of two types of sensor arrays. (a) The sensitivity profiles of sensors in X-ray computed tomography. The sensitivity profile of each sensor is the line connecting the sensor to the X-ray source. (b) The profiles of sensors for biomagnetic measurements.

Many methods successfully used in digital image processing can be proven to be equivalent to this minimum-norm method. For example, the filtered-back-projection algorithm used in X-ray computerized tomography(CT) is one such method[20][21]. The success of the minimum-norm method strongly depends on the property of the gram matrix, which in turn depends on the sensitivity profile of the sensor array. Figure 3.1 schematically illustrates the sensitivity profiles of the sensors for X-ray CT and for the biomagnetic sensor array. In the sensor array for X-ray CT, the sensitivity profile of each sensor consists of the line connecting the sensor to the X-ray source as shown in Fig. 3.1(a). Therefore, the overlap between the sensitivity profiles of different sensors is very small, and the gram matrix is close to the identity matrix. In such a case, the gram matrix is numerically stable since it has a low condition number. On the contrary, the sensors for the biomagnetic measurements have wide-spread sensitivity profiles and the sensitivity profiles of different sensors are more or less similar, as depicted in Fig. 3.1(b). In this case, the elements of $\boldsymbol{G}$ have similar numerical values and $\boldsymbol{G}$ is close to a singular matrix. Thus, calculating $\boldsymbol{G}^{-1}$ generally introduces some errors.

The errors generated when calculating $\boldsymbol{G}^{-1}$ can be reduced by applying a technique called regularization. The simplest form of regularization, called Tikhonov regularization[22][23], calculates $(\boldsymbol{G}+\epsilon\boldsymbol{I})^{-1}$, instead of directly calculating $\boldsymbol{G}^{-1}$. Here ϵ is a user-controllable scalar constant. Using Tikhonov regularization, the solution is expressed as

$$\widehat{\boldsymbol{\nu}}_{vox}(t) = \boldsymbol{L}_V^T(\boldsymbol{G}+\epsilon\boldsymbol{I})^{-1}\boldsymbol{b}(t). \tag{3.11}$$

Interestingly, this solution minimizes the cost function, expressed as

$$\mathcal{F} = \|\boldsymbol{b}(t) - \boldsymbol{L}_V\widehat{\boldsymbol{\nu}}_{vox}(t)\|^2 + \epsilon\|\widehat{\boldsymbol{\nu}}_{vox}(t)\|^2. \tag{3.12}$$

That is, Eq. (3.11) is the solution that minimizes both the norm of the solution vector, $\|\widehat{\boldsymbol{\nu}}_{vox}(t)\|^2$, as well as the squared error term $\|\boldsymbol{b}(t)-\boldsymbol{L}_V\widehat{\boldsymbol{\nu}}_{vox}(t)\|^2$. Because of this property, the solution obtained from Eq. (3.11) is called the minimum-norm solution.

Although the minimum-norm method is commonly used for electromagnetic brain imaging, its performance is known to be poor when applied to reconstructing three-dimensional source spatial distributions[24]. This poor performance problem can be overcome to some extent by using more sophisticated regularization techniques. That is, the general form of the least-squares cost function can be expressed as

$$\mathcal{F} = [\boldsymbol{b}(t) - \boldsymbol{L}_V\widehat{\boldsymbol{\nu}}_{vox}(t)]^T\boldsymbol{\Upsilon}_A^{-1}[\boldsymbol{b}(t) - \boldsymbol{L}_V\widehat{\boldsymbol{\nu}}_{vox}(t)] + \epsilon\widehat{\boldsymbol{\nu}}_{vox}(t)^T\boldsymbol{\Upsilon}_B^{-1}\widehat{\boldsymbol{\nu}}_{vox}(t). \tag{3.13}$$

The solution that minimizes the cost function above is given by:

$$\widehat{\boldsymbol{\nu}}_{vox}(t) = \boldsymbol{\Upsilon}_B\boldsymbol{L}_V^T[\boldsymbol{L}_V\boldsymbol{\Upsilon}_B\boldsymbol{L}_V^T + \epsilon\boldsymbol{\Upsilon}_A]^{-1}\boldsymbol{b}(t). \tag{3.14}$$

It is easy to see that the minimum-norm solution can be obtained by using $\boldsymbol{\Upsilon}_A = \boldsymbol{I}$ and $\boldsymbol{\Upsilon}_B = \boldsymbol{I}$ in Eq. (3.14). The equation above uses the modified gram matrix, $\widetilde{\boldsymbol{G}}$, which is defined as

$$\widetilde{\boldsymbol{G}} = \boldsymbol{L}_V\boldsymbol{\Upsilon}_B\boldsymbol{L}_V^T + \epsilon\boldsymbol{\Upsilon}_A. \tag{3.15}$$

We then have more freedom to improve the numerical stability of the gram matrix, compared to the simple Tikhonov regularization. Therefore, if we use appropriate choices for matrices $\boldsymbol{\Upsilon}_A$ and $\boldsymbol{\Upsilon}_B$, we can improve the quality of the solution. Generally, an appropriate $\boldsymbol{\Upsilon}_A$ should be chosen so as to reflect the property of the noise contained in measurements and an appropriate $\boldsymbol{\Upsilon}_B$ should be chosen so as to reflect the desired property of the source distribution.

3.1.2 Nonadaptive spatial-filter formulation

The minimum-norm method described in the preceding section can also be formulated as a non-adaptive spatial filter. Let us rewrite Eq. (3.10) as

$$\begin{bmatrix} \widehat{s}(\boldsymbol{r}_1,t) \\ \widehat{s}(\boldsymbol{r}_2,t) \\ \vdots \\ \widehat{s}(\boldsymbol{r}_N,t) \end{bmatrix} = \boldsymbol{L}_V^T\boldsymbol{G}^{-1}\boldsymbol{b}(t) = \begin{bmatrix} \boldsymbol{L}^T(\boldsymbol{r}_1) \\ \boldsymbol{L}^T(\boldsymbol{r}_2) \\ \vdots \\ \boldsymbol{L}^T(\boldsymbol{r}_N) \end{bmatrix} \boldsymbol{G}^{-1}\boldsymbol{b}(t). \tag{3.16}$$

Therefore, at each voxel location $\boldsymbol{r}_n$, the relationship,

$$\widehat{s}(\boldsymbol{r}_n,t) = \boldsymbol{L}^T(\boldsymbol{r}_n)\boldsymbol{G}^{-1}\boldsymbol{b}(t), \tag{3.17}$$

holds. The equation above is equivalent to the vector spatial-filter formulation in Eq. (2.74) where the weight matrix $\boldsymbol{W}(\boldsymbol{r})$ is given by:

$$\boldsymbol{W}(\boldsymbol{r}) = \boldsymbol{G}^{-1}\boldsymbol{L}(\boldsymbol{r}). \tag{3.18}$$

When the source orientation at each voxel is predetermined by some means, (for example by using the subject's MRI, or by using the optimal orientation described in Section 2.5.2,) we can derive the scalar form of the filter such that

$$\boldsymbol{w}(\boldsymbol{r}) = \boldsymbol{G}^{-1}\boldsymbol{l}(\boldsymbol{r}). \tag{3.19}$$

The formula to compute the optimum orientation for the scalar spatial-filter weight above is given in Eq. (13.18) in the Appendix. The spatial filter in Eqs. (3.18) or (3.19) is called the minimum-norm filter[1].

The general form of the solution in Eq. (3.14) is expressed as

$$\begin{bmatrix} \widehat{s}(\boldsymbol{r}_1,t) \\ \widehat{s}(\boldsymbol{r}_2,t) \\ \vdots \\ \widehat{s}(\boldsymbol{r}_N,t) \end{bmatrix} = \boldsymbol{\Upsilon}_B\boldsymbol{L}_V^T[\boldsymbol{L}_V\boldsymbol{\Upsilon}_B\boldsymbol{L}_V^T + \epsilon\boldsymbol{\Upsilon}_A]^{-1}\boldsymbol{b}(t) = \begin{bmatrix} \widetilde{\boldsymbol{L}}^T(\boldsymbol{r}_1) \\ \widetilde{\boldsymbol{L}}^T(\boldsymbol{r}_2) \\ \vdots \\ \widetilde{\boldsymbol{L}}^T(\boldsymbol{r}_N) \end{bmatrix} \widetilde{\boldsymbol{G}}^{-1}\boldsymbol{b}(t), \tag{3.20}$$

where $\widetilde{\boldsymbol{L}}^T(\boldsymbol{r}_n)$ is an $M \times 3$ matrix whose three columns are equal to the $3(n-1)+1$, $3(n-1)+2$, and $3n$th rows of the matrix $\boldsymbol{\Upsilon}_B\boldsymbol{L}_V^T$. Also, $\widetilde{\boldsymbol{G}}$ is defined in Eq. (3.15). The vector spatial-filter formulation of Eq. (3.20) is given by:

$$\widehat{s}(\boldsymbol{r}_n,t) = \widetilde{\boldsymbol{L}}^T(\boldsymbol{r}_n)\widetilde{\boldsymbol{G}}^{-1}\boldsymbol{b}(t) = \boldsymbol{W}^T(\boldsymbol{r}_n)\boldsymbol{b}(t), \tag{3.21}$$

where the weight matrix is expressed as

$$\boldsymbol{W}(\boldsymbol{r}) = \widetilde{\boldsymbol{G}}^{-1}\widetilde{\boldsymbol{L}}(\boldsymbol{r}). \tag{3.22}$$

[1] In Eqs. (3.18) and (3.19), the use of regularization is not explicitly indicated. However, Tikhonov regularization is generally applied when implementing these spatial filters.

The corresponding weight for the scalar spatial filter is expressed as

$$w(r) = \widetilde{G}^{-1}\widetilde{l}(r), \tag{3.23}$$

where $\widetilde{l}(r) = \widetilde{L}(r)\eta(r)$ and $\eta(r)$ is the predetermined source orientation at r.

3.2 Variants of the minimum-norm filter

3.2.1 Weight-normalized minimum-norm filter

The weight-normalized minimum-norm filter has been proposed by Dale *et al.* [25]. The idea is to normalize the minimum-norm weight with its norm to ensure that the spatial distribution of the noise is uniform. The scalar form weight is thus given by:

$$w(r) = \frac{G^{-1}l(r)}{\|G^{-1}l(r)\|} = \frac{G^{-1}l(r)}{\sqrt{l^T(r)G^{-2}l(r)}}. \tag{3.24}$$

The formula to compute the optimum orientation for the scalar spatial-filter weight above is given in Eq. (13.20) in the Appendix. The idea of weight normalization can be extended to derive the vector-form weight matrix, such that

$$W(r) = \frac{G^{-1}L(r)}{\sqrt{\mathrm{tr}\{L^T(r)G^{-2}L(r)\}}}, \tag{3.25}$$

where $\mathrm{tr}\{\,\cdot\,\}$ indicates the trace operation. Using this weight matrix, the source vector can be derived as

$$\widehat{s}(r,t) = W^T(r)b(t) = \frac{L^T(r)G^{-1}b(t)}{\sqrt{\mathrm{tr}\{L^T(r)G^{-2}L(r)\}}}. \tag{3.26}$$

3.2.2 sLORETA filter

A method called standardized low-resolution electromagnetic tomography (sLORETA) has been proposed by Pascual-Marqui[26]. This method can be reformulated as a non-adaptive spatial filter. In this method, the minimum-norm reconstruction results are normalized by the square root of the resolution kernel's peak value. The peak value of the resolution kernel of the minimum norm filter is expressed as

$$\mathbb{R}(r,r) = w^T(r)l(r) = l^T(r)G^{-1}l(r), \tag{3.27}$$

and the scalar version of sLORETA spatial filter is expressed as

$$w(r) = \frac{G^{-1}l(r)}{\sqrt{l^T(r)G^{-1}l(r)}}. \tag{3.28}$$

The formula to compute the optimum orientation for the scalar spatial-filter weight above is given in Eq. (13.22) in the Appendix. The resolution kernel of the vector minimum-norm filter in Eq. (3.18) is obtained as the 3×3 matrix,

$$\boldsymbol{W}^T(\boldsymbol{r})\boldsymbol{L}(\boldsymbol{r}) = \boldsymbol{L}^T(\boldsymbol{r})\boldsymbol{G}^{-1}\boldsymbol{L}(\boldsymbol{r}), \tag{3.29}$$

and the weight matrix for the vector sLORETA filter is expressed as

$$\boldsymbol{W}(\boldsymbol{r}) = \boldsymbol{G}^{-1}\boldsymbol{L}(\boldsymbol{r})[\boldsymbol{L}^T(\boldsymbol{r})\boldsymbol{G}^{-1}\boldsymbol{L}(\boldsymbol{r})]^{-1/2}. \tag{3.30}$$

Thus, the source estimate $\widehat{\boldsymbol{s}}(\boldsymbol{r},t)$ is obtained as

$$\widehat{\boldsymbol{s}}(\boldsymbol{r},t) = \boldsymbol{W}^T(\boldsymbol{r})\boldsymbol{b}(t) = [\boldsymbol{L}^T(\boldsymbol{r})\boldsymbol{G}^{-1}\boldsymbol{L}(\boldsymbol{r})]^{-1/2}\boldsymbol{L}^T(\boldsymbol{r})\boldsymbol{G}^{-1}\boldsymbol{b}(t). \tag{3.31}$$

The rationale of the sLORETA weight can be explained in the following manner. We consider a family of spatial-filter weights, with a form expressed as

$$\boldsymbol{w}(\boldsymbol{r}) = \zeta(\boldsymbol{r})\boldsymbol{G}^{-1}\boldsymbol{l}(\boldsymbol{r}), \tag{3.32}$$

where $\zeta(\boldsymbol{r})$ is a scalar coefficient that depends on the filter-pointing location $\boldsymbol{r}$. In this book, we refer to this family of spatial filters as minimum-norm-based spatial filters. We are looking for a coefficient $\zeta(\boldsymbol{r})$ that eliminates source location bias. That is, designating two spatial locations as $\boldsymbol{r}$ and $\boldsymbol{r}'$, when a single unit-magnitude source exists at $\boldsymbol{r}$, the filter weight should satisfy the relationship,

$$\boldsymbol{w}^T(\boldsymbol{r})\boldsymbol{l}(\boldsymbol{r}) \geq \boldsymbol{w}^T(\boldsymbol{r}')\boldsymbol{l}(\boldsymbol{r}), \tag{3.33}$$

for all $\boldsymbol{r}'$ not equal to $\boldsymbol{r}$. This relationship is equal to

$$\zeta(\boldsymbol{r})\boldsymbol{l}^T(\boldsymbol{r})\boldsymbol{G}^{-1}\boldsymbol{l}(\boldsymbol{r}) \geq \zeta(\boldsymbol{r}')\boldsymbol{l}^T(\boldsymbol{r}')\boldsymbol{G}^{-1}\boldsymbol{l}(\boldsymbol{r}). \tag{3.34}$$

When a single unit-magnitude source exists at $\boldsymbol{r}'$, the filter weight should satisfy the relationship,

$$\zeta(\boldsymbol{r}')\boldsymbol{l}^T(\boldsymbol{r}')\boldsymbol{G}^{-1}\boldsymbol{l}(\boldsymbol{r}') \geq \zeta(\boldsymbol{r})\boldsymbol{l}^T(\boldsymbol{r})\boldsymbol{G}^{-1}\boldsymbol{l}(\boldsymbol{r}'), \tag{3.35}$$

Combining Eqs (3.34) and (3.35), we can derive a pair of bounds on the ratio $\zeta(\boldsymbol{r})/\zeta(\boldsymbol{r}')$, such that

$$\frac{\boldsymbol{l}^T(\boldsymbol{r}')\boldsymbol{G}^{-1}\boldsymbol{l}(\boldsymbol{r}')}{\boldsymbol{l}^T(\boldsymbol{r})\boldsymbol{G}^{-1}\boldsymbol{l}(\boldsymbol{r}')} \geq \frac{\zeta(\boldsymbol{r})}{\zeta(\boldsymbol{r}')} \geq \frac{\boldsymbol{l}^T(\boldsymbol{r}')\boldsymbol{G}^{-1}\boldsymbol{l}(\boldsymbol{r})}{\boldsymbol{l}^T(\boldsymbol{r})\boldsymbol{G}^{-1}\boldsymbol{l}(\boldsymbol{r})}. \tag{3.36}$$

One natural choice for $\zeta(\boldsymbol{r})/\zeta(\boldsymbol{r}')$ that satisfies the above relationship is the geometric mean of the bounds, in which

$$\frac{\zeta(\boldsymbol{r})}{\zeta(\boldsymbol{r}')} = \sqrt{\frac{\boldsymbol{l}^T(\boldsymbol{r}')\boldsymbol{G}^{-1}\boldsymbol{l}(\boldsymbol{r}')}{\boldsymbol{l}^T(\boldsymbol{r})\boldsymbol{G}^{-1}\boldsymbol{l}(\boldsymbol{r}')}\frac{\boldsymbol{l}^T(\boldsymbol{r}')\boldsymbol{G}^{-1}\boldsymbol{l}(\boldsymbol{r})}{\boldsymbol{l}^T(\boldsymbol{r})\boldsymbol{G}^{-1}\boldsymbol{l}(\boldsymbol{r})}} = \sqrt{\frac{\boldsymbol{l}^T(\boldsymbol{r}')\boldsymbol{G}^{-1}\boldsymbol{l}(\boldsymbol{r}')}{\boldsymbol{l}^T(\boldsymbol{r})\boldsymbol{G}^{-1}\boldsymbol{l}(\boldsymbol{r})}}. \tag{3.37}$$

This can be achieved if the coefficient has the form

$$\zeta(\boldsymbol{r}) = 1/\sqrt{\boldsymbol{l}^T(\boldsymbol{r})\boldsymbol{G}^{-1}\boldsymbol{l}(\boldsymbol{r})}. \tag{3.38}$$

The resultant weight is exactly the same as in Eq. (3.28). The derivation here shows that the sLORETA filter has no localization bias. This property is further discussed in Section 5.1. The derivation also shows that the sLORETA weight is not the only choice for an unbiased solution and that there are other values of $\zeta(\boldsymbol{r})$ that satisfy Eq. (3.36).

As mentioned, the gram matrix $\boldsymbol{G}$ generally has a large condition number for a typical sensor array used in bioelectromagnetic measurements, and this large condition number can cause a performance degradation or unstable performance particularly when the input data is noisy. This performance degradation can, to some extent, be reduced by applying the regularization technique when calculating $\boldsymbol{G}^{-1}$. The regularized versions of the weight-normalized minimum-norm filter and the sLORETA filter are given, respectively, by:

$$\boldsymbol{w}(\boldsymbol{r}) = \frac{(\boldsymbol{G}+\epsilon\boldsymbol{I})^{-1}\boldsymbol{l}(\boldsymbol{r})}{\sqrt{\boldsymbol{l}^T(\boldsymbol{r})(\boldsymbol{G}+\epsilon\boldsymbol{I})^{-2}\boldsymbol{l}(\boldsymbol{r})}}, \tag{3.39}$$

and

$$\boldsymbol{w}(\boldsymbol{r}) = \frac{(\boldsymbol{G}+\epsilon\boldsymbol{I})^{-1}\boldsymbol{l}(\boldsymbol{r})}{\sqrt{\boldsymbol{l}^T(\boldsymbol{r})(\boldsymbol{G}+\epsilon\boldsymbol{I})^{-1}\boldsymbol{l}(\boldsymbol{r})}}. \tag{3.40}$$

When the regularization constant ϵ is large, both of these weights become

$$\boldsymbol{w}(\boldsymbol{r}) \approx \frac{\boldsymbol{l}(\boldsymbol{r})}{\|\boldsymbol{l}(\boldsymbol{r})\|}. \tag{3.41}$$

The above weight is exactly equal to the spatial matched filter described in the following section. Therefore, the regularized versions of the weight-normalized minimum-norm and sLORETA filters give intermediate performance between their non-regularized versions and the spatial matched filter. When the regularization constant ϵ is increased, their performances approach that of the spatial matched filter.

3.3 Spatial matched filter

The simplest non-adaptive spatial filter can be derived by simply using $\boldsymbol{L}(\boldsymbol{r})$ for the weight matrix $\boldsymbol{W}(\boldsymbol{r})$, i.e.,

$$\boldsymbol{W}(\boldsymbol{r}) = \boldsymbol{L}(\boldsymbol{r})/\|\boldsymbol{L}(\boldsymbol{r})\|. \tag{3.42}$$

The scalar form is expressed as

$$\boldsymbol{w}(\boldsymbol{r}) = \boldsymbol{l}(\boldsymbol{r})/\|\boldsymbol{l}(\boldsymbol{r})\|. \tag{3.43}$$

The formula to compute the optimum orientation for the scalar spatial-filter weight above is given in Eq. (13.24) in the Appendix. This type of spatial filter is called the spatial matched filter or the conventional spatial filter[27].

This spatial filter is almost equivalent to the technique called single-dipole scanning or single-dipole search. In single-dipole search, the least-squares error function defined as

$$\mathcal{F} = \| \left[\boldsymbol{I} - \boldsymbol{l}(\boldsymbol{r})[\boldsymbol{l}^T(\boldsymbol{r})\boldsymbol{l}(\boldsymbol{r})]^{-1}\boldsymbol{l}^T(\boldsymbol{r}) \right] \boldsymbol{b}(t)\|^2 = \|\boldsymbol{b}(t)\|^2 - [\boldsymbol{l}^T(\boldsymbol{r})\boldsymbol{b}(t)]^2/\|\boldsymbol{l}(\boldsymbol{r})\|^2$$

is computed at all voxel locations and the location that minimizes this $\mathcal{F}$ is selected as the source location. (In actual implementation, the above-mentioned exhaustive search is replaced by a more efficient search method to detect the source location.) The cost function above obviously shows that the location that minimizes the cost function is equal to the location that maximizes the inner product $|\boldsymbol{l}^T(\boldsymbol{r})\boldsymbol{b}(t)|/\|\boldsymbol{l}(\boldsymbol{r})\|$. Namely, the peak location in the spatial-matched-filter map is equal to the dipole location found by minimizing the least-squares cost function.

In the following chapters, we show that the spatial matched filter has some desirable properties such as no location bias even in the presence of noise, and no SNR degradation in the reconstruction process. Although its spatial resolution is significantly lower than that of the other spatial filters, methods equivalent to the spatial matched filter have been developed in various fields due to its simplicity. Such methods include the delay-and-sum beamformer[28] used in radar applications and the backprojection operation used in X-ray CT image reconstruction[20].

3.4 Deriving the minimum-norm-based filters using leakage minimization

In Section 3.1, we derive the minimum-norm and related filter methods based on the least-squares principle. In Section 3.2, the weight-normalized minimum-norm and sLORETA filters are derived along with some modifications of the minimum-norm filter. In this section, we present different derivations for these spatial filters based on leakage minimization[29][18]. The derivations here have a close similarity to the derivations of the adaptive spatial filters described in Chapter 4. Therefore, we recommend that readers first read Chapter 4, and then come back to this section.

For a spatial filter pointing at $\boldsymbol{r}$, the leakage from a source located at $\boldsymbol{r}_1$ is expressed using the beam response $\mathcal{H}(\boldsymbol{r}_1) = \boldsymbol{w}^T(\boldsymbol{r})\boldsymbol{l}(\boldsymbol{r}_1)$, as mentioned in Section 2.5.3. We then wish to derive a spatial filter that only passes the signal from a source at the pointing location $\boldsymbol{r}$ and suppresses the leakage from sources at other locations. Such a spatial filter may be derived by imposing a delta-function-like property on the beam response. That is, we define a cost function $\mathcal{F}$ such that

$$\mathcal{F} = \int_{\Omega} [\boldsymbol{w}^T(\boldsymbol{r})\boldsymbol{l}(\boldsymbol{r}_1) - \delta(\boldsymbol{r} - \boldsymbol{r}_1)]^2 \, d^3 r_1. \tag{3.44}$$

We wish to find the $\boldsymbol{w}(\boldsymbol{r})$ that minimizes this $\mathcal{F}$, such that

$$\boldsymbol{w}(\boldsymbol{r}) = \underset{\boldsymbol{w}(\boldsymbol{r})}{\arg\min}\, \mathcal{F}. \tag{3.45}$$

To obtain the weight, we calculate the derivative $\partial\mathcal{F}/\partial\boldsymbol{w}(\boldsymbol{r})$ and set it to zero, i.e.,

$$\begin{aligned}\frac{\partial \mathcal{F}}{\partial \boldsymbol{w}(\boldsymbol{r})} &= 2\int_\Omega \left[\boldsymbol{w}^T(\boldsymbol{r})\boldsymbol{l}(\boldsymbol{r}_1) - \delta(\boldsymbol{r}-\boldsymbol{r}_1)\right]\boldsymbol{l}^T(\boldsymbol{r}_1)\, d^3r_1 \\ &= 2[\boldsymbol{w}^T(\boldsymbol{r})\int_\Omega \boldsymbol{l}(\boldsymbol{r})\boldsymbol{l}^T(\boldsymbol{r})\, d^3r - \boldsymbol{l}^T(\boldsymbol{r})] = 2[\boldsymbol{w}^T(\boldsymbol{r})\boldsymbol{G} - \boldsymbol{l}^T(\boldsymbol{r})] = 0. \end{aligned} \tag{3.46}$$

From this, we can derive the weight vector expressed as $\boldsymbol{w}(\boldsymbol{r}) = \boldsymbol{G}^{-1}\boldsymbol{l}(\boldsymbol{r})$, which is exactly the same as the minimum-norm filter shown in Eq. (3.19).

Other minimum-norm-based filters presented in Section 3.2 can also be derived by a similar minimization formulation. The derivations here are according to Greenblatt *et al.* [30]. In these derivations, we first calculate the total leakage. The total leakage for the filter pointing at $\boldsymbol{r}$ is given by:

$$\int_\Omega \mathbb{R}(\boldsymbol{r},\boldsymbol{r}_1)^2\, d^3r_1 = \boldsymbol{w}^T(\boldsymbol{r})\left[\int_\Omega \boldsymbol{l}(\boldsymbol{r}_1)\boldsymbol{l}^T(\boldsymbol{r}_1)\, d^3r_1\right]\boldsymbol{w}(\boldsymbol{r}) = \boldsymbol{w}^T(\boldsymbol{r})\boldsymbol{G}\boldsymbol{w}(\boldsymbol{r}). \tag{3.47}$$

The weight-normalized minimum-norm filter mentioned in Section 3.2 is derived using

$$\boldsymbol{w}(\boldsymbol{r}) = \underset{\boldsymbol{w}(\boldsymbol{r})}{\arg\min}\, \boldsymbol{w}^T(\boldsymbol{r})\boldsymbol{G}\boldsymbol{w}(\boldsymbol{r}), \quad \text{subject to} \quad \boldsymbol{w}^T(\boldsymbol{r})\boldsymbol{w}(\boldsymbol{r}) = 1. \tag{3.48}$$

Readers may notice a close similarity between the above formulation and that for the unit-noise-gain minimum-variance filter in Eq. (4.11). The only difference is that the former uses the gram matrix, $\boldsymbol{G}$, in Eq. (3.48), instead of using the covariance matrix, $\boldsymbol{R}$, in Eq. (4.11). Therefore, using exactly the same derivation as that for Eq. (4.15), we derive the weight in Eq. (3.24). Despite the superficial similarity between these two types of spatial filters, their performance greatly differs. The unit-noise-gain minimum-variance filter is able to obtain unbiased estimates of the source locations, whereas the weight-normalized minimum-norm filter cannot, as shown in Chapter 5.

The weight for sLORETA in Eq. (3.28) can also be derived using the minimization:

$$\begin{aligned}\boldsymbol{w}(\boldsymbol{r}) = \underset{\boldsymbol{w}(\boldsymbol{r})}{\arg\min}\, \boldsymbol{w}^T(\boldsymbol{r})\boldsymbol{G}\boldsymbol{w}(\boldsymbol{r}), \quad &\text{subject to} \quad \boldsymbol{w}^T(\boldsymbol{r})\boldsymbol{l}(\boldsymbol{r}) = \tau, \\ &\text{and} \quad \boldsymbol{w}^T(\boldsymbol{r})\boldsymbol{G}\boldsymbol{w}(\boldsymbol{r}) = 1.\end{aligned} \tag{3.49}$$

In the above formulation, the minimization problem is solved with the first constraint, $\boldsymbol{w}^T(\boldsymbol{r})\boldsymbol{l}(\boldsymbol{r}) = \tau$, and the scalar constant τ is determined by the second constraint, $\boldsymbol{w}^T(\boldsymbol{r})\boldsymbol{G}\boldsymbol{w}(\boldsymbol{r}) = 1$. It is easy to see that, using the derivation similar to that in Section 4.1.3, the weight for the sLORETA filter in Eq. (3.28) can be derived. However, in this derivation, the constraint $\boldsymbol{w}^T(\boldsymbol{r})\boldsymbol{G}\boldsymbol{w}(\boldsymbol{r}) = 1$ is not easily understood so that the meaning of this constraint is not entirely clear.

Chapter 4

Adaptive spatial filters

This chapter introduces adaptive spatial filters used in electromagnetic brain imaging. We first describe basic formulations to derive minimum-variance spatial filters. We then discuss prerequisites for formulating minimum-variance spatial filters, concerned with the source correlation and signal rank. The scalar and the vector formulations of minimum-variance spatial filters are also described.

4.1 Deriving weights for adaptive spatial filters

4.1.1 Minimum-variance spatial filter with the unit-gain constraint

Let us derive the weight of the minimum-variance spatial filter [28][31][7], which is the best-known adaptive spatial filter. The weight vector $\boldsymbol{w}(\boldsymbol{r})$ is derived as the one which minimizes the output power $\boldsymbol{w}^T(\boldsymbol{r})\boldsymbol{R}\boldsymbol{w}(\boldsymbol{r})$ under the constraint that $\boldsymbol{w}^T(\boldsymbol{r})\boldsymbol{l}(\boldsymbol{r}) = 1$, i.e.,

$$\boldsymbol{w}(\boldsymbol{r}) = \underset{\boldsymbol{w}(\boldsymbol{r})}{\arg\min}\, \boldsymbol{w}^T(\boldsymbol{r})\boldsymbol{R}\boldsymbol{w}(\boldsymbol{r}), \quad \text{subject to} \quad \boldsymbol{w}^T(\boldsymbol{r})\boldsymbol{l}(\boldsymbol{r}) = 1, \tag{4.1}$$

where $\boldsymbol{R}$ is the covariance matrix of the measurements. The inner product $\boldsymbol{w}^T(\boldsymbol{r})\boldsymbol{l}(\boldsymbol{r})$ represents the spatial filter output from a unit-magnitude source located at $\boldsymbol{r}$. Therefore, setting $\boldsymbol{w}^T(\boldsymbol{r})\boldsymbol{l}(\boldsymbol{r})$ equal to one guarantees that the spatial filter passes the signal from $\boldsymbol{r}$ with the gain equal to one (i.e., a unit gain). The output power of the spatial filter $\boldsymbol{w}^T(\boldsymbol{r})\boldsymbol{R}\boldsymbol{w}(\boldsymbol{r})$ generally contains not only the noise contributions but also unwanted contributions such as the influence of sources at locations other than $\boldsymbol{r}$. Accordingly, by minimizing the output power with this unit-gain constraint, we can derive a weight that minimizes such unwanted influence without affecting the signal coming from $\boldsymbol{r}$, the pointing location of the spatial filter.

Let us derive an explicit form of the weight vector by solving the minimization problem in Eq. (4.1). This constrained minimization problem can be solved using the method of the Lagrange multiplier. We define the Lagrange multiplier as a scalar κ, and the Lagrangian as $\mathbb{L}$, such that

$$\mathbb{L}(\boldsymbol{w},\kappa) = \boldsymbol{w}^T\boldsymbol{R}\boldsymbol{w} + \kappa(\boldsymbol{w}^T\boldsymbol{l}(\boldsymbol{r}) - 1), \tag{4.2}$$

where the explicit notation of $(\boldsymbol{r})$ is omitted from $\boldsymbol{w}(\boldsymbol{r})$ for simplicity. Note that because $\boldsymbol{R}$ is a positive definite matrix and vectors $\boldsymbol{w}$ and $\boldsymbol{l}(\boldsymbol{r})$ are real-valued, the Lagrangian $\mathbb{L}(\boldsymbol{w},\kappa)$ is a real-valued function. The weight vector satisfying Eq. (4.1) minimizes the Lagrangian $\mathbb{L}(\boldsymbol{w},\kappa)$ in Eq. (4.2) with no constraints.

The derivative of $\mathbb{L}(\boldsymbol{w},\kappa)$ with respect to $\boldsymbol{w}$ is given by:

$$\frac{\partial \mathbb{L}(\boldsymbol{w},\kappa)}{\partial \boldsymbol{w}} = 2\boldsymbol{R}\boldsymbol{w} + \kappa\boldsymbol{l}(\boldsymbol{r}). \tag{4.3}$$

By setting the right-hand side of the above equation to zero, we obtain

$$\boldsymbol{w} = -\kappa\boldsymbol{R}^{-1}\boldsymbol{l}(\boldsymbol{r})/2. \tag{4.4}$$

Thus, substituting this relationship back into the constraint equation $\boldsymbol{w}^T\boldsymbol{l}(\boldsymbol{r}) = 1$, we get $\kappa = -2/[\boldsymbol{l}^T(\boldsymbol{r})\boldsymbol{R}^{-1}\boldsymbol{l}(\boldsymbol{r})]$. Substituting this κ into Eq. (4.4), the weight vector satisfying Eq. (4.1) is given by:

$$\boldsymbol{w}(\boldsymbol{r}) = \frac{\boldsymbol{R}^{-1}\boldsymbol{l}(\boldsymbol{r})}{[\boldsymbol{l}^T(\boldsymbol{r})\boldsymbol{R}^{-1}\boldsymbol{l}(\boldsymbol{r})]}. \tag{4.5}$$

Using the above weight expression and Eq. (2.65), the spatial filter output is expressed as

$$\widehat{s}(\boldsymbol{r},t) = \frac{\boldsymbol{l}^T(\boldsymbol{r})\boldsymbol{R}^{-1}\boldsymbol{b}(t)}{[\boldsymbol{l}^T(\boldsymbol{r})\boldsymbol{R}^{-1}\boldsymbol{l}(\boldsymbol{r})]}. \tag{4.6}$$

Using Eq. (2.66), the output power of this spatial filter is expressed as

$$\langle \widehat{s}(\boldsymbol{r},t)^2 \rangle = \boldsymbol{w}^T(\boldsymbol{r})\boldsymbol{R}\boldsymbol{w}(\boldsymbol{r}) = \frac{1}{[\boldsymbol{l}^T(\boldsymbol{r})\boldsymbol{R}^{-1}\boldsymbol{l}(\boldsymbol{r})]}. \tag{4.7}$$

Substituting the weight expression in Eq. (4.5) into (2.68), the output SNR, $\mathcal{Z}$, is expressed as

$$\mathcal{Z} = \frac{1}{\sigma_0^2}\frac{\boldsymbol{l}^T(\boldsymbol{r})\boldsymbol{R}^{-1}\boldsymbol{l}(\boldsymbol{r})}{[\boldsymbol{l}^T(\boldsymbol{r})\boldsymbol{R}^{-2}\boldsymbol{l}(\boldsymbol{r})]}. \tag{4.8}$$

The adaptive spatial filter in Eq. (4.5) is derived with the unit-gain constraint. In this book, this type of spatial filter is referred to as the unit-gain (constraint) minimum-variance spatial filter. This spatial filter is sometimes called the minimum-variance distortionless spatial filter in the signal-processing community.

4.1.2 Minimum-variance spatial filter with the array-gain constraint

The constraint $\boldsymbol{w}^T(\boldsymbol{r})\boldsymbol{l}(\boldsymbol{r}) = 1$ is somewhat ad-hoc, and there may be other possibilities, depending on the characteristics of the problem to be solved. For bioelectromagnetic imaging, the norm of the lead-field vector $\|\boldsymbol{l}(\boldsymbol{r})\|$ has a spatial dependence. When the spherical homogeneous conductor model[1] is used for deriving the lead field, $\|\boldsymbol{l}(\boldsymbol{r})\|$ is zero at the center of the sphere. This causes a false intensity increase around the center of the sphere in the source reconstruction results obtained using the unit-gain constraint minimum-variance filter, because the weight becomes infinity at the center of the sphere. We refer to such false intensity increases caused by the non-uniformity of $\|\boldsymbol{l}(\boldsymbol{r})\|$ as lead-field-norm artifacts. We will show an example of these artifacts in Fig. 4.4 of our numerical experiments.

When $\|\boldsymbol{l}(\boldsymbol{r})\|$ has a spatial dependence, it is more reasonable to use the constraint $\boldsymbol{w}^T(\boldsymbol{r})\boldsymbol{l}(\boldsymbol{r}) = \|\boldsymbol{l}(\boldsymbol{r})\|$, instead of using $\boldsymbol{w}^T(\boldsymbol{r})\boldsymbol{l}(\boldsymbol{r}) = 1$. Because $\|\boldsymbol{l}(\boldsymbol{r})\|$ represents the gain of the sensor array, we derive, by using $\boldsymbol{w}^T(\boldsymbol{r})\boldsymbol{l}(\boldsymbol{r}) = \|\boldsymbol{l}(\boldsymbol{r})\|$, a spatial filter whose gain exactly matches the gain of the sensor array. Using exactly the same derivation for Eq. (4.5), the weight vector in this case is obtained as

$$\boldsymbol{w}(\boldsymbol{r}) = \frac{\boldsymbol{R}^{-1}\widetilde{\boldsymbol{l}}(\boldsymbol{r})}{[\widetilde{\boldsymbol{l}}^T(\boldsymbol{r})\boldsymbol{R}^{-1}\widetilde{\boldsymbol{l}}(\boldsymbol{r})]}, \tag{4.9}$$

where $\widetilde{\boldsymbol{l}}(\boldsymbol{r})$ is the normalized lead-field vector defined as $\widetilde{\boldsymbol{l}}(\boldsymbol{r}) = \boldsymbol{l}(\boldsymbol{r})/\|\boldsymbol{l}(\boldsymbol{r})\|$. In Eq. (4.9), the weight is independent of the norm of the lead field, and we can avoid the lead-field-norm artifacts. In this book, this type of spatial filter is referred to as the array-gain (constraint) minimum-variance spatial filter . The output power of this spatial filter is given by[2]:

$$\langle \widehat{s}(\boldsymbol{r},t)^2 \rangle = \frac{1}{[\widetilde{\boldsymbol{l}}^T(\boldsymbol{r})\boldsymbol{R}^{-1}\widetilde{\boldsymbol{l}}(\boldsymbol{r})]} = \frac{\boldsymbol{l}^T(\boldsymbol{r})\boldsymbol{l}(\boldsymbol{r})}{[\boldsymbol{l}^T(\boldsymbol{r})\boldsymbol{R}^{-1}\boldsymbol{l}(\boldsymbol{r})]}. \tag{4.10}$$

The $\mathcal{Z}$ value is expressed in exactly the same way as in Eq. (4.8) because $\mathcal{Z}$ is independent of the norm of the lead field.

4.1.3 Minimum-variance spatial filter with the unit-noise-gain constraint

Another possible constraint is the unit-noise-gain constraint expressed as $\boldsymbol{w}^T(\boldsymbol{r})\boldsymbol{w}(\boldsymbol{r}) = 1$. That is, the filter weight is obtained using

$$\begin{aligned} \boldsymbol{w}(\boldsymbol{r}) = \arg\min_{\boldsymbol{w}(\boldsymbol{r})} \boldsymbol{w}^T(\boldsymbol{r})\boldsymbol{R}\boldsymbol{w}(\boldsymbol{r}), \quad \text{subject to} \quad & \boldsymbol{w}^T(\boldsymbol{r})\boldsymbol{l}(\boldsymbol{r}) = \tau, \\ \text{and} \quad & \boldsymbol{w}^T(\boldsymbol{r})\boldsymbol{w}(\boldsymbol{r}) = 1, \end{aligned} \tag{4.11}$$

[1]The spherical homogeneous conductor model is described in Section 13.7 in the Appendix.
[2]The output power obtained in Eq. (4.10) is sometimes called the neural activity index[9].

where the minimization problem is solved with the first constraint, $\boldsymbol{w}^T(\boldsymbol{r})\boldsymbol{l}(\boldsymbol{r}) = \tau$, and the scalar constant τ is determined by the second constraint, $\boldsymbol{w}^T(\boldsymbol{r})\boldsymbol{w}(\boldsymbol{r}) = 1$. To obtain the weight vector derived from the above minimization, we first calculate the weight using

$$\boldsymbol{w}(\boldsymbol{r}) = \arg\min_{\boldsymbol{w}(\boldsymbol{r})} \boldsymbol{w}^T(\boldsymbol{r})\boldsymbol{R}\boldsymbol{w}(\boldsymbol{r}) \quad \text{subject to} \quad \boldsymbol{w}^T(\boldsymbol{r})\boldsymbol{l}(\boldsymbol{r}) = \tau. \tag{4.12}$$

Following the same exact steps from Eq. (4.2) to (4.5), the weight satisfying Eq. (4.12) is obtained as

$$\boldsymbol{w}(\boldsymbol{r}) = \tau \frac{\boldsymbol{R}^{-1}\boldsymbol{l}(\boldsymbol{r})}{[\boldsymbol{l}^T(\boldsymbol{r})\boldsymbol{R}^{-1}\boldsymbol{l}(\boldsymbol{r})]}. \tag{4.13}$$

Substituting this expression to $\boldsymbol{w}^T(\boldsymbol{r})\boldsymbol{w}(\boldsymbol{r}) = 1$, then, leads to

$$\tau = \frac{\boldsymbol{l}^T(\boldsymbol{r})\boldsymbol{R}^{-1}\boldsymbol{l}(\boldsymbol{r})}{\sqrt{\boldsymbol{l}^T(\boldsymbol{r})\boldsymbol{R}^{-2}\boldsymbol{l}(\boldsymbol{r})}}, \tag{4.14}$$

with the weight satisfying Eq. (4.11) given by:

$$\boldsymbol{w}(\boldsymbol{r}) = \frac{\boldsymbol{R}^{-1}\boldsymbol{l}(\boldsymbol{r})}{\sqrt{\boldsymbol{l}^T(\boldsymbol{r})\boldsymbol{R}^{-2}\boldsymbol{l}(\boldsymbol{r})}}. \tag{4.15}$$

This weight vector again does not depend on the norm of the lead field $\|\boldsymbol{l}(\boldsymbol{r})\|$. The output power of this spatial filter is given by

$$\langle \widehat{s}(\boldsymbol{r},t)^2 \rangle = \frac{\boldsymbol{l}^T(\boldsymbol{r})\boldsymbol{R}^{-1}\boldsymbol{l}(\boldsymbol{r})}{[\boldsymbol{l}^T(\boldsymbol{r})\boldsymbol{R}^{-2}\boldsymbol{l}(\boldsymbol{r})]}. \tag{4.16}$$

Comparing the above equation with Eq. (4.8), one can see that this spatial filter gives $\sigma_0^2\mathcal{Z}$ in Eq. (4.8) as its output power. This spatial filter was first proposed by Borgiotti and Kaplan[32] and, in this book, it is referred to as the unit-noise-gain (constraint) minimum-variance spatial filter .

4.2 Prerequisites for the adaptive spatial-filter formulation

4.2.1 Uncorrelated source time courses

As described in the preceding section, the weight vectors of the minimum-variance spatial filters are obtained by minimizing the output power $\boldsymbol{w}^T(\boldsymbol{r})\boldsymbol{R}\boldsymbol{w}(\boldsymbol{r})$ with a constraint. Let us look at this minimization process in detail. We assume that a total of Q sources are located at $\boldsymbol{r}_1, \boldsymbol{r}_2, \ldots, \boldsymbol{r}_Q$, and that their time courses are denoted $s(\boldsymbol{r}_1,t), s(\boldsymbol{r}_2,t), \ldots, s(\boldsymbol{r}_Q,t)$. Ignoring the noise term in Eq. (2.27) and assuming that the spatial filter is pointing at $\boldsymbol{r}_p$, the location of the pth source, the output power of the spatial filter is given by:

$$\boldsymbol{w}^T(\boldsymbol{r}_p)\boldsymbol{R}\boldsymbol{w}(\boldsymbol{r}_p) = \langle[\boldsymbol{w}^T(\boldsymbol{r}_p)\sum_{q=1}^{Q} s(\boldsymbol{r}_q,t)\boldsymbol{l}(\boldsymbol{r}_q)][\boldsymbol{w}^T(\boldsymbol{r}_p)\sum_{q=1}^{Q} s(\boldsymbol{r}_q,t)\boldsymbol{l}(\boldsymbol{r}_q)]^T\rangle. \quad (4.17)$$

Considering the unit-gain constraint $\boldsymbol{w}^T(\boldsymbol{r}_p)\boldsymbol{l}(\boldsymbol{r}_p) = 1$, we have

$$\begin{aligned}\boldsymbol{w}^T(\boldsymbol{r}_p)\boldsymbol{R}\boldsymbol{w}(\boldsymbol{r}_p) &= \langle s(\boldsymbol{r}_p,t)^2\rangle + \sum_{q\neq p}\langle s(\boldsymbol{r}_q,t)^2\rangle\|\boldsymbol{w}^T(\boldsymbol{r}_p)\boldsymbol{l}(\boldsymbol{r}_q)\|^2 \\ &+ \sum_{q_1\neq q_2}\langle s(\boldsymbol{r}_{q_1},t)s(\boldsymbol{r}_{q_2},t)\rangle\boldsymbol{w}^T(\boldsymbol{r}_p)\boldsymbol{l}(\boldsymbol{r}_{q_1})\boldsymbol{l}^T(\boldsymbol{r}_{q_2})\boldsymbol{w}(\boldsymbol{r}_p), \quad (4.18)\end{aligned}$$

where the notation $\sum_{i\neq j}$ indicates the summation over all combinations of the indices i and j except for the case of $i = j$. We assume that source activities are uncorrelated with each other, i.e.,

$$\langle s(\boldsymbol{r}_{p_1},t)s(\boldsymbol{r}_{p_2},t)\rangle = 0 \quad \text{for} \quad p_1 \neq p_2.$$

Then, the third term on the right-hand side of Eq. (4.18) becomes zero, and we have

$$\boldsymbol{w}^T(\boldsymbol{r}_p)\boldsymbol{R}\boldsymbol{w}(\boldsymbol{r}_p) = \langle s(\boldsymbol{r}_p,t)^2\rangle + \sum_{q\neq p}\langle s(\boldsymbol{r}_q,t)^2\rangle\|\boldsymbol{w}^T(\boldsymbol{r}_p)\boldsymbol{l}(\boldsymbol{r}_q)\|^2. \quad (4.19)$$

Therefore, the weight vector that minimizes the output power $\boldsymbol{w}^T(\boldsymbol{r}_p)\boldsymbol{R}\boldsymbol{w}(\boldsymbol{r}_p)$ satisfies the relationship $\boldsymbol{w}^T(\boldsymbol{r}_p)\boldsymbol{l}(\boldsymbol{r}_q) = 0$ where $q \neq p$. Using such weight vectors, we have

$$\boldsymbol{w}^T(\boldsymbol{r}_p)\boldsymbol{R}\boldsymbol{w}(\boldsymbol{r}_p) = \langle s(\boldsymbol{r}_p,t)^2\rangle. \quad (4.20)$$

In summary, the weight vector obtained by minimizing the output power with the unit-gain constraint has the property

$$\boldsymbol{w}^T(\boldsymbol{r}_p)\boldsymbol{l}(\boldsymbol{r}_q) = \delta_{p,q}, \quad (4.21)$$

where $\delta_{p,q}$ is Kronecker's delta, i.e., $\delta_{p,q} = 1$ when $p = q$ and $\delta_{p,q} = 0$ when $p \neq q$. Equation (4.21) indicates that the weight vector does not pass signals from sources at a location other than the filter pointing location, although it passes a signal from the pointing location with a gain equal to 1. The weight vectors of the adaptive spatial filter attain such performance without explicit information regarding the locations of other sources. This is because the covariance matrix $\boldsymbol{R}$ contains this information, and the adaptive spatial filters automatically utilize it.

We next consider the case where the spatial filter's pointing location $\boldsymbol{r}$ does not equal any of the source locations. Using the same derivation, the output power is expressed as

$$\boldsymbol{w}^T(\boldsymbol{r})\boldsymbol{R}\boldsymbol{w}(\boldsymbol{r}) = \langle s(\boldsymbol{r},t)^2\rangle + \sum_{q=1}^{Q}\langle s(\boldsymbol{r}_q,t)^2\rangle\|\boldsymbol{w}^T(\boldsymbol{r})\boldsymbol{l}(\boldsymbol{r}_q)\|^2, \quad (4.22)$$

and therefore the weight vector obtained by minimizing $\boldsymbol{w}^T(\boldsymbol{r})\boldsymbol{R}\boldsymbol{w}(\boldsymbol{r})$ satisfies the condition

$$\boldsymbol{w}^T(\boldsymbol{r})\boldsymbol{l}(\boldsymbol{r}_q) = 0 \quad (q = 1, \ldots, Q). \tag{4.23}$$

The output power obtained from this minimization is given by:

$$\boldsymbol{w}^T(\boldsymbol{r})\boldsymbol{R}\boldsymbol{w}(\boldsymbol{r}) = \langle s(\boldsymbol{r}, t)^2 \rangle = 0, \tag{4.24}$$

because there is no source at $\boldsymbol{r}$. In summary, the weight vector of the unit-gain-constraint minimum-variance filter has the following property:

$$\begin{aligned} \boldsymbol{w}^T(\boldsymbol{r})\boldsymbol{l}(\boldsymbol{r}_q) &= 1 \quad \text{when} \quad \boldsymbol{r} = \boldsymbol{r}_q, \\ \text{and} \quad \boldsymbol{w}^T(\boldsymbol{r})\boldsymbol{l}(\boldsymbol{r}_q) &= 0 \quad \text{when} \quad \boldsymbol{r} \neq \boldsymbol{r}_q, \end{aligned} \tag{4.25}$$

where $q = 1, \ldots, Q$, and $\boldsymbol{r}_q$ represents one of the source locations. Including the cases of other constraints, Eq. (4.21) changes to

$$\boldsymbol{w}^T(\boldsymbol{r}_p)\boldsymbol{l}(\boldsymbol{r}_q) = \tau\delta_{p,q}, \tag{4.26}$$

and the weight property is summarized as follows:

$$\begin{aligned} \boldsymbol{w}^T(\boldsymbol{r})\boldsymbol{l}(\boldsymbol{r}_q) &= \tau \quad \text{when} \quad \boldsymbol{r} = \boldsymbol{r}_q, \\ \text{and} \quad \boldsymbol{w}^T(\boldsymbol{r})\boldsymbol{l}(\boldsymbol{r}_q) &= 0 \quad \text{when} \quad \boldsymbol{r} \neq \boldsymbol{r}_q, \end{aligned} \tag{4.27}$$

where $\tau = 1$ for the unit-gain constraint, $\tau = \|\boldsymbol{l}(\boldsymbol{r}_q)\|$ for the array-gain constraint, and τ has a value expressed in Eq. (4.14) for the unit-noise-gain constraint.

All these minimum-variance-based filters have null sensitivity on the source locations other than the filter pointing location. The only difference in these adaptive filters is the gain at the pointing location; the gain is determined by the constraint. In other words, in formulating the minimum-variance spatial filters, the value of the inner product $\boldsymbol{w}^T(\boldsymbol{r})\boldsymbol{l}(\boldsymbol{r}')$ is determined only when $\boldsymbol{r}'$ is equal to one of source locations, but no constraints are imposed on the value of $\boldsymbol{w}^T(\boldsymbol{r})\boldsymbol{l}(\boldsymbol{r}')$ when $\boldsymbol{r}'$ is equal to none of source locations, and $\boldsymbol{w}^T(\boldsymbol{r})\boldsymbol{l}(\boldsymbol{r}')$ can have a large non-zero value in such cases. This is the reason why adaptive spatial filters have a strange-looking beam response, which is discussed in Section 5.4.

Also, we should emphasize that the prerequisite for the weight vector to have the property in Eqs. (4.27) is that the source time courses be uncorrelated, i.e.,

$$\langle s(\boldsymbol{r}_{q_1}, t)s(\boldsymbol{r}_{q_2}, t)\rangle = 0 \quad (q_1 \neq q_2). \tag{4.28}$$

This is an essential condition for the formulation of adaptive spatial filters. If this condition is not met, the output of the adaptive spatial filters is affected by the source correlation. The influence of the source correlation is discussed in Chapter 9.

4.2.2 Low-rank signals

We have shown that when source time courses are uncorrelated, the weight vector of the adaptive spatial filter satisfies the condition expressed in Eq. (4.27). Then, the question arises as to whether this condition is always satisfied, regardless of the number of sources. To answer this question, we return to the discussion regarding the signal and the noise subspaces in Section 2.4.1. When the pointing location is not the same as any of the source locations, the weight vector satisfies the condition in Eq. (4.23). Taking a look at this equation in the light of the discussion in Section 2.4.1, it is apparent that the weight vector lies in the noise subspace when the pointing location is equal to none of the source locations, i.e.,

$$\boldsymbol{w}(\boldsymbol{r}) \in \mathcal{E}_N, \quad \text{when} \quad \boldsymbol{r} \neq \boldsymbol{r}_1, \ldots, \boldsymbol{r}_Q. \tag{4.29}$$

Therefore, in order for such a weight vector to exist, the dimension of the noise subspace should be greater than one, and this in turn indicates that the relationship $M > Q$ should hold. In other words, the assumption of the low-rank signal is essential for the adaptive spatial filter formulation.

The necessity of the condition that $M > Q$ can also be discussed in a different manner. We present an alternative way of deriving the weight property in Eq. (4.21), and show that this property cannot be derived unless the condition $M > Q$ is satisfied. Starting from Eq. (2.41), we rewrite this equation as

$$\boldsymbol{R} = \boldsymbol{R}_s + \sigma_0^2 \boldsymbol{I},$$

where the signal covariance matrix $\boldsymbol{R}_s$ is expressed as

$$\boldsymbol{R}_s = \boldsymbol{L}_D \boldsymbol{R}_\nu \boldsymbol{L}_D^T,$$

and $\boldsymbol{L}_D = [\boldsymbol{l}(\boldsymbol{r}_1), \ldots, \boldsymbol{l}(\boldsymbol{r}_Q)]$ (defined in Eq. (2.32)). Let us consider the noiseless limit where the relationship $\boldsymbol{R} = \boldsymbol{R}_s$ holds. In this case, the rank of $\boldsymbol{R}$ is equal to Q, which is less than M, and thus $\boldsymbol{R}$ does not have an inverse. The inverse of $\boldsymbol{R}_s$ can be approximated with the pseudo-inverse of $\boldsymbol{R}_s$. Denoting the non-zero eigenvalues of $\boldsymbol{R}_s$ as λ'_j and the corresponding eigenvectors as $\boldsymbol{e}_j$, where $j = 1, \ldots, Q$, the pseudo-inverse of $\boldsymbol{R}_s$, $\boldsymbol{R}_s^+$, is expressed as

$$\boldsymbol{R}_s^+ = \sum_{j=1}^{Q} \frac{1}{\lambda'_j} \boldsymbol{e}_j \boldsymbol{e}_j^T. \tag{4.30}$$

Then, we can derive

$$\boldsymbol{R}^{-1} \approx \boldsymbol{R}_s^+ = (\boldsymbol{L}_D \boldsymbol{R}_\nu \boldsymbol{L}_D^T)^+ = (\boldsymbol{L}_D^T)^+ \boldsymbol{R}_\nu^+ \boldsymbol{L}_D^+ = (\boldsymbol{L}_D^+)^T \boldsymbol{R}_\nu^{-1} \boldsymbol{L}_D^+. \tag{4.31}$$

Here, $\boldsymbol{L}_D^+$ indicates the pseudo-inverse of $\boldsymbol{L}_D$. In deriving Eq. (4.31), we assume that $\boldsymbol{R}_\nu$ is nonsingular, and use the fact that the transpose and the pseudo-inverse operations are interchangeable.

When the condition $M > Q$ holds, $\boldsymbol{L}_D^+$ can be expressed as $\boldsymbol{L}_D^+ = (\boldsymbol{L}_D^T\boldsymbol{L}_D)^{-1}\boldsymbol{L}_D^T$. Therefore, we have the relationship $\boldsymbol{L}_D^+\boldsymbol{L}_D = \boldsymbol{I}$ and we get the formula

$$\boldsymbol{L}_D^+\boldsymbol{l}(\boldsymbol{r}_q) = \mathbf{1}_q, \tag{4.32}$$

where $\boldsymbol{r}_q$ is one of the source locations and $\mathbf{1}_q$ is the $Q \times 1$ vector that has all elements equal to zero except for the qth element, which is equal to 1. Using this formula and Eq. (4.31), we can obtain

$$\boldsymbol{w}^T(\boldsymbol{r}_p)\boldsymbol{l}(\boldsymbol{r}_q) \approx \frac{\boldsymbol{l}^T(\boldsymbol{r}_p)[\boldsymbol{L}_D^+]^T\boldsymbol{R}_\nu^{-1}\boldsymbol{L}_D^+\boldsymbol{l}(\boldsymbol{r}_q)}{\boldsymbol{l}^T(\boldsymbol{r}_p)[\boldsymbol{L}_D^+]^T\boldsymbol{R}_\nu^{-1}\boldsymbol{L}_D^+\boldsymbol{l}(\boldsymbol{r}_p)} = \frac{\mathbf{1}_p^T\boldsymbol{R}_\nu^{-1}\mathbf{1}_q}{\mathbf{1}_p^T\boldsymbol{R}_\nu^{-1}\mathbf{1}_p} = \frac{[\boldsymbol{R}_\nu^{-1}]_{p,q}}{[\boldsymbol{R}_\nu^{-1}]_{p,p}}, \tag{4.33}$$

where $[\boldsymbol{R}_\nu^{-1}]_{p,q}$ indicates the (p,q) element of the matrix $\boldsymbol{R}_\nu^{-1}$. The above equation was first derived by Zoltowski[33], and will be used for analyzing the influence of the source correlation in Chapter 9. In this section, we assume that the sources are uncorrelated. Thus, we have

$$\boldsymbol{R}_\nu^{-1} = \begin{bmatrix} 1/\sigma_1^2 & 0 & \cdots & 0 \\ 0 & 1/\sigma_2^2 & 0 & 0 \\ & & & \\ 0 & 0 & \ddots & 0 \\ 0 & 0 & \cdots & 1/\sigma_Q^2 \end{bmatrix}, \tag{4.34}$$

where σ_q^2 is the power of the qth source. Using this matrix, we obtain

$$[\boldsymbol{R}_\nu^{-1}]_{p,p}=1/\sigma_p^2, \tag{4.35}$$

$$[\boldsymbol{R}_\nu^{-1}]_{p,q}=0 \quad \text{for} \quad p \neq q, \tag{4.36}$$

and we can derive the relationship

$$\boldsymbol{w}^T(\boldsymbol{r}_p)\boldsymbol{l}(\boldsymbol{r}_q) = \delta_{p,q},$$

which is exactly the same as the relationship in Eq. (4.21). Note that the condition needed to derive this relationship is that the signal be low-rank, i.e., that $M > Q$. If this is not the case, the pseudo-inverse of $\boldsymbol{L}_D$, $\boldsymbol{L}_D^+$, has the form, $\boldsymbol{L}_D^+ = \boldsymbol{L}_D^T(\boldsymbol{L}_D\boldsymbol{L}_D^T)^{-1}$, and we cannot derive the weight property expressed in Eq. (4.21) because $\boldsymbol{L}_D^+\boldsymbol{l}(\boldsymbol{r}_q)$ is not equal to $\mathbf{1}_q$.

4.3 Scalar adaptive spatial filter: deriving the optimum source orientation

Up to this point, we have presented adaptive spatial filters with the assumption that the source orientation is known or predetermined at each location. However, it is generally difficult to predetermine the source orientation and it must be estimated from the measured data. The source orientation can be estimated in the following manner for the scalar-type adaptive spatial filter.

The weight vector of the minimum-variance spatial filter that depends both on the location and the source orientation is obtained using the following optimization[34]:

$$\boldsymbol{w}(\boldsymbol{r},\boldsymbol{\eta}) = \arg\min_{\boldsymbol{w}(\boldsymbol{r},\boldsymbol{\eta})} \boldsymbol{w}^T(\boldsymbol{r},\boldsymbol{\eta})\boldsymbol{R}\boldsymbol{w}(\boldsymbol{r},\boldsymbol{\eta}) \quad \text{subject to} \quad \boldsymbol{w}^T(\boldsymbol{r},\boldsymbol{\eta})\boldsymbol{L}(\boldsymbol{r})\boldsymbol{\eta} = 1. \tag{4.37}$$

The above optimization is the same as Eq. (4.1), except that the source orientation $\boldsymbol{\eta}$ is now explicitly included. The resulting weight vector is given by:

$$\boldsymbol{w}(\boldsymbol{r},\boldsymbol{\eta}) = \frac{\boldsymbol{R}^{-1}\boldsymbol{L}(\boldsymbol{r})\boldsymbol{\eta}}{[\boldsymbol{\eta}^T\boldsymbol{L}^T(\boldsymbol{r})\boldsymbol{R}^{-1}\boldsymbol{L}(\boldsymbol{r})\boldsymbol{\eta}]}, \tag{4.38}$$

and the output power is given by:

$$\langle \widehat{s}(\boldsymbol{r},\boldsymbol{\eta},t)^2 \rangle = \frac{1}{\boldsymbol{\eta}^T[\boldsymbol{L}^T(\boldsymbol{r})\boldsymbol{R}^{-1}\boldsymbol{L}(\boldsymbol{r})]\boldsymbol{\eta}}. \tag{4.39}$$

Therefore, according to the Rayleigh-Ritz formula in Section 13.3, the orientation $\boldsymbol{\eta}_{opt}(\boldsymbol{r})$ that gives the maximum spatial-filter output power is obtained as[35][36]

$$\begin{aligned}\boldsymbol{\eta}_{opt}(\boldsymbol{r}) &= \arg\max_{\boldsymbol{\eta}(\boldsymbol{r})} \left[\frac{1}{\boldsymbol{\eta}^T(\boldsymbol{r})\boldsymbol{L}^T(\boldsymbol{r})\boldsymbol{R}^{-1}\boldsymbol{L}(\boldsymbol{r})\boldsymbol{\eta}(\boldsymbol{r})}\right] \\ &= \arg\min_{\boldsymbol{\eta}(\boldsymbol{r})} \left[\boldsymbol{\eta}^T(\boldsymbol{r})[\boldsymbol{L}^T(\boldsymbol{r})\boldsymbol{R}^{-1}\boldsymbol{L}(\boldsymbol{r})]\boldsymbol{\eta}(\boldsymbol{r})\right] \\ &= \boldsymbol{\vartheta}_{min}\{\boldsymbol{L}^T(\boldsymbol{r})\boldsymbol{R}^{-1}\boldsymbol{L}(\boldsymbol{r})\},\end{aligned} \tag{4.40}$$

where $\boldsymbol{\vartheta}_{min}\{\cdot\}$ indicates the eigenvector corresponding to the minimum eigenvalue of the matrix in the curly braces. That is, the optimum orientation $\boldsymbol{\eta}_{opt}(\boldsymbol{r})$ is given by the eigenvector corresponding to the minimum eigenvalue of $\boldsymbol{L}^T(\boldsymbol{r})\boldsymbol{R}^{-1}\boldsymbol{L}(\boldsymbol{r})$[13][37]. The explicit form of the weight vector for the scalar version of the unit-gain minimum-variance spatial filter is expressed as

$$\boldsymbol{w}(\boldsymbol{r}) = \frac{\boldsymbol{R}^{-1}\boldsymbol{L}(\boldsymbol{r})\boldsymbol{\eta}_{opt}(\boldsymbol{r})}{[\boldsymbol{\eta}_{opt}^T(\boldsymbol{r})\boldsymbol{L}^T(\boldsymbol{r})\boldsymbol{R}^{-1}\boldsymbol{L}(\boldsymbol{r})\boldsymbol{\eta}_{opt}(\boldsymbol{r})]}. \tag{4.41}$$

The output power of this scalar spatial filter is given by

$$\widehat{P}_s(\boldsymbol{r}) = \langle \widehat{s}(\boldsymbol{r},t)^2 \rangle = \frac{1}{[\boldsymbol{\eta}_{opt}^T(\boldsymbol{r})\boldsymbol{L}^T(\boldsymbol{r})\boldsymbol{R}^{-1}\boldsymbol{L}(\boldsymbol{r})\boldsymbol{\eta}_{opt}(\boldsymbol{r})]} = \frac{1}{\lambda_{min}\{\boldsymbol{L}^T(\boldsymbol{r})\boldsymbol{R}^{-1}\boldsymbol{L}(\boldsymbol{r})\}} \tag{4.42}$$

where $\lambda_{min}\{\cdot\}$ is the minimum eigenvalue of the matrix in the curly braces.

For the array-gain constraint minimum-variance spatial filter in Eq. (4.9), the optimum orientation is obtained from

$$\boldsymbol{\eta}_{opt}(\boldsymbol{r}) = \arg\max_{\boldsymbol{\eta}(\boldsymbol{r})} \left[\frac{\boldsymbol{\eta}^T(\boldsymbol{r})[\boldsymbol{L}^T(\boldsymbol{r})\boldsymbol{L}(\boldsymbol{r})]\boldsymbol{\eta}(\boldsymbol{r})}{\boldsymbol{\eta}^T(\boldsymbol{r})[\boldsymbol{L}^T(\boldsymbol{r})\boldsymbol{R}^{-1}\boldsymbol{L}(\boldsymbol{r})]\boldsymbol{\eta}(\boldsymbol{r})}\right]. \tag{4.43}$$

According to the Rayleigh-Ritz formula in Section 13.3, the optimum orientation $\boldsymbol{\eta}_{opt}(\boldsymbol{r})$ is given by[36]:

$$\boldsymbol{\eta}_{opt}(\boldsymbol{r}) = \boldsymbol{\vartheta}_{min}\{\boldsymbol{L}^T(\boldsymbol{r})\boldsymbol{R}^{-1}\boldsymbol{L}(\boldsymbol{r}), \boldsymbol{L}^T(\boldsymbol{r})\boldsymbol{L}(\boldsymbol{r})\}, \tag{4.44}$$

where $\boldsymbol{\vartheta}_{min}\{\cdot,\cdot\}$ indicates the eigenvector corresponding to the minimum generalized eigenvalue of the matrices in the curly braces[3]. That is, $\boldsymbol{\eta}_{opt}(\boldsymbol{r})$ is equal to the eigenvector corresponding to the minimum generalized eigenvalue of the matrix $[\boldsymbol{L}^T(\boldsymbol{r})\boldsymbol{R}^{-1}\boldsymbol{L}(\boldsymbol{r})]$ with the metric $[\boldsymbol{L}^T(\boldsymbol{r})\boldsymbol{L}(\boldsymbol{r})]$. Once $\boldsymbol{\eta}_{opt}(\boldsymbol{r})$ is obtained, the weight vector is calculated using Eq. (4.9) with $\boldsymbol{l}(\boldsymbol{r}) = \boldsymbol{L}(\boldsymbol{r})\boldsymbol{\eta}_{opt}(\boldsymbol{r})$. The output power of this scalar spatial filter is given by:

$$\widehat{P}_s(\boldsymbol{r}) = \frac{1}{\lambda_{min}\{\boldsymbol{L}^T(\boldsymbol{r})\boldsymbol{R}^{-1}\boldsymbol{L}(\boldsymbol{r}), \boldsymbol{L}^T(\boldsymbol{r})\boldsymbol{L}(\boldsymbol{r})\}}, \tag{4.45}$$

where $\lambda_{min}\{\cdot,\cdot\}$ is the minimum generalized eigenvalue of the matrices in the curly braces.

For the unit-noise-gain minimum variance spatial filter, the optimum orientation is given by:

$$\boldsymbol{\eta}_{opt} = \arg\max_{\boldsymbol{\eta}} \left[\frac{\boldsymbol{\eta}^T[\boldsymbol{L}^T(\boldsymbol{r})\boldsymbol{R}^{-1}\boldsymbol{L}(\boldsymbol{r})]\boldsymbol{\eta}}{\boldsymbol{\eta}^T[\boldsymbol{L}^T(\boldsymbol{r})\boldsymbol{R}^{-2}\boldsymbol{L}(\boldsymbol{r})]\boldsymbol{\eta}} \right]. \tag{4.46}$$

The orientation $\boldsymbol{\eta}_{opt}$ is obtained from[36]:

$$\boldsymbol{\eta}_{opt}(\boldsymbol{r}) = \boldsymbol{\vartheta}_{min}\{\boldsymbol{L}^T(\boldsymbol{r})\boldsymbol{R}^{-2}\boldsymbol{L}(\boldsymbol{r}), \boldsymbol{L}^T(\boldsymbol{r})\boldsymbol{R}^{-1}\boldsymbol{L}(\boldsymbol{r})\}, \tag{4.47}$$

i.e., $\boldsymbol{\eta}_{opt}(\boldsymbol{r})$ is equal to the eigenvector corresponding to the minimum generalized eigenvalue of the matrix $[\boldsymbol{L}^T(\boldsymbol{r})\boldsymbol{R}^{-2}\boldsymbol{L}(\boldsymbol{r})]$ with the metric $[\boldsymbol{L}^T(\boldsymbol{r})\boldsymbol{R}^{-1}\boldsymbol{L}(\boldsymbol{r})]$. Once $\boldsymbol{\eta}_{opt}(\boldsymbol{r})$ is obtained, the weight vector is calculated using Eq. (4.15) with $\boldsymbol{l}(\boldsymbol{r}) = \boldsymbol{L}(\boldsymbol{r})\boldsymbol{\eta}_{opt}(\boldsymbol{r})$. The output power of this scalar spatial filter is given by:

$$\widehat{P}_s(\boldsymbol{r}) = \frac{1}{\lambda_{min}\{\boldsymbol{L}^T(\boldsymbol{r})\boldsymbol{R}^{-2}\boldsymbol{L}(\boldsymbol{r}), \boldsymbol{L}^T(\boldsymbol{r})\boldsymbol{R}^{-1}\boldsymbol{L}(\boldsymbol{r})\}}. \tag{4.48}$$

4.4 LCMV spatial filter

Although the adaptive spatial filters described so far are formulated with a single constraint, an adaptive spatial filter can be formulated with multiple linear constraints. An adaptive spatial filter obtained with such multiple constraints is called a linearly-constrained minimum-variance (LCMV) spatial filter[38][7]. The weight vector for an LCMV filter can be derived using

$$\begin{aligned} \boldsymbol{w}(\boldsymbol{r}) = \arg\min_{\boldsymbol{w}(\boldsymbol{r})} \boldsymbol{w}^T(\boldsymbol{r})\boldsymbol{R}\boldsymbol{w}(\boldsymbol{r}), \quad &\text{subject to} \quad \boldsymbol{w}^T(\boldsymbol{r})\boldsymbol{l}(\boldsymbol{r}) = 1, \\ &\qquad\qquad \boldsymbol{w}^T(\boldsymbol{r})\boldsymbol{l}(\boldsymbol{r}_1) = 0, \\ &\qquad\qquad \vdots \\ &\text{and} \qquad \boldsymbol{w}^T(\boldsymbol{r})\boldsymbol{l}(\boldsymbol{r}_d) = 0. \end{aligned} \tag{4.49}$$

[3]Additional arguments concerning the definition are found in Section 13.3 in the Appendix.

The weight vector obtained above passes a signal from $\boldsymbol{r}$ with a unit gain but imposes the null gain at the locations $\boldsymbol{r}_1, \ldots, \boldsymbol{r}_d$. That is, the weight does not pass any signal from these locations. The LCMV spatial filter derived above was originally developed to suppress the influence of highly correlated interferences with known locations. In the formulation above, $\boldsymbol{r}_1, \ldots, \boldsymbol{r}_d$ are the spatial locations of the correlated interferences.

The explicit form of the weight vector for the LCMV spatial filter can be obtained using a derivation similar to that of Eqs. (4.2)–(4.5). That is, to derive the weight to satisfy Eq. (4.49), we define a $(d+1) \times 1$ column vector $\boldsymbol{\kappa}$ whose elements are the Lagrange multipliers, and define the Lagrangian $\mathbb{L}(\boldsymbol{w}, \boldsymbol{\kappa})$ such that

$$\mathbb{L}(\boldsymbol{w}, \boldsymbol{\kappa}) = \boldsymbol{w}^T \boldsymbol{R} \boldsymbol{w} + \boldsymbol{\kappa}^T (\boldsymbol{C}^T \boldsymbol{w} - \boldsymbol{c}), \tag{4.50}$$

where we again omit the explicit notation of $(\boldsymbol{r})$ from the weight vector. In this equation, (the upper-case) $\boldsymbol{C}$ indicates the matrix

$$\boldsymbol{C} = [\boldsymbol{l}(\boldsymbol{r}), \boldsymbol{l}(\boldsymbol{r}_1), \ldots, \boldsymbol{l}(\boldsymbol{r}_d)],$$

and (the lower case) $\boldsymbol{c}$ indicates a vector expressing a response to the constraints, such that

$$\boldsymbol{c} = \begin{bmatrix} 1 \\ 0 \\ \vdots \\ 0 \end{bmatrix}.$$

Calculating the derivative of the Lagrangian in Eq. (4.50) with respect to $\boldsymbol{w}$ and setting the derivative to zero, we obtain

$$\boldsymbol{w} = -\frac{1}{2} \boldsymbol{R}^{-1} \boldsymbol{C} \boldsymbol{\kappa}. \tag{4.51}$$

Substituting this into the constraint equation $\boldsymbol{C}^T \boldsymbol{w} = \boldsymbol{c}$ gives

$$\boldsymbol{\kappa} = -2 [\boldsymbol{C}^T \boldsymbol{R}^{-1} \boldsymbol{C}]^{-1} \boldsymbol{c}, \tag{4.52}$$

and substituting the above equation into Eq. (4.51), we finally obtain the weight vector,

$$\boldsymbol{w}(\boldsymbol{r}) = \boldsymbol{R}^{-1} \boldsymbol{C} [\boldsymbol{C}^T \boldsymbol{R}^{-1} \boldsymbol{C}]^{-1} \begin{bmatrix} 1 \\ 0 \\ \vdots \\ 0 \end{bmatrix}. \tag{4.53}$$

The LCMV spatial filter is used to develop the vector-type adaptive spatial filters in the next section.

4.5 Vector adaptive spatial filter formulation

We next describe the extension of the minimum-variance filters to the vector formulation. As described in Section 2.5.2, the vector spatial filter uses a set of three weight vectors, $\boldsymbol{w}_x(\boldsymbol{r})$, $\boldsymbol{w}_y(\boldsymbol{r})$, $\boldsymbol{w}_z(\boldsymbol{r})$, which detect the x, y, z components of the source vector, respectively. How can we derive such adaptive weights? Let us consider a simple case where a single source exists at $\boldsymbol{r}$ and its magnitude and orientation are denoted $s(\boldsymbol{r},t)$ and $\boldsymbol{\eta} = [\eta_x, \eta_y, \eta_z]^T$, respectively. Ignoring the noise term, the array measurement $\boldsymbol{b}(t)$ is then expressed as

$$\boldsymbol{b}(t) = s(\boldsymbol{r},t)\eta_x \boldsymbol{l}_x(\boldsymbol{r}) + s(\boldsymbol{r},t)\eta_y \boldsymbol{l}_y(\boldsymbol{r}) + s(\boldsymbol{r},t)\eta_z \boldsymbol{l}_z(\boldsymbol{r}). \tag{4.54}$$

One important point here is that the array measurement is equal to that obtained from three perfectly correlated sources whose time courses are equal to $s(\boldsymbol{r},t)\eta_x$, $s(\boldsymbol{r},t)\eta_y$, and $s(\boldsymbol{r},t)\eta_z$. Therefore, to derive the weight for the adaptive vector spatial filter, this virtual source correlation should be taken into consideration, and this can be done by using the LCMV spatial filter described in the preceding section.

4.5.1 Unit-gain constraint spatial filter

The weight vector to detect the x component of the source vector, $\boldsymbol{w}_x$, can be derived using

$$\boldsymbol{w}_x = \arg\min_{\boldsymbol{w}_x} \boldsymbol{w}_x^T \boldsymbol{R} \boldsymbol{w}_x, \quad \text{subject to} \quad \boldsymbol{w}_x^T \boldsymbol{l}_x(\boldsymbol{r}) = 1, \quad \boldsymbol{w}_x^T \boldsymbol{l}_y(\boldsymbol{r}) = 0, \quad \text{and} \quad \boldsymbol{w}_x^T \boldsymbol{l}_z(\boldsymbol{r}) = 0. \tag{4.55}$$

That is, the weight $\boldsymbol{w}_x$ passes a signal from a source at $\boldsymbol{r}$ directed in the x direction with a gain of unity, but does not pass a signal from a source directed in the y or z direction. In exactly the same manner, we can derive $\boldsymbol{w}_y$, such that

$$\boldsymbol{w}_y = \arg\min_{\boldsymbol{w}_y} \boldsymbol{w}_y^T \boldsymbol{R} \boldsymbol{w}_y, \quad \text{subject to} \quad \boldsymbol{w}_y^T \boldsymbol{l}_x(\boldsymbol{r}) = 0, \quad \boldsymbol{w}_y^T \boldsymbol{l}_y(\boldsymbol{r}) = 1, \quad \text{and} \quad \boldsymbol{w}_y^T \boldsymbol{l}_z(\boldsymbol{r}) = 0. \tag{4.56}$$

We can derive $\boldsymbol{w}_z$, such that

$$\boldsymbol{w}_z = \arg\min_{\boldsymbol{w}_z} \boldsymbol{w}_z^T \boldsymbol{R} \boldsymbol{w}_z, \quad \text{subject to} \quad \boldsymbol{w}_z^T \boldsymbol{l}_x(\boldsymbol{r}) = 0, \quad \boldsymbol{w}_z^T \boldsymbol{l}_y(\boldsymbol{r}) = 0, \quad \text{and} \quad \boldsymbol{w}_z^T \boldsymbol{l}_z(\boldsymbol{r}) = 1. \tag{4.57}$$

In the equations above, we omit the explicit notation of $(\boldsymbol{r})$ from the weight vector expressions for simplicity.

Defining the weight matrix $\boldsymbol{W}(\boldsymbol{r})$ such that $\boldsymbol{W}(\boldsymbol{r}) = [\boldsymbol{w}_x(\boldsymbol{r}), \boldsymbol{w}_y(\boldsymbol{r}), \boldsymbol{w}_z(\boldsymbol{r})]$, Equations (4.55)–(4.57) can be rewritten in a compact form as[9][39]

$$\boldsymbol{W}(\boldsymbol{r}) = \underset{\boldsymbol{W}(\boldsymbol{r})}{\arg\min} \, \mathrm{tr}\{\boldsymbol{W}^T(\boldsymbol{r})\boldsymbol{R}\boldsymbol{W}(\boldsymbol{r})\}, \quad \text{subject to} \quad \boldsymbol{W}^T(\boldsymbol{r})\boldsymbol{L}(\boldsymbol{r}) = \boldsymbol{I}. \tag{4.58}$$

It is easy to show that the solution for the weight matrix $\boldsymbol{W}(\boldsymbol{r})$ can be obtained by substituting $\boldsymbol{C}$ in Eq. (4.53) with $\boldsymbol{L}(\boldsymbol{r})$, resulting in[9][10]

$$\boldsymbol{W}(\boldsymbol{r}) = \boldsymbol{R}^{-1}\boldsymbol{L}(\boldsymbol{r})[\boldsymbol{L}^T(\boldsymbol{r})\boldsymbol{R}^{-1}\boldsymbol{L}(\boldsymbol{r})]^{-1}. \tag{4.59}$$

Thus, the output of this vector-type minimum-variance spatial filter is given by:

$$\widehat{\boldsymbol{s}}(\boldsymbol{r},t) = [\widehat{s}_x(\boldsymbol{r},t), \widehat{s}_y(\boldsymbol{r},t), \widehat{s}_z(\boldsymbol{r},t)]^T = [\boldsymbol{L}^T(\boldsymbol{r})\boldsymbol{R}^{-1}\boldsymbol{L}(\boldsymbol{r})]^{-1}\boldsymbol{L}^T(\boldsymbol{r})\boldsymbol{R}^{-1}\boldsymbol{b}(t). \tag{4.60}$$

Using the above equation, the estimated source power matrix defined in Eq. (2.75) is expressed as

$$\begin{aligned}\widehat{\boldsymbol{\Sigma}}_s(\boldsymbol{r}) = \langle \widehat{\boldsymbol{s}}(\boldsymbol{r},t)\widehat{\boldsymbol{s}}^T(\boldsymbol{r},t)\rangle = \\ [\boldsymbol{L}^T(\boldsymbol{r})\boldsymbol{R}^{-1}\boldsymbol{L}(\boldsymbol{r})]^{-1}\boldsymbol{L}^T(\boldsymbol{r})\boldsymbol{R}^{-1}\langle \boldsymbol{b}(t)\boldsymbol{b}^T(t)\rangle \boldsymbol{R}^{-1}\boldsymbol{L}(\boldsymbol{r})[\boldsymbol{L}^T(\boldsymbol{r})\boldsymbol{R}^{-1}\boldsymbol{L}(\boldsymbol{r})]^{-1} \\ = [\boldsymbol{L}^T(\boldsymbol{r})\boldsymbol{R}^{-1}\boldsymbol{L}(\boldsymbol{r})]^{-1}.\end{aligned} \tag{4.61}$$

The source power estimate of the first kind is obtained by computing the trace of the source power matrix

$$\widehat{P}_V^{(\mathrm{I})}(\boldsymbol{r}) = \langle \widehat{s}(\boldsymbol{r},t)^2\rangle = \mathrm{tr}\{[\boldsymbol{L}^T(\boldsymbol{r})\boldsymbol{R}^{-1}\boldsymbol{L}(\boldsymbol{r})]^{-1}\}. \tag{4.62}$$

4.5.2 Array-gain constraint spatial filter

When using the array-gain constraint, we use a formulation for the weight derivation in which

$$\begin{aligned}\boldsymbol{w}_x = \underset{\boldsymbol{w}_x}{\arg\min}\, \boldsymbol{w}_x^T\boldsymbol{R}\boldsymbol{w}_x, \quad & \text{subject to} \quad \boldsymbol{w}_x^T\boldsymbol{l}_x(\boldsymbol{r}) = \|\boldsymbol{l}_x(\boldsymbol{r})\|, \\ & \boldsymbol{w}_x^T\boldsymbol{l}_y(\boldsymbol{r}) = 0, \\ & \text{and} \quad \boldsymbol{w}_x^T\boldsymbol{l}_z(\boldsymbol{r}) = 0,\end{aligned} \tag{4.63}$$

$$\begin{aligned}\boldsymbol{w}_y = \underset{\boldsymbol{w}_y}{\arg\min}\, \boldsymbol{w}_y^T\boldsymbol{R}\boldsymbol{w}_y, \quad & \text{subject to} \quad \boldsymbol{w}_y^T\boldsymbol{l}_x(\boldsymbol{r}) = 0, \\ & \boldsymbol{w}_y^T\boldsymbol{l}_y(\boldsymbol{r}) = \|\boldsymbol{l}_y(\boldsymbol{r})\|, \\ & \text{and} \quad \boldsymbol{w}_y^T\boldsymbol{l}_z(\boldsymbol{r}) = 0,\end{aligned} \tag{4.64}$$

and

$$\begin{aligned}\boldsymbol{w}_z = \underset{\boldsymbol{w}_z}{\arg\min}\, \boldsymbol{w}_z^T\boldsymbol{R}\boldsymbol{w}_z, \quad & \text{subject to} \quad \boldsymbol{w}_z^T\boldsymbol{l}_x(\boldsymbol{r}) = 0, \\ & \boldsymbol{w}_z^T\boldsymbol{l}_y(\boldsymbol{r}) = 0, \\ & \text{and} \quad \boldsymbol{w}_z^T\boldsymbol{l}_z(\boldsymbol{r}) = \|\boldsymbol{l}_z(\boldsymbol{r})\|.\end{aligned} \tag{4.65}$$

In the equations above, we again omit the explicit notation of $(\boldsymbol{r})$ from the weight vector expressions. Using the weight matrix $\boldsymbol{W}(\boldsymbol{r})$, Equations (4.63)–(4.65) can be rewritten as

$$\boldsymbol{W}(\boldsymbol{r}) = \underset{\boldsymbol{W}(\boldsymbol{r})}{\arg\min}\, \mathrm{tr}\{\boldsymbol{W}^T(\boldsymbol{r})\boldsymbol{R}\boldsymbol{W}(\boldsymbol{r})\},$$

$$\text{subject to} \quad \boldsymbol{W}^T(\boldsymbol{r})\boldsymbol{L}(\boldsymbol{r}) = \begin{bmatrix} \|\boldsymbol{l}_x(\boldsymbol{r})\| & 0 & 0 \\ 0 & \|\boldsymbol{l}_y(\boldsymbol{r})\| & 0 \\ 0 & & \|\boldsymbol{l}_z(\boldsymbol{r})\| \end{bmatrix}. \tag{4.66}$$

Using exactly the same derivation as used to derive Eq. (4.59), the vector-version of the array-gain minimum-variance spatial filter is expressed as

$$\boldsymbol{W}(\boldsymbol{r}) = \boldsymbol{R}^{-1}\boldsymbol{L}(\boldsymbol{r})[\boldsymbol{L}^T(\boldsymbol{r})\boldsymbol{R}^{-1}\boldsymbol{L}(\boldsymbol{r})]^{-1} \begin{bmatrix} \|\boldsymbol{l}_x(\boldsymbol{r})\| & 0 & 0 \\ 0 & \|\boldsymbol{l}_y(\boldsymbol{r})\| & 0 \\ 0 & & \|\boldsymbol{l}_z(\boldsymbol{r})\| \end{bmatrix}. \tag{4.67}$$

The source-power matrix is given by:

$$\widehat{\boldsymbol{\Sigma}}_s(\boldsymbol{r}) = \langle \widehat{\boldsymbol{s}}(\boldsymbol{r},t)\widehat{\boldsymbol{s}}^T(\boldsymbol{r},t)\rangle =$$
$$\begin{bmatrix} \|\boldsymbol{l}_x(\boldsymbol{r})\| & 0 & 0 \\ 0 & \|\boldsymbol{l}_y(\boldsymbol{r})\| & 0 \\ 0 & 0 & \|\boldsymbol{l}_z(\boldsymbol{r})\| \end{bmatrix} [\boldsymbol{L}^T(\boldsymbol{r})\boldsymbol{R}^{-1}\boldsymbol{L}(\boldsymbol{r})]^{-1} \begin{bmatrix} \|\boldsymbol{l}_x(\boldsymbol{r})\| & 0 & 0 \\ 0 & \|\boldsymbol{l}_y(\boldsymbol{r})\| & 0 \\ 0 & 0 & \|\boldsymbol{l}_z(\boldsymbol{r})\| \end{bmatrix}. \tag{4.68}$$

This source-power matrix is expressed in a compact form as

$$\widehat{\boldsymbol{\Sigma}}_s(\boldsymbol{r}) = [\widetilde{\boldsymbol{L}}^T(\boldsymbol{r})\boldsymbol{R}^{-1}\widetilde{\boldsymbol{L}}(\boldsymbol{r})]^{-1}, \tag{4.69}$$

where $\widetilde{\boldsymbol{L}}(\boldsymbol{r})$ is a matrix consisting of the normalized lead-field columns, i.e.,

$$\widetilde{\boldsymbol{L}}(\boldsymbol{r}) = \left[\frac{\boldsymbol{l}_x(\boldsymbol{r})}{\|\boldsymbol{l}_x(\boldsymbol{r})\|}, \frac{\boldsymbol{l}_y(\boldsymbol{r})}{\|\boldsymbol{l}_y(\boldsymbol{r})\|}, \frac{\boldsymbol{l}_z(\boldsymbol{r})}{\|\boldsymbol{l}_z(\boldsymbol{r})\|}\right]. \tag{4.70}$$

There is another possibility for formulating the vector-version of the array-gain minimum-variance spatial filter, as shown below.

$$\boldsymbol{w}_x = \underset{\boldsymbol{w}_x}{\arg\min}\, \boldsymbol{w}_x^T\boldsymbol{R}\boldsymbol{w}_x, \quad \text{subject to} \quad \boldsymbol{w}_x^T\boldsymbol{l}_x(\boldsymbol{r}) = \|\boldsymbol{L}(\boldsymbol{r})\|,$$
$$\boldsymbol{w}_x^T\boldsymbol{l}_y(\boldsymbol{r}) = 0, \tag{4.71}$$
$$\text{and} \quad \boldsymbol{w}_x^T\boldsymbol{l}_z(\boldsymbol{r}) = 0,$$

$$\boldsymbol{w}_y = \underset{\boldsymbol{w}_y}{\arg\min}\, \boldsymbol{w}_y^T\boldsymbol{R}\boldsymbol{w}_y, \quad \text{subject to} \quad \boldsymbol{w}_y^T\boldsymbol{l}_x(\boldsymbol{r}) = 0,$$
$$\boldsymbol{w}_y^T\boldsymbol{l}_y(\boldsymbol{r}) = \|\boldsymbol{L}(\boldsymbol{r})\|, \tag{4.72}$$
$$\text{and} \quad \boldsymbol{w}_y^T\boldsymbol{l}_z(\boldsymbol{r}) = 0,$$

and

$$\boldsymbol{w}_z = \arg\min_{\boldsymbol{w}_z} \boldsymbol{w}_z^T \boldsymbol{R} \boldsymbol{w}_z, \qquad \text{subject to} \qquad \begin{aligned} &\boldsymbol{w}_z^T \boldsymbol{l}_x(\boldsymbol{r}) = 0, \\ &\boldsymbol{w}_z^T \boldsymbol{l}_y(\boldsymbol{r}) = 0, \\ \text{and} \qquad &\boldsymbol{w}_z^T \boldsymbol{l}_z(\boldsymbol{r}) = \|\boldsymbol{L}(\boldsymbol{r})\|. \end{aligned} \tag{4.73}$$

The compact form of the formulation is

$$\boldsymbol{W}(\boldsymbol{r}) = \arg\min_{\boldsymbol{W}(\boldsymbol{r})} \mathrm{tr}[\boldsymbol{W}^T(\boldsymbol{r}) \boldsymbol{R} \boldsymbol{W}(\boldsymbol{r})],$$
$$\text{subject to} \quad \boldsymbol{W}^T(\boldsymbol{r}) \boldsymbol{L}(\boldsymbol{r}) = \|\boldsymbol{L}(\boldsymbol{r})\| \boldsymbol{I}. \tag{4.74}$$

The weight matrix, in this case, is expressed as

$$\boldsymbol{W}(\boldsymbol{r}) = \|\boldsymbol{L}(\boldsymbol{r})\| \boldsymbol{R}^{-1} \boldsymbol{L}(\boldsymbol{r}) [\boldsymbol{L}^T(\boldsymbol{r}) \boldsymbol{R}^{-1} \boldsymbol{L}(\boldsymbol{r})]^{-1} \tag{4.75}$$

The source-power matrix is given by:

$$\widehat{\boldsymbol{\Sigma}}_s(\boldsymbol{r}) = \langle \widehat{\boldsymbol{s}}(\boldsymbol{r}, t) \widehat{\boldsymbol{s}}^T(\boldsymbol{r}, t) \rangle = \|\boldsymbol{L}(\boldsymbol{r})\|^2 [\boldsymbol{L}^T(\boldsymbol{r}) \boldsymbol{R}^{-1} \boldsymbol{L}(\boldsymbol{r})]^{-1} \tag{4.76}$$

4.5.3 Unit-noise-gain constraint spatial filter

The vector-version of the unit-noise-gain constraint spatial filter is obtained using

$$\boldsymbol{w}_x = \arg\min_{\boldsymbol{w}_x} \boldsymbol{w}_x^T \boldsymbol{R} \boldsymbol{w}_x, \qquad \text{subject to} \qquad \begin{aligned} &\boldsymbol{w}_x^T \boldsymbol{w}_x = 1, \\ &\boldsymbol{w}_x^T \boldsymbol{l}_y(\boldsymbol{r}) = 0, \\ \text{and} \qquad &\boldsymbol{w}_x^T \boldsymbol{l}_z(\boldsymbol{r}) = 0, \end{aligned} \tag{4.77}$$

$$\boldsymbol{w}_y = \arg\min_{\boldsymbol{w}_y} \boldsymbol{w}_y^T \boldsymbol{R} \boldsymbol{w}_y, \qquad \text{subject to} \qquad \begin{aligned} &\boldsymbol{w}_y^T \boldsymbol{l}_x(\boldsymbol{r}) = 0, \\ &\boldsymbol{w}_y^T \boldsymbol{w}_y = 1, \\ \text{and} \qquad &\boldsymbol{w}_y^T \boldsymbol{l}_z(\boldsymbol{r}) = 0, \end{aligned} \tag{4.78}$$

and

$$\boldsymbol{w}_z = \arg\min_{\boldsymbol{w}_z} \boldsymbol{w}_z^T \boldsymbol{R} \boldsymbol{w}_z, \qquad \text{subject to} \qquad \begin{aligned} &\boldsymbol{w}_z^T \boldsymbol{l}_x(\boldsymbol{r}) = 0, \\ &\boldsymbol{w}_z^T \boldsymbol{l}_y(\boldsymbol{r}) = 0, \\ \text{and} \qquad &\boldsymbol{w}_z^T \boldsymbol{w}_z = 1. \end{aligned} \tag{4.79}$$

We can derive the explicit forms of these weight vectors using a derivation similar to that in Section 4.1.3. That is, to derive the weight vector $\boldsymbol{w}_x$, let us first define the scalar constant θ such that $\boldsymbol{w}_x^T \boldsymbol{l}_x(\boldsymbol{r}) = \theta$, and rewrite Eq. (4.77) as

$$\boldsymbol{w}_x = \arg\min_{\boldsymbol{w}_x} \boldsymbol{w}_x^T \boldsymbol{R} \boldsymbol{w}_x, \quad \text{subject to} \quad \boldsymbol{L}^T(\boldsymbol{r}) \boldsymbol{w}_x = \theta \begin{bmatrix} 1 \\ 0 \\ 0 \end{bmatrix}. \tag{4.80}$$

The solution for the above optimization problem is expressed as

$$\boldsymbol{w}_x = \theta \boldsymbol{R}^{-1}\boldsymbol{L}(\boldsymbol{r})[\boldsymbol{L}^T(\boldsymbol{r})\boldsymbol{R}^{-1}\boldsymbol{L}(\boldsymbol{r})]^{-1}\begin{bmatrix} 1 \\ 0 \\ 0 \end{bmatrix}, \tag{4.81}$$

where

$$\boldsymbol{w}_x^T \boldsymbol{w}_x = \theta^2 \Upsilon_{1,1} \tag{4.82}$$

and $\Upsilon_{1,1}$ is the $(1,1)$ component of a matrix defined as

$$\boldsymbol{\Upsilon} = [\boldsymbol{L}^T(\boldsymbol{r})\boldsymbol{R}^{-1}\boldsymbol{L}(\boldsymbol{r})]^{-1}\boldsymbol{L}^T(\boldsymbol{r})\boldsymbol{R}^{-2}\boldsymbol{L}(\boldsymbol{r})[\boldsymbol{L}^T(\boldsymbol{r})\boldsymbol{R}^{-1}\boldsymbol{L}(\boldsymbol{r})]^{-1}. \tag{4.83}$$

Therefore, to impose the constraint $\boldsymbol{w}_x^T \boldsymbol{w}_x = 1$, the scalar constant θ should be

$$\theta = 1/\sqrt{\Upsilon_{1,1}}.$$

We can derive $\boldsymbol{w}_y$ and $\boldsymbol{w}_z$ in a similar fashion, so that the solution for the weight vectors is given by[40]:

$$\begin{aligned}
\boldsymbol{w}_x(\boldsymbol{r}) &= \frac{\boldsymbol{R}^{-1}\boldsymbol{L}(\boldsymbol{r})[\boldsymbol{L}^T(\boldsymbol{r})\boldsymbol{R}^{-1}\boldsymbol{L}(\boldsymbol{r})]^{-1}}{\sqrt{\Upsilon_{1,1}}}\begin{bmatrix} 1 \\ 0 \\ 0 \end{bmatrix}, \\
\boldsymbol{w}_y(\boldsymbol{r}) &= \frac{\boldsymbol{R}^{-1}\boldsymbol{L}(\boldsymbol{r})[\boldsymbol{L}^T(\boldsymbol{r})\boldsymbol{R}^{-1}\boldsymbol{L}(\boldsymbol{r})]^{-1}}{\sqrt{\Upsilon_{2,2}}}\begin{bmatrix} 0 \\ 1 \\ 0 \end{bmatrix}, \\
\boldsymbol{w}_z(\boldsymbol{r}) &= \frac{\boldsymbol{R}^{-1}\boldsymbol{L}(\boldsymbol{r})[\boldsymbol{L}^T(\boldsymbol{r})\boldsymbol{R}^{-1}\boldsymbol{L}(\boldsymbol{r})]^{-1}}{\sqrt{\Upsilon_{3,3}}}\begin{bmatrix} 0 \\ 0 \\ 1 \end{bmatrix},
\end{aligned} \tag{4.84}$$

and the weight matrix is expressed in a compact form

$$\boldsymbol{W}(\boldsymbol{r}) = \boldsymbol{R}^{-1}\boldsymbol{L}(\boldsymbol{r})[\boldsymbol{L}^T(\boldsymbol{r})\boldsymbol{R}^{-1}\boldsymbol{L}(\boldsymbol{r})]^{-1}\begin{bmatrix} 1/\sqrt{\Upsilon_{1,1}} & 0 & 0 \\ 0 & 1/\sqrt{\Upsilon_{2,2}} & 0 \\ 0 & 0 & 1/\sqrt{\Upsilon_{3,3}} \end{bmatrix}. \tag{4.85}$$

Then, we can derive the expression for the estimated source power matrix such that

$$\widehat{\boldsymbol{\Sigma}}_s(\boldsymbol{r}) = \langle \widehat{\boldsymbol{s}}(\boldsymbol{r},t)\widehat{\boldsymbol{s}}^T(\boldsymbol{r},t)\rangle = [\breve{\boldsymbol{L}}^T(\boldsymbol{r})\boldsymbol{R}^{-1}\breve{\boldsymbol{L}}(\boldsymbol{r})]^{-1}, \tag{4.86}$$

where $\breve{\boldsymbol{L}}(\boldsymbol{r})$ is the modified lead-field matrix, which is expressed as

$$\breve{\boldsymbol{L}}(\boldsymbol{r}) = [\sqrt{\Upsilon_{1,1}}\boldsymbol{l}_x(\boldsymbol{r}), \sqrt{\Upsilon_{2,2}}\boldsymbol{l}_y(\boldsymbol{r}), \sqrt{\Upsilon_{3,3}}\boldsymbol{l}_z(\boldsymbol{r})]. \tag{4.87}$$

4.5.4 Equivalence between the adaptive scalar and vector formulations

We have presented general discussion regarding the equivalence between the scalar and vector formulations in Section 2.5.2. These two formulations are generally not equivalent for adaptive spatial filters. However, for the unit-gain minimum-variance filter, we can show that the two formulations are equivalent, i.e., that the following two conditions are met:

$$\bar{\boldsymbol{\eta}}_{opt} = \boldsymbol{\eta}_{opt}, \tag{4.88}$$

and

$$\widehat{P}_V^{(\mathrm{II})}(\boldsymbol{r}) = \widehat{P}_s(\boldsymbol{r}). \tag{4.89}$$

For the following discussion, we define the eigendecomposition of $\boldsymbol{L}^T(\boldsymbol{r})\boldsymbol{R}^{-1}\boldsymbol{L}(\boldsymbol{r})$, such that

$$\boldsymbol{L}^T(\boldsymbol{r})\boldsymbol{R}^{-1}\boldsymbol{L}(\boldsymbol{r}) = \sum_{j=1}^{3} \phi_j \boldsymbol{z}_j \boldsymbol{z}_j^T, \tag{4.90}$$

where ϕ_j and $\boldsymbol{z}_j$ are the jth eigenvalue and its corresponding eigenvector of the matrix $\boldsymbol{L}^T(\boldsymbol{r})\boldsymbol{R}^{-1}\boldsymbol{L}(\boldsymbol{r})$. Here, the eigenvalues are numbered in decreasing order. Then, using Eq. (4.61), we have

$$\widehat{\boldsymbol{\Sigma}}_s(\boldsymbol{r}) = [\boldsymbol{L}^T(\boldsymbol{r})\boldsymbol{R}^{-1}\boldsymbol{L}(\boldsymbol{r})]^{-1} = \sum_{j=1}^{3} \frac{1}{\phi_j} \boldsymbol{z}_j \boldsymbol{z}_j^T. \tag{4.91}$$

We first show that Eq. (4.88) holds for the unit-gain minimum-variance filter. The optimum orientation defined in Eq. (2.78), $\bar{\boldsymbol{\eta}}_{opt}$, is obtained as

$$\bar{\boldsymbol{\eta}}_{opt} = \arg\max_{\boldsymbol{\eta}} \boldsymbol{\eta}^T \widehat{\boldsymbol{\Sigma}}_s(\boldsymbol{r})\boldsymbol{\eta}. \tag{4.92}$$

According to the Rayleigh-Ritz formula, the right-hand side of this equation is equal to the eigenvector corresponding to the maximum eigenvalue of the matrix $\widehat{\boldsymbol{\Sigma}}_s(\boldsymbol{r})$. Using Eq. (4.91) and recalling the relationship, $\phi_1 \geq \phi_2 \geq \phi_3$, the maximum eigenvalue of $\widehat{\boldsymbol{\Sigma}}_s(\boldsymbol{r})$ is $1/\phi_3$ and we have

$$\bar{\boldsymbol{\eta}}_{opt} = \boldsymbol{z}_3 = \boldsymbol{\vartheta}_{min}\{\boldsymbol{L}^T(\boldsymbol{r})\boldsymbol{R}^{-1}\boldsymbol{L}(\boldsymbol{r})\}. \tag{4.93}$$

Comparing the above equation with Eq. (4.40), we can thus derive

$$\bar{\boldsymbol{\eta}}_{opt} = \boldsymbol{\eta}_{opt}.$$

To show that Eq. (4.89) holds, we calculate $\widehat{P}_V^{(\mathrm{II})}(\boldsymbol{r})$ such that

$$\widehat{P}_V^{(\mathrm{II})}(\boldsymbol{r}) = \max_{\boldsymbol{\eta}} \left[\boldsymbol{\eta}^T \widehat{\boldsymbol{\Sigma}}_s(\boldsymbol{r})\boldsymbol{\eta}\right] = \lambda_{max}\{\widehat{\boldsymbol{\Sigma}}_s(\boldsymbol{r})\} = \frac{1}{\phi_3}. \tag{4.94}$$

On the other hand, ϕ_3 is equal to $\lambda_{min}\{[\boldsymbol{L}^T(\boldsymbol{r})\boldsymbol{R}^{-1}\boldsymbol{L}(\boldsymbol{r})]\}$, and thus we have

$$\widehat{P}_V^{(\mathrm{II})}(\boldsymbol{r}) = \frac{1}{\lambda_{min}\{[\boldsymbol{L}^T(\boldsymbol{r})\boldsymbol{R}^{-1}\boldsymbol{L}(\boldsymbol{r})]\}}. \tag{4.95}$$

Comparing the above equation with Eq. (4.42), we can derive

$$\widehat{P}_V^{(\mathrm{II})}(\boldsymbol{r}) = \widehat{P}_s(\boldsymbol{r}).$$

Thus, we can see that the scalar and vector formulations are equivalent for the unit-gain minimum-variance spatial filter.

For the minimum-variance filters with the other constraints, Eq. (4.88) does not hold. This fact can easily be shown, for example, for the array-gain minimum-variance spatial filter. For this filter, the source-power matrix $\widehat{\boldsymbol{\Sigma}}_s$ is given in Eq. (4.69), and thus the optimum orientation for the vector formulation, $\bar{\boldsymbol{\eta}}_{opt}$, is obtained as

$$\bar{\boldsymbol{\eta}}_{opt} = \boldsymbol{\vartheta}_{min}\{\widetilde{\boldsymbol{L}}^T(\boldsymbol{r})\boldsymbol{R}^{-1}\widetilde{\boldsymbol{L}}(\boldsymbol{r})\}. \tag{4.96}$$

Apparently, this $\bar{\boldsymbol{\eta}}_{opt}$ is different from the $\boldsymbol{\eta}_{opt}$ obtained in Eq. (4.44). For the unit-noise-gain spatial filter, in the same manner, we can show that Eq. (4.88) does not hold. Additionally, using the arguments in this section, for the two kinds of the power estimates $\widehat{P}_V^{(\mathrm{I})}$ and $\widehat{P}_V^{(\mathrm{II})}$, we can show the relationship

$$\widehat{P}_V^{(\mathrm{II})} = \frac{1}{\phi_3} < \sum_{j=1}^{3}\frac{1}{\phi_j} = \widehat{P}_V^{(\mathrm{I})}. \tag{4.97}$$

It is clear that this relationship holds for the array-gain and the unit-noise-gain minimum-variance filters.

4.6 Frequency-domain implementation

In this section, we describe an implementation of the adaptive spatial filter in the frequency domain, and derive the frequency-specific weight. We first define the Fourier transform of the measurement vector $\boldsymbol{b}(t)$ as a vector

$$\boldsymbol{g}(f) = \begin{bmatrix} g_1(f) \\ g_2(f) \\ \vdots \\ g_M(f) \end{bmatrix}, \tag{4.98}$$

where $g_m(f)$ is the Fourier spectrum of the mth-channel recording $b_m(t)$, i.e.,

$$g_m(f) = \int_{-\infty}^{\infty} b_m(t)e^{-2\pi f t}dt. \tag{4.99}$$

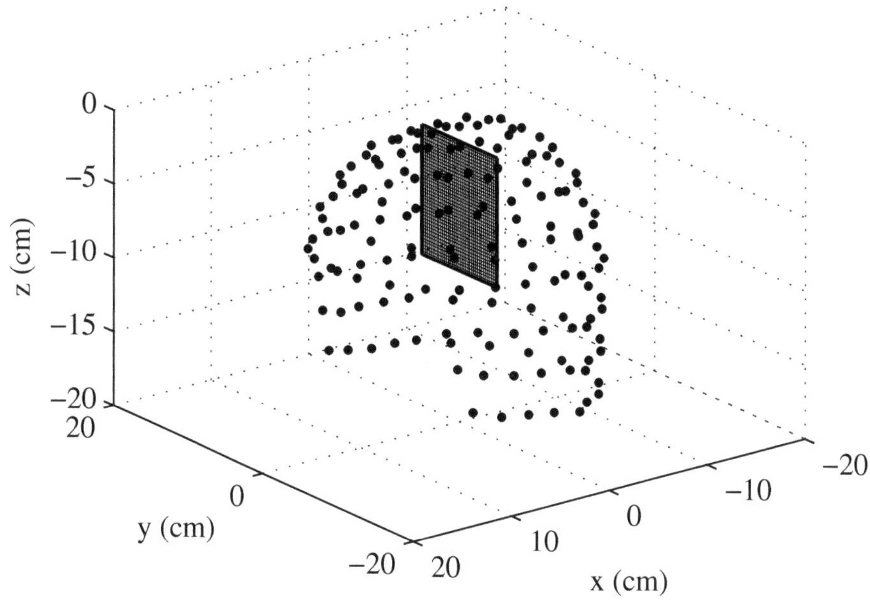

Figure 4.1: Three-dimensional view of the locations of the 148 sensors for the whole-head sensor array used in the numerical experiments in this book. The hatched plane indicates the plane $x = 0$ cm where sources are assumed to exist. The posterior to anterior direction is defined as that from negative to positive x coordinates, and the direction from the left to right hemispheres is defined as that from negative to positive y coordinates.

We define the cross-spectrum matrix of the measured data as $\boldsymbol{\Gamma}(f)$:

$$\boldsymbol{\Gamma}(f) = \langle \boldsymbol{g}(f)\boldsymbol{g}^H(f) \rangle, \tag{4.100}$$

where the superscript H indicates the Hermitian transpose, which is the matrix transpose with the complex conjugate operation. Then, the frequency-selective weight $\boldsymbol{w}(\boldsymbol{r}, f)$ is obtained using,

$$\boldsymbol{w}(\boldsymbol{r}, f) = \underset{\boldsymbol{w}(\boldsymbol{r},f)}{\arg\min}\, \boldsymbol{w}^H(\boldsymbol{r}, f)\boldsymbol{\Gamma}(f)\boldsymbol{w}(\boldsymbol{r}, f), \quad \text{subject to} \quad \boldsymbol{w}^H(\boldsymbol{r}, f)\boldsymbol{l}(\boldsymbol{r}) = 1. \tag{4.101}$$

The resultant weight is expressed as

$$\boldsymbol{w}(\boldsymbol{r}, f) = \frac{\boldsymbol{\Gamma}^{-1}(f)\boldsymbol{l}(\boldsymbol{r})}{\boldsymbol{l}^T(\boldsymbol{r})\boldsymbol{\Gamma}^{-1}(f)\boldsymbol{l}(\boldsymbol{r})}. \tag{4.102}$$

The estimated source activity at $\boldsymbol{r}$ and for the frequency f, $\widehat{s}(\boldsymbol{r}, f)$, is obtained as

$$\widehat{s}(\boldsymbol{r}, f) = \boldsymbol{w}^H(\boldsymbol{r}, f)\boldsymbol{g}(f). \tag{4.103}$$

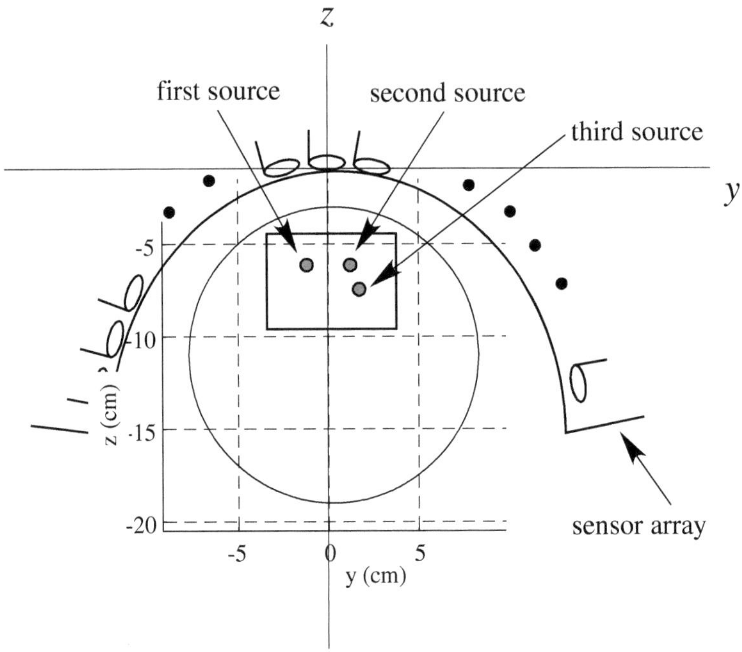

Figure 4.2: The source and sensor configuration used in the numerical experiments. The plane $x = 0$ cm is shown. The three filled circles show the locations of the three sources, and a large circle shows the boundary of the sphere used for the forward calculation. The square shows the reconstruction region used in the experiments of Section 4.7.

The cross-spectrum matrix can be calculated for a specific frequency window F_w, instead of a particular frequency f and in that case, the matrix is given by

$$\boldsymbol{\Gamma}(F_w) = \sum_{f \in F_w} \langle \boldsymbol{g}(f) \boldsymbol{g}^H(f) \rangle, \tag{4.104}$$

where $\sum_{f \in F_w}$ indicates the summation over the frequency window F_w. The corresponding weight is expressed as

$$\boldsymbol{w}(\boldsymbol{r}, F_w) = \frac{\boldsymbol{\Gamma}^{-1}(F_w)\boldsymbol{l}(\boldsymbol{r})}{\boldsymbol{l}^T(\boldsymbol{r})\boldsymbol{\Gamma}^{-1}(F_w)\boldsymbol{l}(\boldsymbol{r})}. \tag{4.105}$$

The output power at $\boldsymbol{r}$ in this case, $\langle \widehat{s}(\boldsymbol{r}, F_w)^2 \rangle$, is given by

$$\langle \widehat{s}(\boldsymbol{r}, F_w)^2 \rangle = \boldsymbol{w}^T(\boldsymbol{r}, F_w)\boldsymbol{\Gamma}(F_w)\boldsymbol{w}(\boldsymbol{r}, F_w) = \frac{1}{\boldsymbol{l}^T(\boldsymbol{r})\boldsymbol{\Gamma}^{-1}(F_w)\boldsymbol{l}(\boldsymbol{r})}. \tag{4.106}$$

Using the array-gain constraint $\boldsymbol{w}^H(\boldsymbol{r}, f)\boldsymbol{l}(\boldsymbol{r}) = \|\boldsymbol{l}(\boldsymbol{r})\|$, we can derive the weight for the array-gain-constraint frequency-domain minimum-variance filter, such that

$$\boldsymbol{w}(\boldsymbol{r}, F_w) = \frac{\boldsymbol{\Gamma}^{-1}(F_w)\widetilde{\boldsymbol{l}}(\boldsymbol{r})}{\widetilde{\boldsymbol{l}}^T(\boldsymbol{r})\boldsymbol{\Gamma}^{-1}(F_w)\widetilde{\boldsymbol{l}}(\boldsymbol{r})}, \tag{4.107}$$

where $\widetilde{\boldsymbol{l}}(\boldsymbol{r})$ is again the normalized lead-field vector. The output power in this case is expressed as

$$\langle \widehat{s}(\boldsymbol{r}, F_w)^2 \rangle = \frac{1}{\widetilde{\boldsymbol{l}}^T(\boldsymbol{r})\boldsymbol{\Gamma}^{-1}(F_w)\widetilde{\boldsymbol{l}}(\boldsymbol{r})}. \tag{4.108}$$

Using the unit-noise-gain constraint $\boldsymbol{w}^T\boldsymbol{w}(\boldsymbol{r}) = 1$, we can derive the weight for the unit-noise-gain constraint frequency-domain minimum-variance filter, such that

$$\boldsymbol{w}(\boldsymbol{r}, F_w) = \frac{\boldsymbol{\Gamma}^{-1}(F_w)\boldsymbol{l}(\boldsymbol{r})}{\sqrt{\boldsymbol{l}^T(\boldsymbol{r})\boldsymbol{\Gamma}^{-2}(F_w)\boldsymbol{l}(\boldsymbol{r})}}, \tag{4.109}$$

and the output power in this case is expressed as

$$\langle \widehat{s}(\boldsymbol{r}, F_w)^2 \rangle = \frac{\boldsymbol{l}^T(\boldsymbol{r})\boldsymbol{\Gamma}^{-1}(F_w)\boldsymbol{l}(\boldsymbol{r})}{\boldsymbol{l}^T(\boldsymbol{r})\boldsymbol{\Gamma}^{-2}(F_w)\boldsymbol{l}(\boldsymbol{r})}. \tag{4.110}$$

4.7 Numerical examples

Numerical experiments were conducted to illustrate some results of our arguments in this chapter. For the numerical experiments throughout this book, (except the last experiments in Section 8.5,) we assumed a whole-head sensor array where 148 sensors were arranged on a helmet-shaped surface as shown in Fig. 4.1. This sensor array is the one used in Magnes 2500^{TM} (4D Neuroimaging Inc., San Diego, CA). The coordinate origin was chosen as the center of the sensor array, and the x, y, and z directions are defined as shown in Fig. 4.1. The values of (x, y, z) are expressed in centimeters. We assumed a single vertical plane ($x = 0$), which is shown by a hatched plane in this figure. For the numerical experiments, we assumed that three point sources existed on this plane of $x = 0$, as depicted in Fig. 4.2. The locations and orientations of the three sources are shown in Table I.

Table I. Source parameter values used in the numerical experiments

Source number	Location (cm)	Orientation
1	(0, −1.0, −6.0)	(1.0, 0.0, 0.0)
2	(0, 1.0, −6.0)	(0.7, 0.7, 0.0)
3	(0, 1.6, −7.2)	(0.0, 0.7, 0.7)

The assumed time courses of the three sources, $s_1(t)$, $s_2(t)$, $s_3(t)$, are shown in Fig. 4.3(a). The sensor-array output, $\boldsymbol{b}(t)$, was then calculated at 400 time points using

$$\boldsymbol{b}(t) = \sum_{q=1}^{3} s_q(t)\boldsymbol{l}(\boldsymbol{r}_q) + \boldsymbol{n}(t), \tag{4.111}$$

where $\boldsymbol{l}(\boldsymbol{r}_q)$, $(q = 1, 2, 3)$, is the lead-field vector at the qth source location and orientation. These lead-field vectors were calculated using the formula for the spherical homogeneous conductor in Eq. (13.137), with the center of the sphere set at $(0, 0, -11)$. The noise, $\boldsymbol{n}(t)$, was generated using a Gaussian random-number generator. The noise variance and the powers of the three sources were set so that the SNR defined in Eq. (2.45) for the three sources was 8. (The corresponding α values in Eq. (2.46), α_1, α_2, and α_3, were all equal to $8M$.) The calculated sensor-array measurements are shown in Fig. 4.3(b). The theoretical covariance matrix was obtained using

$$\boldsymbol{R} = \sum_{q=1}^{3} \sigma_q^2 \boldsymbol{l}(\boldsymbol{r}_q)\boldsymbol{l}^T(\boldsymbol{r}_q) + \sigma_0^2 \boldsymbol{I} \propto \sum_{q=1}^{3} \frac{\alpha_q}{\|\boldsymbol{l}(\boldsymbol{r}_q)\|^2} \boldsymbol{l}(\boldsymbol{r}_q)\boldsymbol{l}^T(\boldsymbol{r}_q) + \boldsymbol{I}, \tag{4.112}$$

Using this covariance matrix, we first applied two types of scalar minimum-variance spatial filters: the unit-gain and the array-gain minimum-variance spatial filters. The results of (the square-root of) the source-power reconstruction are shown in Fig. 4.4. The results of the unit-gain spatial filter are shown in (a), and those of the array-gain spatial filter are shown in (b). Here, the three sources are accurately reconstructed in both sets of results. However, the results in (a) show that a false intensity increase, called an l-norm artifact, occurs around the center of the sphere. The l-norm artifact is completely removed in the results of (b).

We next compare the array-gain and the unit-noise gain scalar spatial filters. The results from the array-gain spatial filter are shown in Fig. 4.5. Here, the three sources and their time courses are accurately reconstructed. The results from the unit-noise gain spatial filter are shown in Fig. 4.6. The three sources and their time courses are also accurately reconstructed. Moreover, compared to the results in Fig. 4.5, we can see that the spatial resolution is much higher than that of the results of the array-gain minimum-variance spatial filter. We will discuss the spatial resolution of the adaptive spatial filters in Chapter 5.

The results of the array-gain constraint vector spatial filter are shown in Fig. 4.7 and those of the unit-noise-gain vector spatial filter are in Fig. 4.8. In these reconstruction results, because the spherical homogeneous conductor was used, the two tangential components of the source vector were obtained. We then calculated $\widehat{s}_{\parallel}(\boldsymbol{r}, t)$ and $\widehat{s}_{\perp}(\boldsymbol{r}, t)$, which respectively represent the source component in the $\bar{\boldsymbol{\eta}}_{opt}$ direction and in the direction perpendicular to $\bar{\boldsymbol{\eta}}_{opt}$. In Figs. 4.7 and 4.8, we used $\widehat{s}_{\parallel}(\boldsymbol{r}, t)$ and $\widehat{s}_{\perp}(\boldsymbol{r}, t)$ when displaying the estimated source time courses. These results show that the scalar and the corresponding vector version of spatial filters give almost identical results.

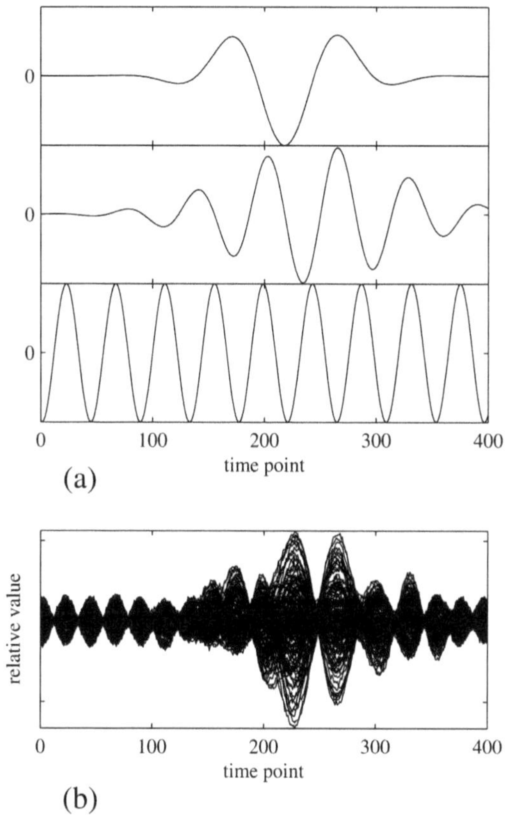

Figure 4.3: (a) The time courses of the three sources assumed in the numerical experiments. Time courses from the first to third sources are shown from the top to bottom row, respectively. Each time course is normalized by its maximum value. (b) The simulated sensor-array measurements calculated at 400 time points using the source parameters in Table I.

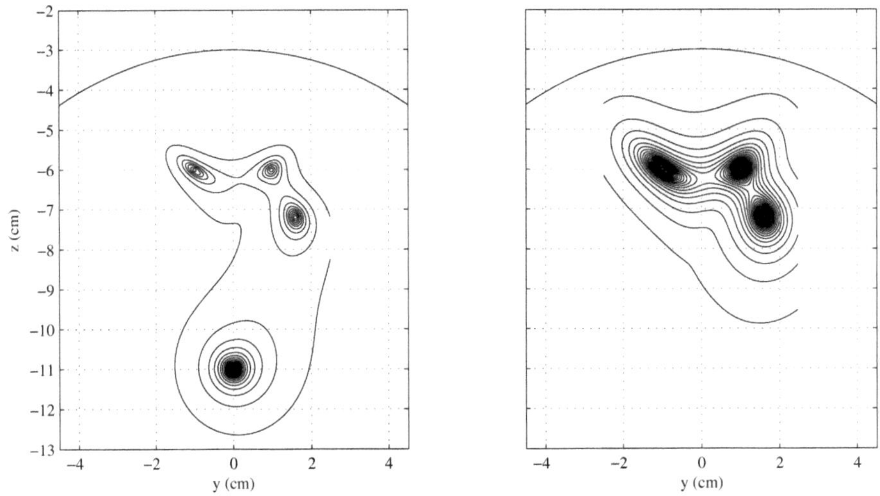

(a) (b)

Figure 4.4: Square root of the source-power reconstruction, $\sqrt{\widehat{P}_s(\boldsymbol{r})}$, from the scalar minimum-variance spatial filter. (a) Unit-gain-constraint spatial filter used. (b) Array-gain-constraint spatial filter used.

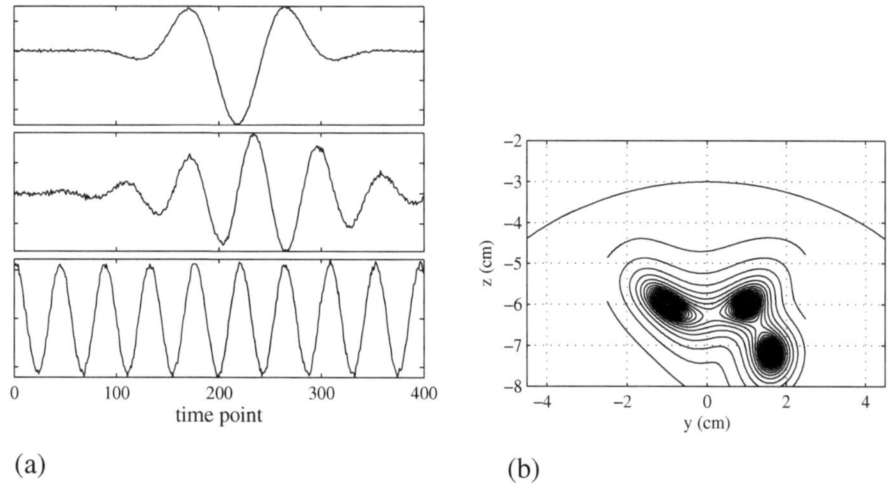

Figure 4.5: Results of the scalar array-gain minimum-variance spatial filter source reconstruction. (a) The time courses of the three sources obtained as the spatial filter outputs at the first to third source locations. (b) Square root of the source-power reconstruction, $\sqrt{\widehat{P}_s(\boldsymbol{r})}$.

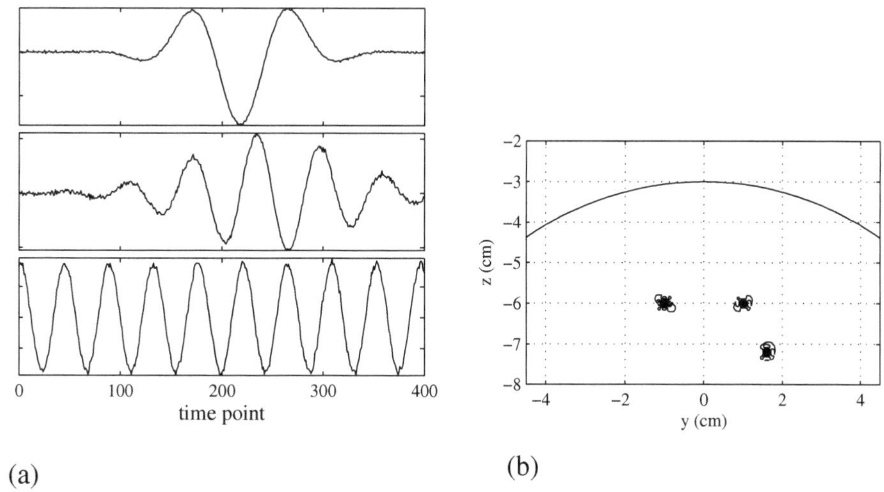

Figure 4.6: Results of the scalar unit-noise-gain minimum-variance spatial filter source reconstruction. (a) The time courses of the three sources obtained as the spatial-filter outputs at the first to third source locations. (b) Square root of the source-power reconstruction $\sqrt{\widehat{P}_s(\boldsymbol{r})}$.

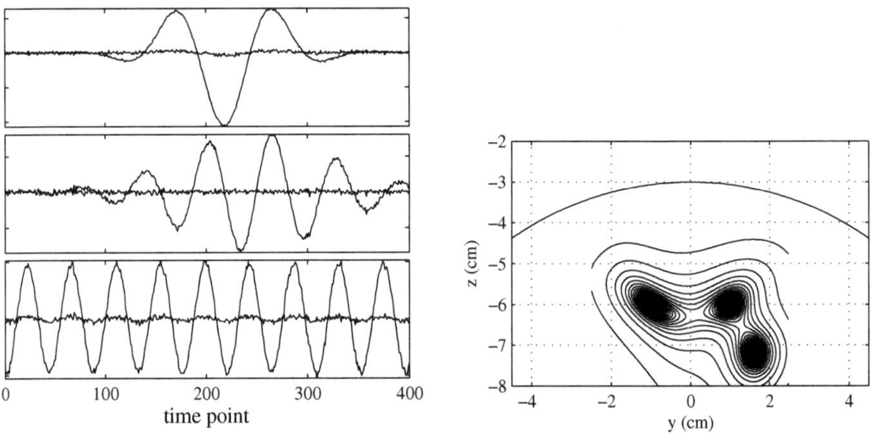

(a) (b)

Figure 4.7: Results of the vector array-gain minimum-variance spatial filter source reconstruction. (a) The time courses of the three sources obtained as the spatial-filter outputs at the first to third source locations. The two time courses correspond to $\widehat{s}_{\|}(\boldsymbol{r},t)$ and $\widehat{s}_{\perp}(\boldsymbol{r},t)$. (b) Square root of the source-power reconstruction $\sqrt{\widehat{P}_V^{(\mathrm{I})}(\boldsymbol{r})}$.

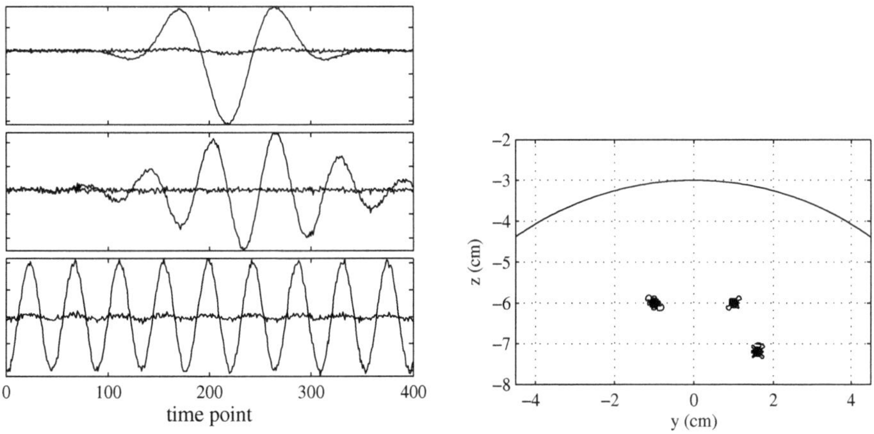

(a) (b)

Figure 4.8: Results of the vector unit-noise-gain minimum-variance spatial filter source reconstruction. (a) The time courses of the three sources obtained as the spatial filter outputs at the first to third source locations. The two time courses correspond to $\widehat{s}_{\parallel}(\boldsymbol{r}, t)$ and $\widehat{s}_{\perp}(\boldsymbol{r}, t)$. (b) Square root of the source-power reconstruction $\sqrt{\widehat{P}_V^{(\mathrm{I})}(\boldsymbol{r})}$.

Chapter 5

Location bias, spatial resolution, and beam response

In this chapter, we discuss the location bias and the spatial resolution of adaptive and non-adaptive spatial filters introduced in the previous chapters. We first analyze the location bias for several representative adaptive and non-adaptive spatial filters, and then compare the spatial resolution for several spatial filters having no location bias. The point-spread function plays a key role in these analyses. We then discuss the beam response of the spatial filters, which characterizes the gain on the leakage of sources located elsewhere from the filter pointing location.

5.1 Bias properties of various spatial filters

5.1.1 Definition of source location bias

Here we consider whether the various types of non-adaptive and adaptive spatial filters introduced in the previous chapters have a bias in their estimated source locations. As mentioned in Section 2.5.3, assuming that a single point source exists at $\boldsymbol{r}_1$, the point-spread function is defined in Eq. (2.84), which expresses the reconstruction of the point source located at $\boldsymbol{r}_1$. The most desirable property of the point-spread function is that it has a maximum value at the source location $\boldsymbol{r}_1$. If this condition is not met, the spatial filter will reconstruct a source at a location different from the true source location. The difference between the true and the reconstructed source locations is called the location bias. A conceptual view of a point-spread function is given in Fig. 5.1. A case where the reconstruction suffers from source location bias is given in Fig. 5.1(a). A case where a point-spread function has its maximum at the true source location $\boldsymbol{r}_1$ is shown in Fig. 5.1(b). Naturally, when the bias is large, the reconstruction results become meaningless.

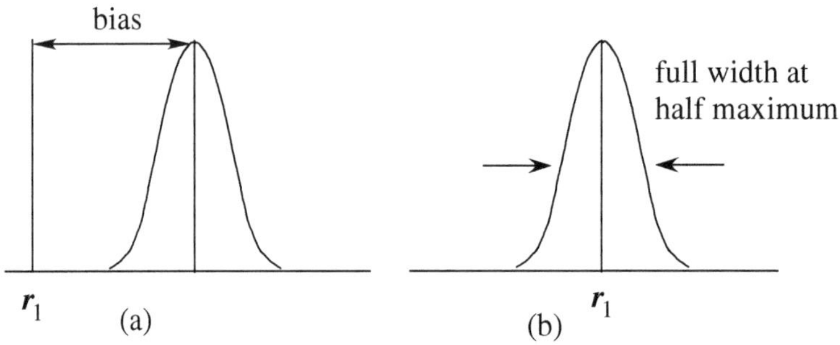

Figure 5.1: Conceptual view of two point-spread functions. (a) The case where the point-spread function does not have its maximum at $\boldsymbol{r}_1$, and the spatial filter output has a location bias. (b) The case where the point-spread function has its maximum at $\boldsymbol{r}_1$, and the filter output has no location bias. In this case, the full-width-at-half-maximum is used as a measure of the spatial resolution of the reconstruction results.

Therefore, in this section, the bias for the estimated source location is evaluated by checking whether the point-spread function has its maximum at the source location $\boldsymbol{r}_1$, i.e., whether the condition

$$\mathcal{F}(\boldsymbol{r}_1) > \mathcal{F}(\boldsymbol{r}) \tag{5.1}$$

holds for any $\boldsymbol{r}$ $(\boldsymbol{r} \neq \boldsymbol{r}_1)$ for various spatial filters. Note that this is equivalent to evaluating the source location bias using a simple scenario where a single source exists in a noiseless environment.

When a point-spread function has its maximum at the true source location, the reconstructed results are the smoothed version of the true source distribution. The width of the point-spread function's main lobe is a measure of the spatial resolution. To evaluate the spatial resolution, we introduce the normalized point spread function $\mathcal{F}_N(\boldsymbol{r})$, which is defined as

$$\mathcal{F}_N(\boldsymbol{r}) = \mathcal{F}(\boldsymbol{r})/\mathcal{F}(\boldsymbol{r}_1). \tag{5.2}$$

5.1.2 Bias for the spatial matched filter

We first check whether the relationship in Eq. (5.1) holds for the spatial matched filter, which is the most basic non-adaptive spatial filter. We denote the lead-field vector at the source location and the source orientation as $\boldsymbol{f}$, i.e., $\boldsymbol{l}(\boldsymbol{r}_1) = \boldsymbol{f}$. The point-spread function is expressed using the weight in Eq. (3.43) as

$$\mathcal{F}(\boldsymbol{r}) = \boldsymbol{w}^T(\boldsymbol{r})\boldsymbol{l}(\boldsymbol{r}_1) = \frac{\boldsymbol{l}^T(\boldsymbol{r})\boldsymbol{l}(\boldsymbol{r}_1)}{\|\boldsymbol{l}(\boldsymbol{r})\|} = \|\boldsymbol{f}\| \cos(\boldsymbol{l}, \boldsymbol{f}), \tag{5.3}$$

where the explicit notation of $(\boldsymbol{r})$ is omitted from $\boldsymbol{l}(\boldsymbol{r})$. The generalized cosine is defined in Eq. (13.39) in the Appendix, and we have $\digamma(\boldsymbol{r}_1) = \|\boldsymbol{f}\|$. Thus, from Eq. (5.1), the condition for the spatial matched filter to have no bias is given by:

$$\cos(\boldsymbol{l}, \boldsymbol{f}) < 1. \tag{5.4}$$

According to the arguments in Section 13.4, the relationship in Eq. (5.4) holds for any pair $\boldsymbol{l}$ and $\boldsymbol{f}$ (where $\boldsymbol{l} \neq \boldsymbol{f}$), and the spatial matched filter is found to have no location bias in the noiseless case.

5.1.3 Bias for the minimum-norm filter

For the minimum-norm filter, the point-spread function is given by:

$$\digamma(\boldsymbol{r}) = \boldsymbol{w}^T(\boldsymbol{r})\boldsymbol{f} = \boldsymbol{l}^T(\boldsymbol{r})\boldsymbol{G}^{-1}\boldsymbol{f}, \tag{5.5}$$

and the condition in Eq. (5.1) is expressed as

$$\boldsymbol{f}^T\boldsymbol{G}^{-1}\boldsymbol{f} > \boldsymbol{l}^T\boldsymbol{G}^{-1}\boldsymbol{f}. \tag{5.6}$$

Since the norm of $\boldsymbol{l}$ may become larger than the norm of $\boldsymbol{f}$ in a region close to the sensors, the above inequality obviously does not always hold. Actually, it is well known that the source reconstruction of the minimum-norm filter is severely biased toward the sensor array[24]. We give an example of such a biased reconstruction in our numerical experiments.

5.1.4 Bias for the weight-normalized minimum-norm filter

For the weight-normalized minimum-norm filter in Eq. (3.24), the point-spread function is expressed as

$$\digamma(\boldsymbol{r}) = \frac{\boldsymbol{l}^T\boldsymbol{G}^{-1}\boldsymbol{f}}{\sqrt{\boldsymbol{l}^T\boldsymbol{G}^{-2}\boldsymbol{l}}}. \tag{5.7}$$

Thus, the condition in Eq. (5.1) is expressed as

$$\frac{\boldsymbol{f}^T\boldsymbol{G}^{-1}\boldsymbol{f}}{\sqrt{\boldsymbol{f}^T\boldsymbol{G}^{-2}\boldsymbol{f}}} > \frac{\boldsymbol{l}^T\boldsymbol{G}^{-1}\boldsymbol{f}}{\sqrt{\boldsymbol{l}^T\boldsymbol{G}^{-2}\boldsymbol{l}}}. \tag{5.8}$$

It is not straightforward to see whether this inequality holds for any pair of $\boldsymbol{l}$ and $\boldsymbol{f}$. However, we can theoretically and numerically show that it does not generally hold. The theoretical proof is given in [30], and we give a numerical example of a biased source-reconstruction obtained from this weight-normalized minimum-norm filter in Section 5.5 of this chapter.

5.1.5 Bias for the sLORETA filter

For sLORETA, the point-spread function is given by:

$$F(\boldsymbol{r}) = \frac{\boldsymbol{l}^T\boldsymbol{G}^{-1}\boldsymbol{f}}{\sqrt{\boldsymbol{l}^T\boldsymbol{G}^{-1}\boldsymbol{l}}} = \sqrt{\boldsymbol{f}^T\boldsymbol{G}^{-1}\boldsymbol{f}}\cos(\boldsymbol{l},\boldsymbol{f}|\boldsymbol{G}^{-1}), \tag{5.9}$$

where the generalized cosine with the metric $\boldsymbol{G}^{-1}$ is defined in Eq. (13.37) in the Appendix. We have

$$F(\boldsymbol{r}_1) = \frac{\boldsymbol{f}^T\boldsymbol{G}^{-1}\boldsymbol{f}}{\sqrt{\boldsymbol{f}^T\boldsymbol{G}^{-1}\boldsymbol{f}}} = \sqrt{\boldsymbol{f}^T\boldsymbol{G}^{-1}\boldsymbol{f}}.$$

Thus, the condition in Eq. (5.1) is expressed as

$$\cos(\boldsymbol{l},\boldsymbol{f}|\boldsymbol{G}^{-1}) < 1. \tag{5.10}$$

Since $\boldsymbol{G}^{-1}$ is a positive definite matrix, this condition holds for any pair $\boldsymbol{f}$ and $\boldsymbol{l}$, as mentioned in Section 13.4. It is empirically shown that sLORETA has no location bias for a single source by Pascual-Marqui[26], and the analysis presented here validates such empirical findings.

5.1.6 Bias for the unit-gain minimum-variance spatial filter

For the unit-gain minimum-variance spatial filter, the point-spread function is given by:

$$F(\boldsymbol{r}) = \frac{\boldsymbol{l}^T\boldsymbol{R}^{-1}\boldsymbol{f}}{\boldsymbol{l}^T\boldsymbol{R}^{-1}\boldsymbol{l}}. \tag{5.11}$$

We have

$$F(\boldsymbol{r}_1) = \frac{\boldsymbol{f}^T\boldsymbol{R}^{-1}\boldsymbol{f}}{\boldsymbol{f}^T\boldsymbol{R}^{-1}\boldsymbol{f}} = 1.$$

Thus, the condition for having no location bias is that

$$\frac{\boldsymbol{l}^T\boldsymbol{R}^{-1}\boldsymbol{f}}{\boldsymbol{l}^T\boldsymbol{R}^{-1}\boldsymbol{l}} < 1. \tag{5.12}$$

Using Eqs. (13.46) and (13.47) in Section 13.4 in the Appendix, the above condition can be rewritten as

$$\frac{\|\boldsymbol{f}\|}{\|\boldsymbol{l}\|}\frac{\cos(\boldsymbol{l},\boldsymbol{f})}{1+\alpha[1-\cos^2(\boldsymbol{l},\boldsymbol{f})]} < 1. \tag{5.13}$$

In the above equation, $\alpha = (\sigma_1^2/\sigma_0^2)\|\boldsymbol{f}\|^2$ is the input SNR defined in Eq. (2.46). It can be seen that the inequality in Eq. (5.13) may not hold when $\|\boldsymbol{l}\|$ is small. This can happen in a region near the center of the sphere when the spherical homogeneous conductor model is used to compute the lead field. As a result, severe artifacts appear near the center of the sphere, as shown in Fig. 4.4(a).

5.1.7 Bias for the array-gain minimum-variance spatial filter

For the array-gain-constraint minimum-variance spatial filter, the point-spread function is given by:

$$\mathcal{F}(\boldsymbol{r}) = \frac{\widetilde{\boldsymbol{l}}^T \boldsymbol{R}^{-1} \boldsymbol{f}}{\widetilde{\boldsymbol{l}}^T \boldsymbol{R}^{-1} \widetilde{\boldsymbol{l}}} = \|\boldsymbol{f}\| \frac{\cos(\widetilde{\boldsymbol{l}}, \boldsymbol{f})}{1 + \alpha[1 - \cos^2(\widetilde{\boldsymbol{l}}, \boldsymbol{f})]}. \tag{5.14}$$

We have

$$\mathcal{F}(\boldsymbol{r}_1) = \frac{\widetilde{\boldsymbol{f}}^T \boldsymbol{R}^{-1} \boldsymbol{f}}{\widetilde{\boldsymbol{f}}^T \boldsymbol{R}^{-1} \widetilde{\boldsymbol{f}}} = \|\boldsymbol{f}\|,$$

where $\widetilde{\boldsymbol{f}} = \widetilde{\boldsymbol{f}}/\|\boldsymbol{f}\|$. Thus, the condition for no location bias in this case is expressed as

$$\frac{\cos(\widetilde{\boldsymbol{l}}, \boldsymbol{f})}{1 + \alpha[1 - \cos^2(\widetilde{\boldsymbol{l}}, \boldsymbol{f})]} < 1. \tag{5.15}$$

Since α is positive, it is clear that this inequality is always fulfilled for any $\boldsymbol{f}$ and $\widetilde{\boldsymbol{l}}$ (where $\boldsymbol{f} \neq \widetilde{\boldsymbol{l}}$). Thus, we can conclude that the array-gain constraint spatial filter has no location bias.

5.1.8 Bias for the unit-noise-gain minimum-variance spatial filter

For the unit-noise-gain minimum variance spatial filter, the point-spread function is expressed as

$$\mathcal{F}(\boldsymbol{r}) = \frac{\boldsymbol{l}^T \boldsymbol{R}^{-1} \boldsymbol{f}}{\sqrt{\boldsymbol{l}^T \boldsymbol{R}^{-2} \boldsymbol{l}}}. \tag{5.16}$$

Using Eqs. (13.45) and (13.48), we have

$$\mathcal{F}(\boldsymbol{r}_1) = \frac{\boldsymbol{f}^T \boldsymbol{R}^{-1} \boldsymbol{f}}{\sqrt{\boldsymbol{f}^T \boldsymbol{R}^{-2} \boldsymbol{f}}} = \|\boldsymbol{f}\|, \tag{5.17}$$

and using Eqs. (5.16), (13.47), and (13.49), the condition for the point-spread function having the maximum at the source location is finally given by:

$$\frac{\cos(\boldsymbol{l}, \boldsymbol{f})}{\sqrt{1 + \alpha(\alpha + 2)(1 - \cos^2(\boldsymbol{l}, \boldsymbol{f}))}} < 1. \tag{5.18}$$

Because α is positive, the above condition obviously holds for any $\boldsymbol{l}$ and $\boldsymbol{f}$, and we can conclude that the unit-noise-gain spatial filter does not have source location bias.

5.2 Effects of noise on the location bias

In the preceding section, the source location bias was analyzed using the point-spread function and such analysis should be valid when the SNR is sufficiently high. However, when the SNR is low, the noise may cause bias in the estimated source location even though the point-spread function has its maximum at the true source location. We here investigate the effects of noise on the source location bias, assuming that the noise is uncorrelated white Gaussian noise. The output signal power at a pointing location $\boldsymbol{r}$ is equal to $\sigma_1^2\mathbb{R}(\boldsymbol{r},\boldsymbol{r}_1)^2$ where $\mathbb{R}(\boldsymbol{r},\boldsymbol{r}_1)$ indicates the resolution kernel when a source is located at $\boldsymbol{r}_1$. The output noise power is equal to $\sigma_0^2\|\boldsymbol{w}(\boldsymbol{r})\|^2$. Therefore, when the noise is taken into account, the condition for no location bias is:

$$\sigma_1^2\mathbb{R}(\boldsymbol{r}_1,\boldsymbol{r}_1)^2+\sigma_0^2\|\boldsymbol{w}(\boldsymbol{r}_1)\|^2>\sigma_1^2\mathbb{R}(\boldsymbol{r},\boldsymbol{r}_1)^2+\sigma_0^2\|\boldsymbol{w}(\boldsymbol{r})\|^2. \tag{5.19}$$

For the spatial matched filter, since the weight norm is always equal to 1, i.e., $\|\boldsymbol{w}(\boldsymbol{r})\|=\|\boldsymbol{w}(\boldsymbol{r}_1)\|=1$, by substituting Eq. (5.3) into Eq. (5.19), we derive the condition for no location bias in this case such that

$$\cos^2(\boldsymbol{l},\boldsymbol{f})<1. \tag{5.20}$$

This condition is equivalent to that for the no-noise case in Eq. (5.4), and the above relationship always holds, indicating that the spatial matched filter does not have a source location bias even when the noise cannot be ignored.

For other spatial filters, this condition in Eq. (5.19) can be rewritten as

$$\frac{\mathbb{R}(\boldsymbol{r},\boldsymbol{r}_1)^2}{\mathbb{R}(\boldsymbol{r}_1,\boldsymbol{r}_1)^2}\frac{[1+\Phi(\boldsymbol{r})/\alpha]}{[1+\Phi(\boldsymbol{r}_1)/\alpha]}<1, \tag{5.21}$$

where α is again the input SNR and $\Phi(\boldsymbol{r})$ is given by:

$$\Phi(\boldsymbol{r})=\frac{\|\boldsymbol{f}\|^2\|\boldsymbol{w}(\boldsymbol{r})\|^2}{\mathbb{R}(\boldsymbol{r},\boldsymbol{r}_1)^2}. \tag{5.22}$$

Note that since $\Phi(\boldsymbol{r})$ does not depend on the input noise or signal power, when α is very large, the condition in Eq. (5.21) becomes identical to the condition for no-noise cases in Eq. (5.1).

In the case of sLORETA filter, Eq. (5.21) can be rewritten as

$$\frac{1+\Phi(\boldsymbol{r})/\alpha}{1+\Phi(\boldsymbol{r}_1)/\alpha}\cos^2(\boldsymbol{l},\boldsymbol{f}|\boldsymbol{G}^{-1})<1 \tag{5.23}$$

It can be seen in Eq. (5.23) that when α is very large, this condition becomes identical to Eq. (5.10), and as a result, this condition holds. However, in general, the value of α determines whether the condition in Eq. (5.23) is satisfied for any $\boldsymbol{l}$ and $\boldsymbol{f}$. In our numerical experiments, we present both examples, in which Eq. (5.23) is either fulfilled or not.

For the array-gain minimum-variance spatial filter, noting that $\widetilde{\boldsymbol{l}} = \boldsymbol{l}/\|\boldsymbol{l}\|$, the condition for no location bias in Eq. (5.21) is rewritten as

$$\frac{[1+\Phi(\boldsymbol{r})/\alpha]}{[1+\Phi(\boldsymbol{r}_1)/\alpha]} \frac{\cos^2(\widetilde{\boldsymbol{l}},\boldsymbol{f})}{[1+\alpha(1-\cos^2(\widetilde{\boldsymbol{l}},\boldsymbol{f}))]^2} < 1, \tag{5.24}$$

where $\Phi(\boldsymbol{r})$, in this case, is given by:

$$\Phi(\boldsymbol{r}) = \frac{\|\boldsymbol{f}\|^2[\widetilde{\boldsymbol{l}}^T \boldsymbol{R}^{-2}\widetilde{\boldsymbol{l}}]}{[\widetilde{\boldsymbol{l}}^T \boldsymbol{R}^{-1}\boldsymbol{f}]^2}. \tag{5.25}$$

Substituting Eq. (5.25) into Eq. (5.24), and using Eqs. (13.47) and (13.49), the condition in Eq. (5.24) can be simplified to

$$\frac{1}{1+\alpha\sin^2(\widetilde{\boldsymbol{l}},\boldsymbol{f})} < 1. \tag{5.26}$$

Since $\alpha > 0$, this relationship holds for any $\widetilde{\boldsymbol{l}}$ and $\boldsymbol{f}$, and this fact indicates that the array-gain constraint minimum-variance spatial filter has no location bias even when the SNR is low. For the unit-noise-gain spatial filter, the condition in Eq. (5.21) is simplified to

$$\frac{1+\alpha\sin^2(\boldsymbol{l},\boldsymbol{f})}{[1+\alpha\sin^2(\boldsymbol{l},\boldsymbol{f})+\alpha(\alpha+1)\sin^2(\boldsymbol{l},\boldsymbol{f})]} < 1. \tag{5.27}$$

It is again clear that this condition holds for any $\boldsymbol{l}$ and $\boldsymbol{f}$, and the unit-noise-gain minimum-variance filter does not have a location bias even when the SNR is low.

5.3 Spatial resolution

When the point-spread function has its maximum at the source location, the reconstructed source distribution can be interpreted as the smoothed version of the true source distribution, and the main-lobe width of the point-spread function can be a measure of the spatial resolution. To compare the main-lobe width of the point-spread function, the normalized point-spread function, $\mathcal{F}_N(\boldsymbol{r})$, is derived for the spatial filters with no location bias. For the spatial matched filter, $\mathcal{F}_N(\boldsymbol{r})$ is expressed as

$$\mathcal{F}_N(\boldsymbol{r}) = \cos(\boldsymbol{l},\boldsymbol{f}). \tag{5.28}$$

For sLORETA, we have

$$\mathcal{F}_N(\boldsymbol{r}) = \cos(\boldsymbol{l},\boldsymbol{f}|\boldsymbol{G}^{-1}). \tag{5.29}$$

For either the array-gain or the unit-gain minimum-variance spatial filter, the normalized point-spread function is expressed as

$$\mathcal{F}_N(\boldsymbol{r}) = \frac{\cos(\widetilde{\boldsymbol{l}},\boldsymbol{f})}{1+\alpha[1-\cos^2(\widetilde{\boldsymbol{l}},\boldsymbol{f})]} = \frac{\cos(\boldsymbol{l},\boldsymbol{f})}{1+\alpha[1-\cos^2(\boldsymbol{l},\boldsymbol{f})]}. \tag{5.30}$$

For the unit-noise-gain spatial filter, it is expressed as

$$F_N(\boldsymbol{r}) = \frac{\cos(\boldsymbol{l}, \boldsymbol{f})}{\sqrt{1 + \alpha(\alpha+2)(1 - \cos^2(\boldsymbol{l}, \boldsymbol{f}))}} \approx \frac{\cos(\boldsymbol{l}, \boldsymbol{f})}{\sqrt{1 + \alpha^2(1 - \cos^2(\boldsymbol{l}, \boldsymbol{f}))}}, \tag{5.31}$$

where we use the fact that $\alpha \gg 1$.

The major difference between the point-spread functions of the non-adaptive spatial filters and those of the adaptive spatial filters is that the spread functions of the adaptive spatial filters strongly depend on the input SNR, α. This dependency has been reported previously[11]. Because α usually has a value greater than the number of sensors M, the denominators in Eqs. (5.30) and (5.31) cause a rapid decay, and consequently, the spatial resolution of the adaptive spatial filters is usually much higher than that of the non-adaptive spatial filters. In Section 5.5, numerical examples of the point-spread functions are presented to demonstrate the high spatial resolution of the adaptive spatial filters. Also, we can see that the denominator of Eq. (5.31) contains α^2, so that the unit-noise-gain spatial filter can attain a significantly higher spatial resolution than the array-gain spatial filter. We have already seen this high spatial resolution of the unit-noise-gain filters in the numerical experiments in Section 4.7.

5.4 Spatial-filter beam response

We next discuss the beam response of spatial filters. The beam response of the adaptive spatial filters has a very interesting property. The beam response, as briefly discussed in Section 2.5.3, expresses the sensitivity of a spatial filter to sources located elsewhere from the filter-pointing location. For an ideal spatial filter[1], the beam response of the filter should have the shape of the delta function. When the beam response is different from the delta function, as is typical, the beam response represents the gain for the unwanted leakage signal from sources located elsewhere from the filter-pointing location.

Assuming that $\boldsymbol{r}_0$ is the spatial-filter pointing location, according to Eq. (2.85), the beam response $\mathcal{H}(\boldsymbol{r})$ is defined as

$$\mathcal{H}(\boldsymbol{r}) = \mathbb{R}(\boldsymbol{r}_0, \boldsymbol{r}) = \boldsymbol{w}^T(\boldsymbol{r}_0)\boldsymbol{l}(\boldsymbol{r}). \tag{5.32}$$

For the spatial matched filter, substituting the weight vector in Eq. (3.43) into Eq. (5.32), we derive

$$\mathcal{H}(\boldsymbol{r}) = \frac{\boldsymbol{l}^T(\boldsymbol{r}_0)\boldsymbol{l}(\boldsymbol{r})}{\|\boldsymbol{l}(\boldsymbol{r}_0)\|} = \|\boldsymbol{l}(\boldsymbol{r})\| \cos\left(\boldsymbol{l}(\boldsymbol{r}_0), \boldsymbol{l}(\boldsymbol{r})\right). \tag{5.33}$$

The above equation shows that, in the case of the spatial matched filter, the beam response and the point-spread function have the same form. Consequently, if we implement spatial filter scanning using the spatial matched filter, the reconstructed source distribution has a blur represented by the point-spread function.

[1]Of course, such an ideal spatial filter never exists.

In contrast to the case of the spatial matched filter, the point-spread function and the beam response are generally different for most spatial filters. Let us look into the beam response of adaptive spatial filters, and calculate the beam response of the unit-gain minimum-variance spatial filter. Using Eqs. (4.5) and (5.32), the beam response is obtained as

$$\mathcal{H}(\boldsymbol{r}) = \left[\frac{\boldsymbol{l}^T(\boldsymbol{r}_0)\boldsymbol{R}^{-1}\boldsymbol{l}(\boldsymbol{r})}{[\boldsymbol{l}^T(\boldsymbol{r}_0)\boldsymbol{R}^{-1}\boldsymbol{l}(\boldsymbol{r}_0)]} \right]. \tag{5.34}$$

We assume here the same simple scenario as used in the previous sections in which only a single source with a power of σ_1^2 exists at $\boldsymbol{r}_1$. Substituting Eq. (13.43) into the above expression and letting $\boldsymbol{l}_0 = \boldsymbol{l}(\boldsymbol{r}_0)$, we derive

$$\mathcal{H}(\boldsymbol{r}) = \frac{[\cos(\boldsymbol{l}_0, \boldsymbol{l}) + \alpha(\cos(\boldsymbol{l}_0, \boldsymbol{l}) - \cos(\boldsymbol{l}_0, \boldsymbol{f})\cos(\boldsymbol{l}, \boldsymbol{f}))]}{[1 + \alpha \sin^2(\boldsymbol{l}_0, \boldsymbol{f})]}, \tag{5.35}$$

where $\boldsymbol{l} = \boldsymbol{l}(\boldsymbol{r})$, which is the lead-field vector for an arbitrary location and orientation, $\boldsymbol{l}_0$ is the lead-field vector for the filter pointing location and orientation, and $\boldsymbol{f}$ is the lead-field vector for the source location and orientation, i.e., $\boldsymbol{f} = \boldsymbol{l}(\boldsymbol{r}_1)$.

It is not straightforward to interpret the actual shape of $\mathcal{H}(\boldsymbol{r})$ because of its complex dependency on $\cos(\boldsymbol{l}_0, \boldsymbol{l})$, $\cos(\boldsymbol{l}, \boldsymbol{f})$, and $\cos(\boldsymbol{l}_0, \boldsymbol{f})$. Let us check the sensitivity at the source location $\boldsymbol{r}_1$ while the filter is pointing at $\boldsymbol{r}_0$. Substituting $\boldsymbol{f}$ for $\boldsymbol{l}$ in Eq. (5.35), we have,

$$\mathcal{H}(\boldsymbol{r}_1) = \frac{\cos(\boldsymbol{l}_0, \boldsymbol{f})}{[1 + \alpha \sin^2(\boldsymbol{l}_0, \boldsymbol{f})]}. \tag{5.36}$$

When the pointing location is equal to the source location, by setting $\boldsymbol{l}_0$ equal to $\boldsymbol{f}$, we obtain $\mathcal{H}(\boldsymbol{r}_1) = 1$. This corresponds to the fact that the value of the beam response at a source location is equal to the filter gain. When $\boldsymbol{l}_0 \neq \boldsymbol{f}$, i.e., when the pointing location is different from the source location, from Eq. (5.36) we generally have $\mathcal{H}(\boldsymbol{r}_1) \ll 1$ because $1 + \alpha \sin^2(\boldsymbol{l}_0, \boldsymbol{f}) \gg 1$. (Note that α is generally much greater than M, the total number of sensors.) Thus, in summary, we can see that the weight vector has the property

$$\boldsymbol{w}^T(\boldsymbol{r}_p)\boldsymbol{l}(\boldsymbol{r}_q) \approx \delta_{p,q}.$$

This weight property is consistent with the discussion in Chapter 4. Equation (5.35) also reveals the interesting fact that the beam response generally has a non-zero value when $\boldsymbol{l} \neq \boldsymbol{f}$. That is, the weight vector has non-zero sensitivity and $\boldsymbol{w}^T(\boldsymbol{r}_0)\boldsymbol{l}(\boldsymbol{r}) \neq 0$ at arbitrary $\boldsymbol{r}$ not equal to the source location. This property of the weight vector was already pointed out in Section 4.2.1. Accordingly, $\mathcal{H}(\boldsymbol{r})$ of the minimum-variance filter does not form a peak at the pointing location, as is the case for the beam response for the spatial matched filter. In other words, the minimum-variance filter does not form a virtual sensor with sensitivity localized around the pointing location; an ideal spatial filter forming such a virtual sensor is depicted in Fig. 2.2(b). In our numerical experiments we show, in Fig. 5.10, some numerical examples of the beam response of the minimum-variance spatial filter for a case where three sources exist.

5.5 Numerical examples

We give numerical examples of the results of our arguments in this chapter. We used the same computer simulation scheme in Section 4.7, except that a single point source existed at $\boldsymbol{r}_1 = (0, -1.5, -6)$. The source and the coordinate system are shown schematically in Fig. 5.2.

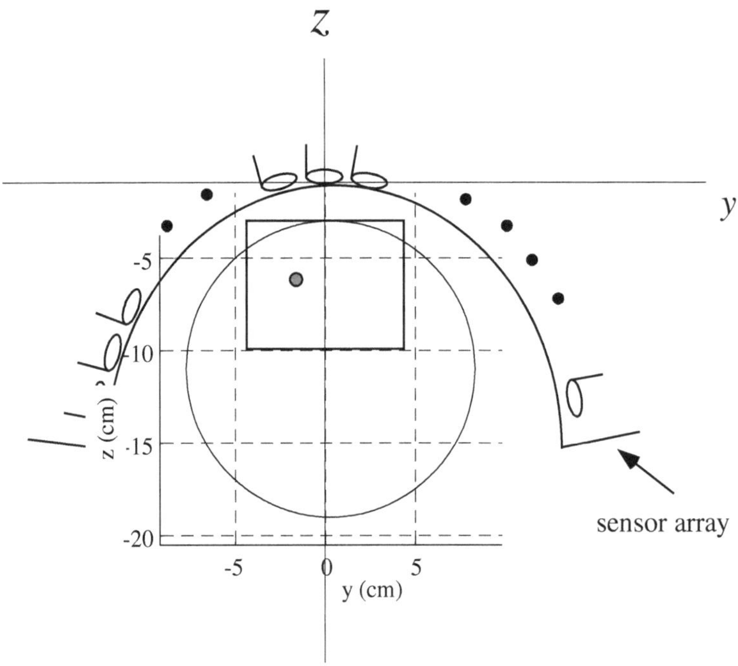

Figure 5.2: The geometry of the numerical experiments used for plotting point-spread functions in Section 5.5. A single source located at $(0, -1.5, -6)$ is shown by a gray circle, and the square shows the region for plotting the point-spread function in Figs. 5.3–5.6.

We first calculated the square of the point-spread function $\mathcal{F}(\boldsymbol{r})^2$ using the four types of non-adaptive spatial filters: (1) the spatial matched filter, (2) the minimum-norm filter, (3) the weight-normalized minimum-norm filter, and (4) sLORETA. The point-spread functions plotted on the plane $x = 0$ are shown in Fig. 5.3. When deriving the weight vectors for these non-adaptive spatial filters, the gram matrix $\boldsymbol{G}$ was obtained by using Eq. (2.10) by numerically integrating over the assumed source space, defined as $-4 \leq x \leq 4$, $-4 \leq y \leq 4$, and $-12 \leq z \leq -3$.

These results show that the point-spread functions of the spatial matched filter and sLORETA have their maximum at the source location. The point-

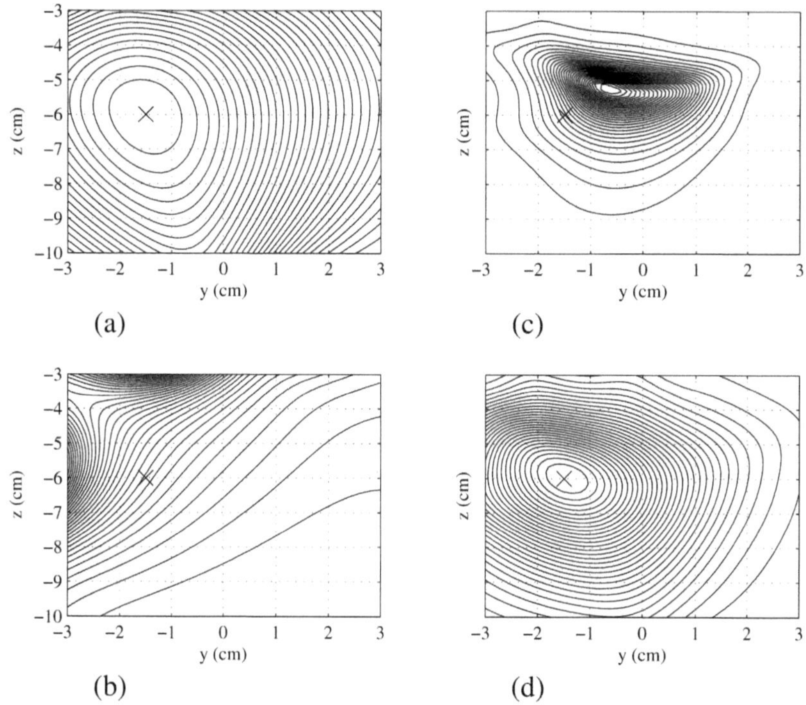

Figure 5.3: The plot of the square of the point-spread function on the plane $x = 0$ for four types of non-adaptive spatial filters. The single source located at $(0, -1.5, -6)$; this location is indicated by the cross mark $\times$. The results of (a) the spatial matched filter, (b) the minimum-norm filter, (c) the weight-normalized minimum-norm filter, and (d) the sLORETA filter.

spread function for the minimum-norm filter shows a severe location bias toward the sensors. The point-spread function for the weight-normalized minimum-norm spatial filter has its maximum at an incorrect location, although the bias is not as large as that of the original minimum-norm filter. These results are consistent with the arguments of Section 5.1.

We next calculated point-spread functions for three types of minimum-variance spatial filters: the unit-gain-constraint filter, the array-gain-constraint filter, and the unit-noise-gain filter. The point-spread functions plotted on the plane $x = 0$ are shown in Fig. 5.4. In these experiments, the covariance matrix $\boldsymbol{R}$ is obtained using

$$\boldsymbol{R} = \sigma_0^2 \boldsymbol{I} + \sigma_1^2 \boldsymbol{f}\boldsymbol{f}^T, \tag{5.37}$$

and the input SNR ($\alpha = (\sigma_1^2/\sigma_0^2)\|\boldsymbol{f}\|^2$) is set to M. (M is the number of sensors and $M = 148$ in these numerical experiments.) The point-spread function for the unit-gain minimum-variance spatial filter has two peaks: one at the source

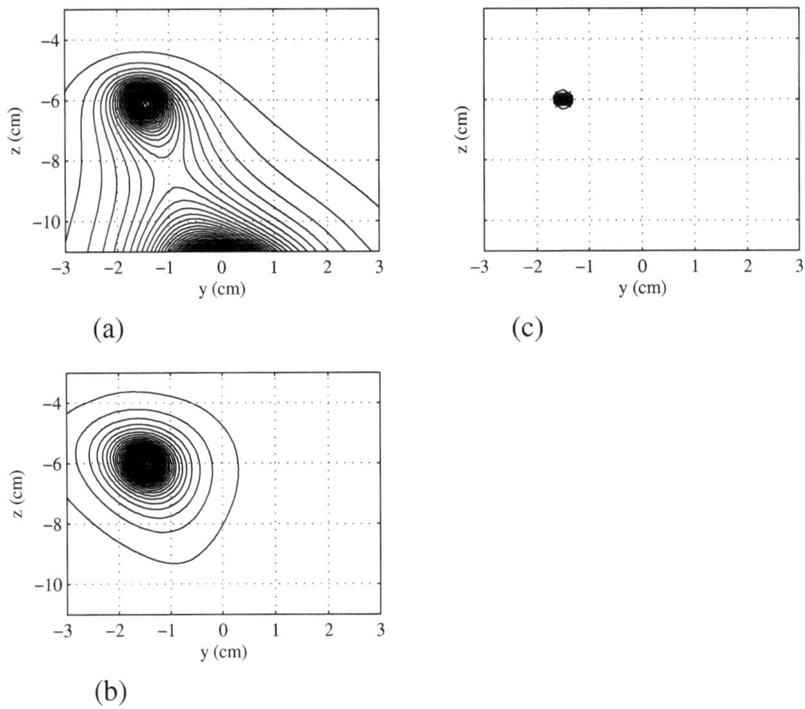

Figure 5.4: The plot of the square of the point-spread function on the plane $x = 0$ for three types of adaptive spatial filters. The single source is located at $(0, -1.5, -6)$. The results of (a) the unit-gain-constraint minimum-variance spatial filter, (b) the array-gain-constraint minimum-variance spatial filter, and (c) the unit-noise-gain-constraint minimum-variance spatial filter.

location and the other at the sphere origin, which is located at $(0, 0, -12)$. The peak intensity at the sphere origin is higher than that at the source location. The point-spread function of the array-gain minimum-variance spatial filter and the unit-noise-gain spatial filter have their maximum at the source location. These results are consistent with the arguments of Section 5.1.

We next show the results of point-source reconstruction when a significant amount of noise exists. Assuming that a point source exists at $(0, -1.5, -6)$, the reconstructed source power, $\langle \widehat{s}(\boldsymbol{r})^2 \rangle$, of the single source is calculated with the three input SNRs: $\alpha = 8M$, $\alpha = 4M$, and $\alpha = M$ The results from the spatial matched filter and sLORETA are shown in Fig. 5.5. Here, sLORETA can reconstruct the source at the correct location for the two higher-SNR cases. However, it cannot reconstruct the source when $\alpha = M$, i.e., when the input SNR is the lowest among the three cases. On the contrary, the spatial matched filter reconstructs the source at the correct location, regardless of the input SNR. We conducted the same experiments using the array-gain minimum-variance spatial filter and unit-noise-gain minimum-variance spatial filter. The results are shown in Fig. 5.6. The results show that these spatial filters reconstruct the source at the correct location, regardless of the input SNR. These results are consistent with the arguments in Section 5.2.

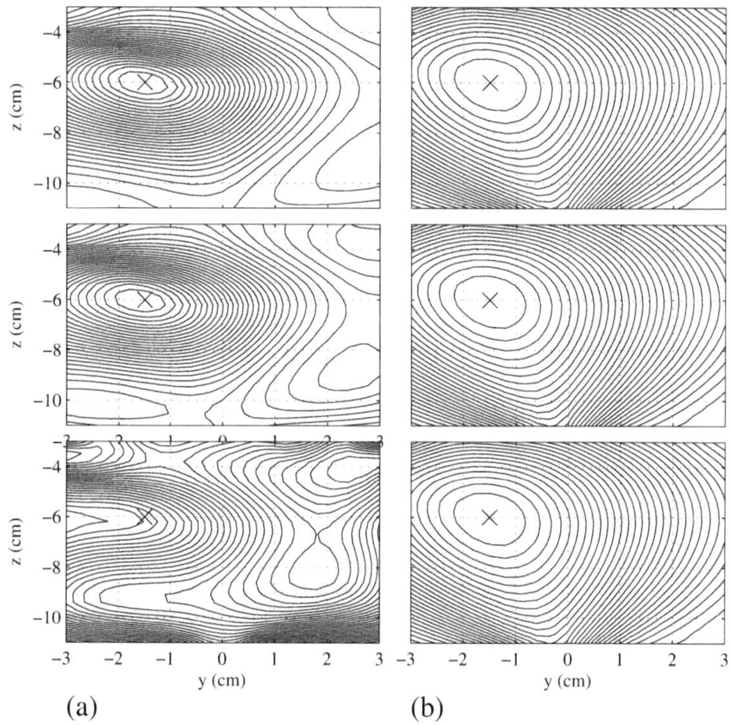

Figure 5.5: The results of experiments on the effects of noise on location bias. Results of point-source reconstruction from (a) the sLORETA filter and (b) the spatial matched filter. The point source is located at $(0, -1.5, -6)$ cm; this location is indicated by the cross mark $\times$. The contour plots from top to bottom show, respectively, the results when the input SNR, α, equals $8M$, $4M$, and M.

We then compare the spatial resolution by plotting the cross-sectional view of the normalized point-spread function. The point-spread functions for the spatial matched filter, sLORETA, the array-gain minimum-variance spatial filter, and the unit-noise-gain spatial filter are plotted in Fig. 5.7. In these plots, the reconstructed results of the point source at $(0, -1.5, -6)$ are plotted with respect to the y direction. We used Eqs. (5.28) – (5.31) to calculate the point-spread functions, and the input SNR, α, was set to M when calculating the adaptive spatial-filter point-spread functions. The plots here clearly demonstrate that the adaptive spa-

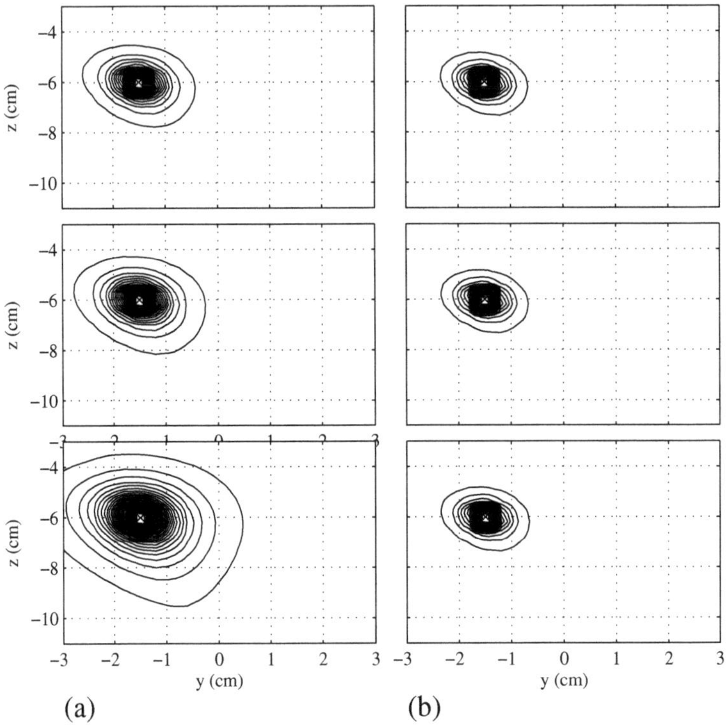

Figure 5.6: The results of experiments on the effects of noise on location bias. Results of point-source reconstruction from (a) the array-gain-constraint minimum-variance spatial filter and (b) the unit-noise-gain minimum-variance spatial filter. The point source is located at $(0, -1.5, -6)$ cm. The contour plots from top to bottom show, respectively, the results when the input SNR, α, equals $8M$, $4M$, and M.

tial filters attain much higher spatial resolution than the non-adaptive spatial filters.

We compare the point-spread functions of the array-gain minimum-variance spatial filter and the unit-noise-gain minimum-variance spatial filter for three SNR cases, $\alpha = 8M$, $4M$, and M. The results are shown in Figs. 5.8(a) and (b). These plots illustrate the fact that the spatial resolution of these minimum-variance filters depends on the input SNR, and that the unit-noise-gain spatial filter attains a significantly higher spatial resolution than that of the array-gain minimum-variance spatial filter.

Numerical examples of the beam response are shown next. We assume three sources located at $(0, -3.5, -4.75)$, $(0, 2.5, -5.5)$, and $(0, -1, -6.5)$ on the plane $x = 0$. We first plot the square of the beam response, $\mathcal{H}(\boldsymbol{r})^2$, of the spatial matched filter in Fig. 5.9. In these results, the spatial-filter pointing locations

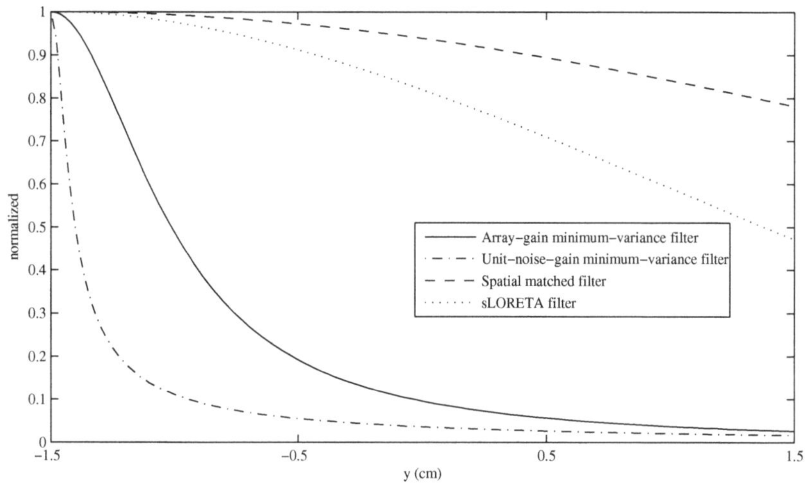

Figure 5.7: Cross-sectional view of the point spread functions. The point-spread functions of the spatial matched filter (broken line), sLORETA (dotted line), the array-gain constraint minimum-variance spatial filter (solid line), and the unit-noise-gain constraint minimum-variance spatial filter (dash-dot line). The input SNR, α, is set to M for the adaptive spatial filters. The point source was assumed to exist at $(0, -1.5, -6)$, and the abscissa indicates the distance from the source location in the y direction.

$\boldsymbol{r}_0$ are indicated by the cross mark $\times$, and the source locations are indicated by the circles. The plotted beam response in (a) shows a case in which the pointing location $\boldsymbol{r}_0$ is not equal to any of the source locations. The beam responses in (b), (c), and (d) show the cases in which the pointing location $\boldsymbol{r}_0$ is equal to the locations of the first, second, and third source, respectively. These plots confirm our arguments that for the spatial matched filter, the beam response and the point spread function have the same form.

Next, we plot the square of the beam response of the unit-gain constraint minimum-variance filter. The square of the beam response, $\mathcal{H}(\boldsymbol{r})^2$, is calculated using Eq. (5.34), and the results are shown in Fig. 5.10. The plotted beam response in (a) shows a case in which the pointing location is not equal to any of the source locations. The beam responses in (b), (c), and (d) show the cases in which the pointing location $\boldsymbol{r}_0$ is equal to the locations of the first, second, and third source, respectively. In these results, the gray-colored area indicates the region in which the beam response is nearly equal to zero and such a region is called the null sensitivity region. In the plot in (a), the null sensitivity region covers the locations of all three sources, whereas in the plots in (b)–(d), the null sensitivity region covers the two sources that are not equal to the filter pointing location. These results indicate the interesting fact that the minimum-variance spatial filters extract the

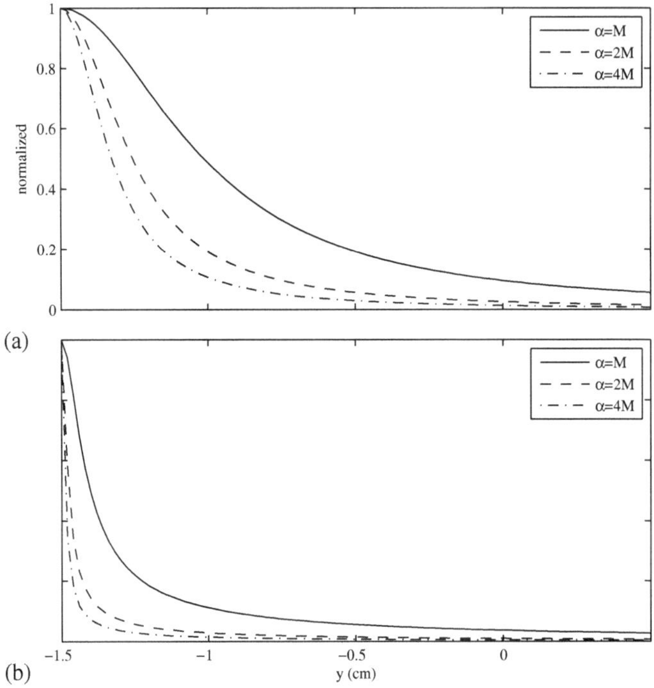

Figure 5.8: Cross-sectional view of the point spread functions. (a) The point-spread functions of the array-gain constraint minimum-variance spatial filter for the three values of α. (b) The point-spread functions of the unit-noise-gain constraint minimum-variance spatial filter for the same three values of α. The point source was assumed to exist at $(0, -1.5, -6)$, and the abscissa indicates the distance from the source location in the y direction.

signal from a source at the filter pointing location by suppressing the signals from other sources, i.e., by enforcing the null sensitivity on the locations of other sources. These plotted beam responses show this property of the minimum-variance spatial filters. The minimum-variance filter can do this without using explicit information on the locations of these sources, because it automatically uses the information on the source locations implicitly contained in the data covariance matrix.

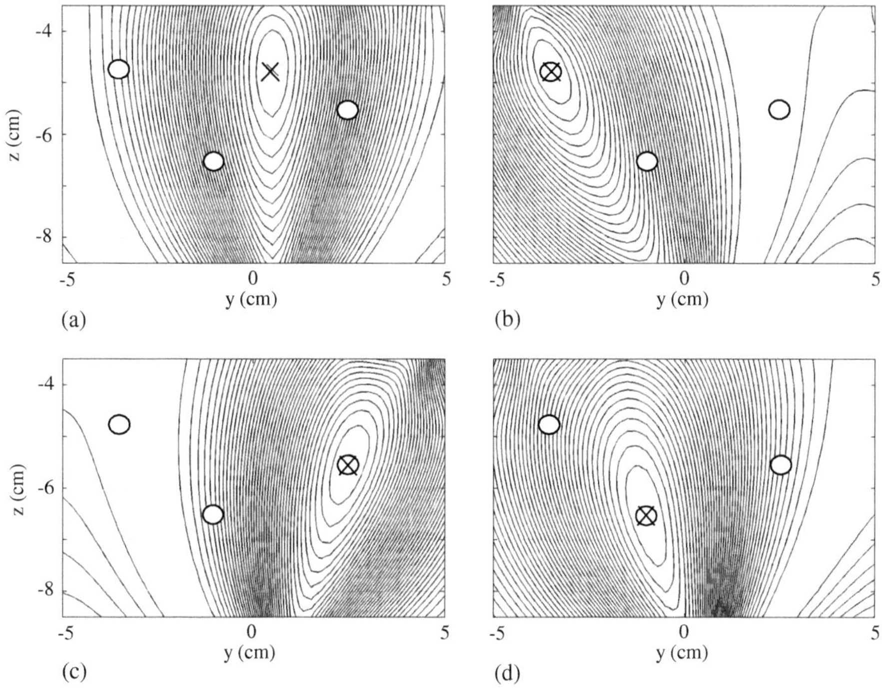

Figure 5.9: The plot of the square of the beam response, $\mathcal{H}(\boldsymbol{r})^2$ of the spatial matched filter. Three sources are assumed to exist at $(0, -3.5, -4.75)$, $(0, 2.5, -5.5)$, and $(0, -1, -6.5)$ on the plane $x = 0$. The spatial-filter pointing locations $\boldsymbol{r}_0$ are indicated by the cross mark $\times$, and the source locations are indicated by the circles. (a) The case where the pointing location is not equal to any of the three source locations. (b) The case where the pointing location is equal to the first source location. (c) The case where the pointing location is equal to the second source location. (b) The case where the pointing location is equal to the third source location.

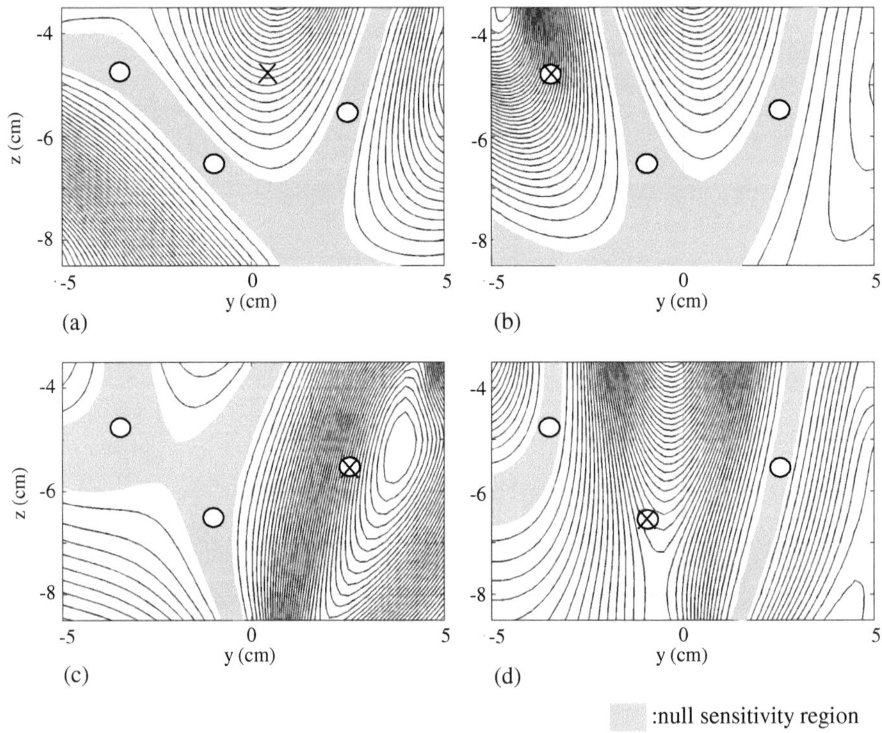

Figure 5.10: The plot of the square of the beam response, $\mathcal{H}(\boldsymbol{r})^2$ of the minimum-variance filter. Three sources are assumed to exist at $(0, -3.5, -4.75)$, $(0, 2.5, -5.5)$, and $(0, -1, -6.5)$ on the plane $x = 0$. The spatial-filter pointing locations $\boldsymbol{r}_0$ are indicated by the cross mark $\times$, and the source locations are indicated by the circles. The gray area indicate the region in which the beam response is nearly equal to zero (null sensitivity region). (a) The case where the pointing location is not equal to any of the three source locations. (b) The case where the pointing location is equal to the first source location. (c) The case where the pointing location is equal to the second source location. (b) The case where the pointing location is equal to the third source location.

Chapter 6

Output SNR and array mismatch

This chapter discusses the signal-to-noise ratio (SNR) in the outputs of spatial filters. We first show that the adaptive spatial filters attain the maximum signal-to-interference-plus-noise ratio among all types of spatial filters. We next derive the SNR transfer factor, which is the ratio between the input and output SNRs, for several representative non-adaptive and adaptive spatial filters. We then show that a significant SNR degradation is caused in adaptive-spatial filter outputs by the array mismatch, which indicates a situation where the lead-field used for computing the spatial filter weight is different from the true lead field. We describe two kinds of techniques, diagonal loading and eigenspace-projection, that can reduce the SNR degradation caused by the array mismatch.

6.1 Output SINR

Let us consider the case where Q uncorrelated sources exist and the measured data contains external interference in addition to the sensor noise. Assuming that this interference is uncorrelated with the signal of interest, the measurement covariance matrix $\boldsymbol{R}$ is given by:

$$\boldsymbol{R} = \sum_{q=1}^{Q} \sigma_q^2 \boldsymbol{l}(\boldsymbol{r}_q)\boldsymbol{l}^T(\boldsymbol{r}_q) + \boldsymbol{R}_{i+n}, \tag{6.1}$$

where $\boldsymbol{R}_{i+n}$ indicates the interference-plus-noise covariance matrix, which is the covariance matrix only for the interference and the sensor noise. Let us next consider the case where, among the Q sources, we are attempting to reconstruct the first source, which is located at $\boldsymbol{r}_1$. In such a case, the source at $\boldsymbol{r}_1$ is the signal source or the source of interest and all other sources should be considered interference sources. Therefore, in such cases, the interference-plus-noise covariance matrix should include the contributions from the sources located at $\boldsymbol{r}_2$ to $\boldsymbol{r}_Q$. The contribution is equal to $\sum_{q=2}^{Q} \sigma_q^2 \boldsymbol{l}(\boldsymbol{r}_q)\boldsymbol{l}^T(\boldsymbol{r}_q)$, and equation (6.1) is rewritten as

$$\boldsymbol{R} = \sigma_1^2 \boldsymbol{l}(\boldsymbol{r}_1)\boldsymbol{l}^T(\boldsymbol{r}_1) + \boldsymbol{R}_{i+n}^{ex}(\boldsymbol{r}_1), \tag{6.2}$$

where

$$\boldsymbol{R}_{i+n}^{ex}(\boldsymbol{r}_1) = \sum_{q=2}^{Q} \sigma_q^2 \boldsymbol{l}(\boldsymbol{r}_q)\boldsymbol{l}^T(\boldsymbol{r}_q) + \boldsymbol{R}_{i+n}. \tag{6.3}$$

Here, $\boldsymbol{R}_{i+n}^{ex}$ is called the extended interference-plus-noise covariance matrix. Note that this $\boldsymbol{R}_{i+n}^{ex}$ depends on the spatial-filter pointing location, which is $\boldsymbol{r}_1$ in Eqs. (6.2) and (6.3), and we write $\boldsymbol{R}_{i+n}^{ex}$ as $\boldsymbol{R}_{i+n}^{ex}(\boldsymbol{r}_1)$ to express this spatial-location dependency explicitly. When the pointing location $\boldsymbol{r}$ is equal to none of the source locations, $\boldsymbol{R}_{i+n}^{ex}(\boldsymbol{r})$ is equal to the covariance matrix $\boldsymbol{R}$.

When a source is located at $\boldsymbol{r}_1$ and the spatial filter is pointing at this location, substituting Eq. (6.2) into (2.66) produces the power of the output of the spatial filter, which is expressed as

$$\langle \widehat{s}(\boldsymbol{r}_1, t)^2 \rangle = \boldsymbol{w}^T(\boldsymbol{r}_1)\boldsymbol{R}\boldsymbol{w}(\boldsymbol{r}_1) = \sigma_1^2 \|\boldsymbol{w}^T(\boldsymbol{r}_1)\boldsymbol{l}(\boldsymbol{r}_1)\|^2 + \boldsymbol{w}^T(\boldsymbol{r}_1)\boldsymbol{R}_{i+n}^{ex}(\boldsymbol{r}_1)\boldsymbol{w}(\boldsymbol{r}_1). \tag{6.4}$$

The first term on the right-hand side represents the power of the signal at $\boldsymbol{r}_1$ and the second term represents the contribution from noise, interference, and other sources. The ratio between the first and the second terms on the right-hand side of Eq. (6.4) is denoted $\mathcal{Z}_0$, i.e.,

$$\mathcal{Z}_0 = \frac{\sigma_1^2 \|\boldsymbol{w}^T(\boldsymbol{r}_1)\boldsymbol{l}(\boldsymbol{r}_1)\|^2}{\boldsymbol{w}^T(\boldsymbol{r}_1)\boldsymbol{R}_{i+n}^{ex}(\boldsymbol{r}_1)\boldsymbol{w}(\boldsymbol{r}_1)}. \tag{6.5}$$

This $\mathcal{Z}_0$ is called the signal-to-interference-plus-noise ratio (SINR), which plays an important role in the performance analysis of adaptive spatial filters. In Section 6.2, we show that the minimum-variance spatial filters maximize this signal-to-interference-plus-noise ratio.

This $\mathcal{Z}_0$ can be used for theoretical analysis. However, we cannot use $\mathcal{Z}_0$ to evaluate the SNR of the spatial filter output in actual measurements, because σ_1^2 and $\boldsymbol{R}_{i+n}^{ex}$ are unknown. The numerator in Eq. (6.5) can be replaced with $\boldsymbol{w}^T(\boldsymbol{r})\boldsymbol{R}\boldsymbol{w}(\boldsymbol{r})$, assuming that $\sigma_1^2 \|\boldsymbol{w}^T(\boldsymbol{r}_1)\boldsymbol{l}(\boldsymbol{r}_1)\|^2 \gg \boldsymbol{w}^T(\boldsymbol{r}_1)\boldsymbol{R}_{i+n}^{ex}\boldsymbol{w}(\boldsymbol{r}_1)$ in Eq. (6.4). In the denominator of Eq. (6.5), the noise and interference covariance matrix, $\boldsymbol{R}_{i+n}^{ex}$, is replaced with $\sigma_0^2 \boldsymbol{I}$ when no information regarding $\boldsymbol{R}_{i+n}^{ex}$ is available. We then derive

$$\mathcal{Z} = \frac{\boldsymbol{w}^T(\boldsymbol{r})\boldsymbol{R}\boldsymbol{w}(\boldsymbol{r})}{\sigma_0^2 \|\boldsymbol{w}(\boldsymbol{r})\|^2},$$

which is equal to Eq. (2.68). This $\mathcal{Z}$ can be computed from measurements, although we should estimate the variance of the input noise σ_0^2.

6.2 Adaptive spatial filters that attain the maximum SINR

The weight vectors in Eqs. (4.5), (4.9), and (4.15) have the form expressed as

$$\boldsymbol{w}(\boldsymbol{r}) = \xi \boldsymbol{R}^{-1}\boldsymbol{l}(\boldsymbol{r}), \tag{6.6}$$

where ξ is a scalar constant, which is equal to $\xi = 1/[\boldsymbol{l}^T(\boldsymbol{r})\boldsymbol{R}^{-1}\boldsymbol{l}(\boldsymbol{r})]$, $\xi = \|\boldsymbol{l}(\boldsymbol{r})\|/[\boldsymbol{l}^T(\boldsymbol{r})\boldsymbol{R}^{-1}\boldsymbol{l}(\boldsymbol{r})]$, and $\xi = 1/[\boldsymbol{l}^T(\boldsymbol{r})\boldsymbol{R}^{-2}\boldsymbol{l}(\boldsymbol{r})]^{-1/2}$ for the unit-gain, array-gain, and unit-noise-gain minimum-variance filters, respectively. We show in this section that this family of weight vectors attains the maximum output signal-to-interference-plus-noise ratio (SINR) at each source location. We assume that a source exists at $\boldsymbol{r}_1$, and derive a weight vector $\boldsymbol{w}(\boldsymbol{r}_1)$ that detects this source with the maximum SINR[11]. For this purpose, we define the modified weight vector $\widetilde{\boldsymbol{w}}(\boldsymbol{r}_1)$ such that $\widetilde{\boldsymbol{w}}(\boldsymbol{r}_1) = (\boldsymbol{R}_{i+n}^{ex})^{1/2}\boldsymbol{w}(\boldsymbol{r}_1)$. Here, $\boldsymbol{R}_{i+n}^{ex}$ is the the extended interference-plus-noise covariance matrix defined in Eq. (6.3), and we omit the explicit notation of $(\boldsymbol{r}_1)$ for simplicity. We decompose it such that

$$\boldsymbol{R}_{i+n}^{ex} = (\boldsymbol{R}_{i+n}^{ex})^{1/2}(\boldsymbol{R}_{i+n}^{ex})^{1/2}. \tag{6.7}$$

Because $\boldsymbol{R}_{i+n}^{ex}$ is a positive definite matrix, we can apply this decomposition to $\boldsymbol{R}_{i+n}^{ex}$. The signal-to-interference-plus-noise ratio, $\mathcal{Z}_0$, defined in Eq. (6.5) is then expressed as

$$\begin{aligned}\mathcal{Z}_0 = \frac{\sigma_1^2\|\boldsymbol{w}^T(\boldsymbol{r}_1)\boldsymbol{l}(\boldsymbol{r}_1)\|^2}{\boldsymbol{w}^T(\boldsymbol{r}_1)\boldsymbol{R}_{i+n}^{ex}\boldsymbol{w}(\boldsymbol{r}_1)} &= \sigma_1^2\frac{\|\widetilde{\boldsymbol{w}}^T(\boldsymbol{r}_1)(\boldsymbol{R}_{i+n}^{ex})^{-1/2}\boldsymbol{l}(\boldsymbol{r}_1)\|^2}{\widetilde{\boldsymbol{w}}^T(\boldsymbol{r}_1)\widetilde{\boldsymbol{w}}(\boldsymbol{r}_1)} \\ &\le \sigma_1^2\frac{\|\widetilde{\boldsymbol{w}}(\boldsymbol{r}_1)\|^2\|(\boldsymbol{R}_{i+n}^{ex})^{-1/2}\boldsymbol{l}(\boldsymbol{r}_1)\|^2}{\|\widetilde{\boldsymbol{w}}(\boldsymbol{r}_1)\|^2} = \sigma_1^2\|(\boldsymbol{R}_{i+n}^{ex})^{-1/2}\boldsymbol{l}(\boldsymbol{r}_1)\|^2,\end{aligned} \tag{6.8}$$

where we use the Schwartz inequality,

$$\|\widetilde{\boldsymbol{w}}^T(\boldsymbol{r}_1)(\boldsymbol{R}_{i+n}^{ex})^{-1/2}\boldsymbol{l}(\boldsymbol{r}_1)\|^2 \le \|\widetilde{\boldsymbol{w}}(\boldsymbol{r}_1)\|^2\|(\boldsymbol{R}_{i+n}^{ex})^{-1/2}\boldsymbol{l}(\boldsymbol{r}_1)\|^2.$$

In Eq. (6.8), the equality holds and $\mathcal{Z}_0$ has its maximum value when the following relationship holds:

$$\widetilde{\boldsymbol{w}}(\boldsymbol{r}_1) = \xi(\boldsymbol{R}_{i+n}^{ex})^{-1/2}\boldsymbol{l}(\boldsymbol{r}_1) \quad \text{or} \quad \boldsymbol{w}(\boldsymbol{r}_1) = \xi(\boldsymbol{R}_{i+n}^{ex})^{-1}\boldsymbol{l}(\boldsymbol{r}_1), \tag{6.9}$$

where ξ is an arbitrary constant.

According to Eq. (6.8), the value of $\mathcal{Z}_0$ does not depend on the norm of the weight vector, i.e., $\mathcal{Z}_0$ does not depend on the value of ξ. Thus, this constant should be determined by some criterion other than maximization of $\mathcal{Z}_0$. Using the unit-gain constraint $\boldsymbol{w}^T(\boldsymbol{r}_1)\boldsymbol{l}(\boldsymbol{r}_1) = 1$, we have

$$\xi = 1/[\boldsymbol{l}^T(\boldsymbol{r}_1)(\boldsymbol{R}_{i+n}^{ex})^{-1}\boldsymbol{l}(\boldsymbol{r}_1)],$$

and the weight vector is given by:

$$\boldsymbol{w}_{i+n}(\boldsymbol{r}_1) = \frac{(\boldsymbol{R}_{i+n}^{ex})^{-1}\boldsymbol{l}(\boldsymbol{r}_1)}{[\boldsymbol{l}^T(\boldsymbol{r}_1)(\boldsymbol{R}_{i+n}^{ex})^{-1}\boldsymbol{l}(\boldsymbol{r}_1)]}. \tag{6.10}$$

Using the array-gain constraint $\boldsymbol{w}^T(\boldsymbol{r}_1)\boldsymbol{l}(\boldsymbol{r}_1) = \|\boldsymbol{l}(\boldsymbol{r}_1)\|$, we get

$$\xi = 1/[\widetilde{\boldsymbol{l}}^T(\boldsymbol{r}_1)(\boldsymbol{R}_{i+n}^{ex})^{-1}\widetilde{\boldsymbol{l}}(\boldsymbol{r}_1)],$$

and the weight vector[1] is given by:

$$\boldsymbol{w}_{i+n}(\boldsymbol{r}_1) = \frac{(\boldsymbol{R}_{i+n}^{ex})^{-1}\widetilde{\boldsymbol{l}}(\boldsymbol{r}_1)}{[\widetilde{\boldsymbol{l}}^T(\boldsymbol{r}_1)(\boldsymbol{R}_{i+n}^{ex})^{-1}\widetilde{\boldsymbol{l}}(\boldsymbol{r}_1)]}. \tag{6.11}$$

Also, using $\boldsymbol{w}^T\boldsymbol{w} = 1$ leads to

$$\xi = 1/\sqrt{\boldsymbol{l}^T(\boldsymbol{r}_1)(\boldsymbol{R}_{i+n}^{ex})^{-2}\boldsymbol{l}(\boldsymbol{r}_1)},$$

and the weight vector is given by:

$$\boldsymbol{w}_{i+n}(\boldsymbol{r}_1) = \frac{(\boldsymbol{R}_{i+n}^{ex})^{-1}\boldsymbol{l}(\boldsymbol{r}_1)}{\sqrt{\boldsymbol{l}^T(\boldsymbol{r}_1)(\boldsymbol{R}_{i+n}^{ex})^{-2}\boldsymbol{l}(\boldsymbol{r}_1)}}. \tag{6.12}$$

The only difference between the weight vectors derived above from those derived in Section 4.1 is that the interference-plus-noise covariance $\boldsymbol{R}_{i+n}^{ex}$ is used instead of the measurement covariance $\boldsymbol{R}$. To indicate this explicitly, we use the notation $\boldsymbol{w}_{i+n}(\boldsymbol{r})$ for the weight vectors in (6.10)–(6.12).

Next, we show that $\boldsymbol{w}(\boldsymbol{r})$ is equal to $\boldsymbol{w}_{i+n}(\boldsymbol{r})$ at each source location. That is, the weight vectors in Eqs. (4.5), (4.9), and (4.15) also maximize the signal to interference-plus-noise ratio, $\mathcal{Z}_0$, at the source locations. This can be shown in the following manner. First, we show that $\boldsymbol{w}(\boldsymbol{r})$ in Eq. (4.5) is equal to $\boldsymbol{w}_{i+n}(\boldsymbol{r})$ in Eq. (6.10) at the source location, $\boldsymbol{r}_1$. Denoting $\boldsymbol{f} = \boldsymbol{l}(\boldsymbol{r}_1)$, Eq. (6.2) can be rewritten as

$$\boldsymbol{R} = \sigma_1^2\boldsymbol{f}\boldsymbol{f}^T + \boldsymbol{R}_{i+n}^{ex}. \tag{6.13}$$

(Here, we assume that the source at $\boldsymbol{r}_1$ is uncorrelated with all the other sources.) Applying the matrix inversion formula in Eq. (13.51), we get

$$\boldsymbol{R}^{-1} = (\boldsymbol{R}_{i+n}^{ex})^{-1} - (\boldsymbol{R}_{i+n}^{ex})^{-1}\frac{\sigma_1^2\boldsymbol{f}\boldsymbol{f}^T}{1+\sigma_1^2\boldsymbol{f}^T(\boldsymbol{R}_{i+n}^{ex})^{-1}\boldsymbol{f}}(\boldsymbol{R}_{i+n}^{ex})^{-1}. \tag{6.14}$$

By multiplying $\boldsymbol{f}$ from the both sides of the equation above, we get

$$\boldsymbol{f}^T\boldsymbol{R}^{-1}\boldsymbol{f} = \frac{\boldsymbol{f}^T(\boldsymbol{R}_{i+n}^{ex})^{-1}\boldsymbol{f}}{1+\sigma_1^2\boldsymbol{f}^T(\boldsymbol{R}_{i+n}^{ex})^{-1}\boldsymbol{f}}. \tag{6.15}$$

[1] $\widetilde{\boldsymbol{l}}(\boldsymbol{r}_1)$ again indicates the normalized lead-field vector defined as $\widetilde{\boldsymbol{l}}(\boldsymbol{r}_1) = \boldsymbol{l}(\boldsymbol{r}_1)/\|\boldsymbol{l}(\boldsymbol{r}_1)\|$.

By multiplying $\boldsymbol{f}$ from the right-hand side of Eq. (6.14), we get

$$\boldsymbol{R}^{-1}\boldsymbol{f} = \frac{(\boldsymbol{R}_{i+n}^{ex})^{-1}\boldsymbol{f}}{1+\sigma_1^2\boldsymbol{f}^T(\boldsymbol{R}_{i+n}^{ex})^{-1}\boldsymbol{f}}. \tag{6.16}$$

Then, using Eqs. (6.15) and (6.16), we have the relationship,

$$\boldsymbol{w}(\boldsymbol{r}_1) = \frac{\boldsymbol{R}^{-1}\boldsymbol{f}}{\boldsymbol{f}^T\boldsymbol{R}^{-1}\boldsymbol{f}} = \frac{(\boldsymbol{R}_{i+n}^{ex})^{-1}\boldsymbol{f}}{\boldsymbol{f}^T(\boldsymbol{R}_{i+n}^{ex})^{-1}\boldsymbol{f}} = \boldsymbol{w}_{i+n}(\boldsymbol{r}_1). \tag{6.17}$$

We can show the equivalence between the weight vectors in Eqs. (4.9) and (6.11) and the equivalence between those in Eqs. (4.15) and (6.12) in exactly the same manner. Although the weight vectors $\boldsymbol{w}(\boldsymbol{r})$ and $\boldsymbol{w}_{i+n}(\boldsymbol{r})$ are equal at each source location, these two weight vectors perform very differently at locations where no sources exist[11]. In the field of electromagnetic brain imaging, the weight vectors $\boldsymbol{w}_{i+n}(\boldsymbol{r})$ in Eqs. (6.10)–(6.12) have never been used because it is difficult to obtain $\boldsymbol{R}_{i+n}^{ex}$. Therefore, we do not discuss the properties of these weight vectors further in this book.

6.3 SNR transfer factor

We next argue how much SNR is maintained in the spatial filter reconstruction process, and compare the performances of the adaptive and non-adaptive spatial filters introduced in the previous chapters in this regard. For this argument, we again assume the same simple scenario in which a single source with its power of σ_1^2 exists at $\boldsymbol{r}_1$ under the uncorrelated Gaussian noise with its variance σ_0^2. Thus, we can write $\boldsymbol{R}_{i+n} = \sigma_0^2\boldsymbol{I}$. Denoting again $\boldsymbol{f}$ as $\boldsymbol{f} = \boldsymbol{l}(\boldsymbol{r}_1)$, the output SNR of the spatial filter[2], $\mathcal{Z}_0$, which was defined in Eq. (6.5), is expressed as

$$\mathcal{Z}_0 = \frac{\sigma_1^2\|\boldsymbol{f}\|^2}{\sigma_0^2}\frac{[\boldsymbol{w}^T(\boldsymbol{r}_1)\boldsymbol{f}]^2}{\|\boldsymbol{f}\|^2\|\boldsymbol{w}(\boldsymbol{r}_1)\|^2} = \alpha\Theta, \tag{6.18}$$

where α is again the input SNR, and Θ is defined as

$$\Theta = \frac{[\boldsymbol{w}^T(\boldsymbol{r}_1)\boldsymbol{f}]^2}{\|\boldsymbol{f}\|^2\|\boldsymbol{w}(\boldsymbol{r}_1)\|^2}. \tag{6.19}$$

It can be seen in Eq. (6.18) that Θ represents the ratio between the input and output SNRs, and it indicates how much SNR is preserved through the spatial filter reconstruction process. Because of this property, Θ is called the SNR transfer factor.

Under this single-source scenario, we can also derive the relationship between Θ and $\mathcal{Z}$ defined in Eq. (2.68). That is, using Eq. (13.42), we have

$$\boldsymbol{w}^T\boldsymbol{R}\boldsymbol{w} = \sigma_1^2\boldsymbol{w}^T\boldsymbol{f}\boldsymbol{f}^T\boldsymbol{w} + \sigma_0^2\boldsymbol{w}^T\boldsymbol{w} = \sigma_1^2[\boldsymbol{w}^T\boldsymbol{f}]^2 + \sigma_0^2\|\boldsymbol{w}\|^2, \tag{6.20}$$

[2] $\mathcal{Z}_0$ is called the output SNR here, because there is no interference in this argument.

and thus we have

$$\mathcal{Z} = \frac{\boldsymbol{w}^T \boldsymbol{R} \boldsymbol{w}}{\sigma_0^2 \|\boldsymbol{w}\|^2} = \frac{\sigma_1^2 \|\boldsymbol{f}\|^2}{\sigma_0^2} \frac{[\boldsymbol{w}^T \boldsymbol{f}]^2}{\|\boldsymbol{f}\|^2 \|\boldsymbol{w}\|^2} + 1 = \alpha\Theta + 1 = \mathcal{Z}_0 + 1. \tag{6.21}$$

Let us first derive a value of Θ for representative non-adaptive spatial filters. For the spatial matched filter, the weight vector should have the following form at the source location: $\boldsymbol{w}(\boldsymbol{r}_1) = \boldsymbol{f}/\|\boldsymbol{f}\|$. Substituting this into Eq. (6.19), we get the relationship

$$\Theta = 1. \tag{6.22}$$

The spatial matched filter preserves SNR, and there is no SNR loss in the reconstruction process.

According to the arguments in Section 3.2, the general form of the minimum-norm-based filters can be expressed as

$$\boldsymbol{w}(\boldsymbol{r}_1) = \zeta \boldsymbol{G}^{-1} \boldsymbol{f}, \tag{6.23}$$

where $\zeta = 1$ for the minimum-norm filter, $\zeta = [\boldsymbol{f}^T \boldsymbol{G}^{-2} \boldsymbol{f}]^{-1/2}$ for the weight-normalized minimum-norm filter, and $\zeta = [\boldsymbol{f}^T \boldsymbol{G}^{-1} \boldsymbol{f}]^{-1/2}$ for sLORETA. Therefore, substituting Eq. (6.23) into Eq. (6.19), we have

$$\Theta = \frac{[\boldsymbol{f}^T \boldsymbol{G}^{-1} \boldsymbol{f}]^2}{\|\boldsymbol{f}\|^2 \|\boldsymbol{G}^{-1} \boldsymbol{f}\|^2}. \tag{6.24}$$

Using the Schwartz inequality,

$$[\boldsymbol{f}^T \boldsymbol{G}^{-1} \boldsymbol{f}]^2 < \|\boldsymbol{f}\|^2 \|\boldsymbol{G}^{-1} \boldsymbol{f}\|^2,$$

we can derive

$$\Theta < 1.$$

Here, note that, since the relationship $\boldsymbol{G}^{-1} \propto \boldsymbol{I}$ never holds, Θ is never equal to 1. Therefore, for the minimum-norm-based non-adaptive spatial filters, the SNR is not preserved and there is an SNR loss in the reconstruction process.

We next check if the adaptive spatial filters preserve the SNR. For all three minimum-variance spatial filters, the weight vectors pointing at $\boldsymbol{r}_1$ have the following form

$$\boldsymbol{w}(\boldsymbol{r}_1) = \xi \boldsymbol{R}^{-1} \boldsymbol{f},$$

We then derive

$$\Theta = \frac{[\boldsymbol{f}^T \boldsymbol{R}^{-1} \boldsymbol{f}]^2}{\|\boldsymbol{f}\|^2 [\boldsymbol{f}^T \boldsymbol{R}^{-2} \boldsymbol{f}]}. \tag{6.25}$$

Substituting Eqs. (13.45) and (13.48) into Eq. (6.25) leads to the relationship

$$\Theta = 1.$$

This result indicates that, for the single source case, the minimum-variance spatial filters preserve the input SNR and there is no SNR loss in the reconstruction process.

We next consider the case where two sources exist at $\boldsymbol{r}_1$ and $\boldsymbol{r}_2$. We denote the power of these sources as σ_1^2 and σ_2^2 and define $\boldsymbol{f}$ and $\boldsymbol{g}$ such that $\boldsymbol{f} = \boldsymbol{l}(\boldsymbol{r}_1)$ and $\boldsymbol{g} = \boldsymbol{l}(\boldsymbol{r}_2)$. We assume that the spatial filter is pointing at the first source location. Then, setting $\boldsymbol{l} = \boldsymbol{f}$ in Eq. (13.57), we can derive

$$\boldsymbol{f}^T\boldsymbol{R}^{-1}\boldsymbol{f} = \frac{\|\boldsymbol{f}\|^2}{\Gamma}[1 - \frac{\alpha_2}{1+\alpha_2}\cos^2(\boldsymbol{f},\boldsymbol{g})], \tag{6.26}$$

where α_2 is the input SNR for the second source, which is defined as $\alpha_2 = (\sigma_2^2/\sigma_0^2)\|\boldsymbol{g}\|^2$, and Γ is defined in Eq. (13.58). Using Eq. (13.65), we have

$$\boldsymbol{f}^T\boldsymbol{R}^{-2}\boldsymbol{f} = \frac{\|\boldsymbol{f}\|^2}{\Gamma^2}\left[1 - \frac{\alpha_2(2+\alpha_2)}{(1+\alpha_2)^2}\cos^2(\boldsymbol{f},\boldsymbol{g})\right]. \tag{6.27}$$

Then, substituting Eqs. (6.27) and (6.26) into Eq. (6.25), we derive

$$\Theta = \frac{[1 - \frac{\alpha_2}{1+\alpha_2}\cos^2(\boldsymbol{f},\boldsymbol{g})]^2}{[1 - \frac{\alpha_2(2+\alpha_2)}{(1+\alpha_2)^2}\cos^2(\boldsymbol{f},\boldsymbol{g})]}. \tag{6.28}$$

If the second source is so weak that $\alpha_2 \approx 0$, we obtain the relationship $\Theta \approx 1$, which is exactly equal to the relationship for the single-source case. However, since in general we can assume $\alpha_2 \gg 1$, equation (6.28) becomes[36][41]

$$\Theta \approx [1 - \cos^2(\boldsymbol{f},\boldsymbol{g})]. \tag{6.29}$$

The above equation shows that the output SNR of the first source is influenced by the generalized cosine between $\boldsymbol{f}$ and $\boldsymbol{g}$, which is often referred to as the spatial correlation between the two sources[42]. This is an important property of the adaptive spatial filters.

6.4 Two types of SNR definitions for the vector minimum-variance spatial filter

We mentioned that, for the vector spatial filter, the definition of the output power is not unique and we present two definitions, $\widehat{P}_V^{(\mathrm{I})}(\boldsymbol{r})$ and $\widehat{P}_V^{(\mathrm{II})}(\boldsymbol{r})$, in Section 2.5.2. Similarly, for the vector spatial filter, the definition of the output SNR is not unique. In this section, we present two different definitions of output SNR and we discuss the relationship between these definitions for the vector minimum-variance spatial filter. We also discuss the relationship between these two SNR definitions and the output-SNR definition for the scalar minimum-variance spatial filter[36].

The two definitions for the output SNR of the vector minimum-variance filter are expressed as

$$\mathcal{Z}_V^{(\mathrm{I})}(\boldsymbol{r}) = \frac{\mathrm{tr}\{\boldsymbol{W}^T\boldsymbol{R}\boldsymbol{W}\}}{\sigma_0^2\mathrm{tr}\{\boldsymbol{W}^T\boldsymbol{W}\}}, \tag{6.30}$$

$$\mathcal{Z}_V^{(\mathrm{II})}(\boldsymbol{r}) = \max_{\boldsymbol{\eta}} \frac{\boldsymbol{\eta}^T\widehat{\boldsymbol{\Sigma}}_s(\boldsymbol{r})\boldsymbol{\eta}}{\sigma_0^2\|\boldsymbol{W}(\boldsymbol{r})\boldsymbol{\eta}\|^2} = \frac{1}{\sigma_0^2}\left[\max_{\boldsymbol{\eta}} \frac{\boldsymbol{\eta}^T\widehat{\boldsymbol{\Sigma}}_s(\boldsymbol{r})\boldsymbol{\eta}}{\boldsymbol{\eta}^T\boldsymbol{\Upsilon}\boldsymbol{\eta}}\right], \tag{6.31}$$

where the matrix $\boldsymbol{\Upsilon}$ has already been defined in Eq. (4.83). Here, $\mathcal{Z}_V^{(\mathrm{I})}(\boldsymbol{r})$ and $\mathcal{Z}_V^{(\mathrm{II})}(\boldsymbol{r})$ are orientation-non-optimized and orientation-optimized definitions, which correspond, respectively, to the definitions of the output power $\widehat{P}_V^{(\mathrm{I})}(\boldsymbol{r})$ and $\widehat{P}_V^{(\mathrm{II})}(\boldsymbol{r})$. The output SNR for the scalar minimum-variance filter is defined in Eq. (4.8), and is expressed as

$$\mathcal{Z}_S(\boldsymbol{r}) = \frac{1}{\sigma_0^2} \frac{\boldsymbol{l}^T(\boldsymbol{r})\boldsymbol{R}^{-1}\boldsymbol{l}(\boldsymbol{r})}{[\boldsymbol{l}^T(\boldsymbol{r})\boldsymbol{R}^{-2}\boldsymbol{l}(\boldsymbol{r})]}. \tag{6.32}$$

In this section, we let σ_0^2 equal 1 in order to simplify the notations.

We first show the relationship $\mathcal{Z}_S(\boldsymbol{r}) = \mathcal{Z}_V^{(\mathrm{II})}(\boldsymbol{r})$. According to Eq. (4.48), $\mathcal{Z}_S(\boldsymbol{r})$ is given by:

$$\begin{aligned}\mathcal{Z}_S(\boldsymbol{r}) &= \frac{1}{\lambda_{min}\{\boldsymbol{L}^T(\boldsymbol{r})\boldsymbol{R}^{-2}\boldsymbol{L}(\boldsymbol{r}), \boldsymbol{L}^T(\boldsymbol{r})\boldsymbol{R}^{-1}\boldsymbol{L}(\boldsymbol{r})\}} \\ &= \frac{1}{\lambda_{min}\{[\boldsymbol{L}^T(\boldsymbol{r})\boldsymbol{R}^{-1}\boldsymbol{L}(\boldsymbol{r})]^{-1}[\boldsymbol{L}^T(\boldsymbol{r})\boldsymbol{R}^{-2}\boldsymbol{L}(\boldsymbol{r})]\}}.\end{aligned} \tag{6.33}$$

Also, according to Eq. (6.31), $\mathcal{Z}_V^{(\mathrm{II})}(\boldsymbol{r})$ is expressed as

$$\mathcal{Z}_V^{(\mathrm{II})}(\boldsymbol{r}) = \frac{1}{\lambda_{min}\{\boldsymbol{\Upsilon}, \widehat{\boldsymbol{\Sigma}}_s(\boldsymbol{r})\}} = \frac{1}{\lambda_{min}\{[\boldsymbol{L}^T(\boldsymbol{r})\boldsymbol{R}^{-2}\boldsymbol{L}(\boldsymbol{r})][\boldsymbol{L}^T(\boldsymbol{r})\boldsymbol{R}^{-1}\boldsymbol{L}(\boldsymbol{r})]^{-1}\}}. \tag{6.34}$$

Since the relationship,

$$\begin{aligned}&\lambda_{min}\{[\boldsymbol{L}^T(\boldsymbol{r})\boldsymbol{R}^{-1}\boldsymbol{L}(\boldsymbol{r})]^{-1}[\boldsymbol{L}^T(\boldsymbol{r})\boldsymbol{R}^{-2}\boldsymbol{L}(\boldsymbol{r})]\} \\ &\qquad = \lambda_{min}\{[\boldsymbol{L}^T(\boldsymbol{r})\boldsymbol{R}^{-2}\boldsymbol{L}(\boldsymbol{r})][\boldsymbol{L}^T(\boldsymbol{r})\boldsymbol{R}^{-1}\boldsymbol{L}(\boldsymbol{r})]^{-1}\},\end{aligned}$$

holds[43], we can derive

$$\mathcal{Z}_S(\boldsymbol{r}) = \mathcal{Z}_V^{(\mathrm{II})}(\boldsymbol{r}). \tag{6.35}$$

We next compare $\mathcal{Z}_V^{(\mathrm{I})}(\boldsymbol{r})$ and $\mathcal{Z}_V^{(\mathrm{II})}(\boldsymbol{r})$. However, since the comparison for the general case is difficult, we again assume the single-source scenario in which a single source with its power of σ_1^2 exists at $\boldsymbol{r}_1$, and the source orientation is denoted $\boldsymbol{\eta}_1$. The lead-field vector at the source location and the source orientation is denoted $\boldsymbol{f}$, i.e., $\boldsymbol{f} = \boldsymbol{L}(\boldsymbol{r}_1)\boldsymbol{\eta}_1$. Under this scenario, we then compare $\mathcal{Z}_V^{(\mathrm{I})}(\boldsymbol{r}_1)$ and $\mathcal{Z}_V^{(\mathrm{II})}(\boldsymbol{r}_1)$. We have

$$\mathcal{Z}_V^{(\mathrm{II})}(\boldsymbol{r}_1) = \mathcal{Z}_S(\boldsymbol{r}_1) = \frac{\boldsymbol{f}^T\boldsymbol{R}^{-1}\boldsymbol{f}}{[\boldsymbol{f}^T\boldsymbol{R}^{-2}\boldsymbol{f}]}. \tag{6.36}$$

Using Eqs. (13.45) and (13.48), it is easy to derive

$$\mathcal{Z}_V^{(\mathrm{II})}(\boldsymbol{r}_1) = 1 + \alpha, \tag{6.37}$$

where $\alpha = (\sigma_1^2/\sigma_0^2)\|\boldsymbol{f}\|^2$ is the input SNR. Therefore, using Eq. (6.21), the value of Θ, the SNR transfer factor, is obtained as

$$\Theta = 1,$$

and we can see that the orientation-optimized definition, $\mathcal{Z}_V^{(\mathrm{II})}$, preserves the input SNR.

Next, we discuss $\mathcal{Z}_V^{(\mathrm{I})}(\boldsymbol{r}_1)$, and for this discussion, we define the eigenvalues and eigenvectors of the 3×3 matrix $\boldsymbol{L}^T(\boldsymbol{r}_1)\boldsymbol{R}^{-1}\boldsymbol{L}(\boldsymbol{r}_1)$ as ϕ_j and $\boldsymbol{z}_j$ where $j = 1, 2, 3$. To formulate $\mathcal{Z}_V^{(\mathrm{I})}(\boldsymbol{r}_1)$, instead of simply using the x, y, z directions, we use the three orthogonal directions represented by $\boldsymbol{z}_1$, $\boldsymbol{z}_2$, and $\boldsymbol{z}_3$, We define $\boldsymbol{l}_j$ such that $\boldsymbol{l}_j = \boldsymbol{L}(\boldsymbol{r}_1)\boldsymbol{z}_j$ $(j = 1, 2, 3)$. Then, omitting the explicit notation $(\boldsymbol{r}_1)$ from $\boldsymbol{L}(\boldsymbol{r}_1)$ for simplicity, the output power at the source location, $\boldsymbol{r}_1$, is expressed as

$$\mathrm{tr}[\boldsymbol{W}^T\boldsymbol{R}\boldsymbol{W}] = \mathrm{tr}\{[\boldsymbol{L}^T\boldsymbol{R}^{-1}\boldsymbol{L}]^{-1}\} = \sum_{j=1}^{3}\frac{1}{\phi_j}. \tag{6.38}$$

Here ϕ_j can be obtained as

$$\phi_j = \boldsymbol{z}_j^T[\boldsymbol{L}^T\boldsymbol{R}^{-1}\boldsymbol{L}]\boldsymbol{z}_j = \boldsymbol{l}_j^T\boldsymbol{R}^{-1}\boldsymbol{l}_j = \|\boldsymbol{l}_j\|^2(1 - \frac{\alpha}{1+\alpha}\cos^2(\boldsymbol{l}_j, \boldsymbol{f})), \tag{6.39}$$

where we use Eq. (13.46) with $\sigma_0^2 = 1$. Considering the relationship, $\cos(\boldsymbol{l}_j, \boldsymbol{f}) = \cos(\boldsymbol{L}\boldsymbol{z}_j, \boldsymbol{L}\boldsymbol{\eta}_1) = \cos(\boldsymbol{z}_j, \boldsymbol{\eta}_1|\boldsymbol{L}^T\boldsymbol{L}))$, we have

$$\mathrm{tr}[\boldsymbol{W}^T\boldsymbol{R}\boldsymbol{W}] = \sum_{j=1}^{3}\frac{1}{\|\boldsymbol{l}_j\|^2[1 - \omega\cos^2(\boldsymbol{z}_j, \boldsymbol{\eta}_1|\boldsymbol{L}^T\boldsymbol{L})]}, \tag{6.40}$$

where $\omega = \alpha/(1+\alpha)$. The noise gain is expressed as

$$\mathrm{tr}\{\boldsymbol{W}^T\boldsymbol{W}\} = \sum_{j=1}^{3}\boldsymbol{z}_j^T\boldsymbol{\Upsilon}\boldsymbol{z}_j = \sum_{j=1}^{3}\frac{[1 - (2\omega - \omega^2)\cos^2(\boldsymbol{z}_j, \boldsymbol{\eta}_1|\boldsymbol{L}^T\boldsymbol{L})]}{\|\boldsymbol{l}_j\|^2[1 - \omega\cos^2(\boldsymbol{z}_j, \boldsymbol{\eta}_1|\boldsymbol{L}^T\boldsymbol{L})]^2}, \tag{6.41}$$

where we use the relationship

$$\begin{aligned}
\boldsymbol{z}_j^T\boldsymbol{\Upsilon}\boldsymbol{z}_j &= \boldsymbol{z}_j^T[\boldsymbol{L}^T(\boldsymbol{r}_1)\boldsymbol{R}^{-1}\boldsymbol{L}(\boldsymbol{r}_1)]^{-1}[\boldsymbol{L}^T(\boldsymbol{r}_1)\boldsymbol{R}^{-2}\boldsymbol{L}(\boldsymbol{r}_1)][\boldsymbol{L}^T(\boldsymbol{r}_1)\boldsymbol{R}^{-1}\boldsymbol{L}(\boldsymbol{r}_1)]^{-1}\boldsymbol{z}_j \\
&= \boldsymbol{z}_j^T[\sum_{i=1}^{3}\frac{1}{\phi_i}\boldsymbol{z}_i\boldsymbol{z}_i^T][\boldsymbol{L}^T(\boldsymbol{r}_1)\boldsymbol{R}^{-2}\boldsymbol{L}(\boldsymbol{r}_1)][\sum_{i=1}^{3}\frac{1}{\phi_i}\boldsymbol{z}_i\boldsymbol{z}_i^T]\boldsymbol{z}_j = \\
&= \frac{\boldsymbol{z}_j^T[\boldsymbol{L}^T(\boldsymbol{r}_1)\boldsymbol{R}^{-2}\boldsymbol{L}(\boldsymbol{r}_1)]\boldsymbol{z}_j}{\phi_j^2} = \frac{[\boldsymbol{l}_j^T\boldsymbol{R}^{-2}\boldsymbol{l}_j]}{\phi_j^2} = \frac{[\boldsymbol{l}_j^T\boldsymbol{R}^{-2}\boldsymbol{l}_j]}{[\boldsymbol{l}_j^T\boldsymbol{R}^{-1}\boldsymbol{l}_j]^2}.
\end{aligned} \tag{6.42}$$

As discussed in Section 4.3, $\boldsymbol{z}_3$, which is the eigenvector corresponding to the minimum eigenvalue ϕ_3, can represent the source orientation and it is generally nearly equal to the source orientation $\boldsymbol{\eta}_1$. Therefore, we can assume that

$\cos^2(\boldsymbol{z}_3, \boldsymbol{\eta}_1|\boldsymbol{L}^T\boldsymbol{L}) \approx 1$. Also, since the directions represented by $\boldsymbol{z}_1$ and $\boldsymbol{z}_2$ are orthogonal to $\boldsymbol{z}_3$, i.e., to $\boldsymbol{\eta}_1$, we can assume that $\cos^2(\boldsymbol{z}_j, \boldsymbol{\eta}_1|\boldsymbol{L}^T\boldsymbol{L}) \ll 1$ for $j = 1, 2$. Thus, by substituting Eqs. (6.38) and (6.41) into (6.30), considering the above-mentioned relationships, we obtain

$$\mathcal{Z}_V^{(\mathrm{I})}(\boldsymbol{r}_1) \approx \frac{\frac{1}{\|\boldsymbol{l}_1\|^2} + \frac{1}{\|\boldsymbol{l}_2\|^2} + \frac{1}{\|\boldsymbol{l}_3\|^2[1-\omega]}}{\frac{1}{\|\boldsymbol{l}_1\|^2} + \frac{1}{\|\boldsymbol{l}_2\|^2} + \frac{1-(2\omega-\omega^2)}{\|\boldsymbol{l}_3\|^2[1-\omega]^2}}. \tag{6.43}$$

Further assuming that $\|\boldsymbol{l}_1\|^2 \approx \|\boldsymbol{l}_2\|^2 \approx \|\boldsymbol{l}_3\|^2$, $\mathcal{Z}_V^{(\mathrm{I})}(\boldsymbol{r}_1)$ is finally expressed as

$$\mathcal{Z}_V^{(\mathrm{I})}(\boldsymbol{r}_1) \approx \frac{2 + \frac{1}{(1-\omega)}}{2 + \frac{1-(2\omega-\omega^2)}{(1-\omega)^2}} = 1 + \frac{1}{3}\alpha, \tag{6.44}$$

and the value of Θ is obtained as

$$\Theta \approx \frac{1}{3}. \tag{6.45}$$

The above analysis indicates that $\mathcal{Z}_V^{(\mathrm{I})}(\boldsymbol{r}_1)$ becomes approximately one third of the input SNR.

When the spherically symmetric homogeneous conductor model is used for the forward calculation, the source vector is expressed along the two tangential components, as explained in Section 13.7. As a result, the lead-field matrix $\boldsymbol{L}(\boldsymbol{r}_1)$ is an $M \times 2$ matrix, and $\boldsymbol{L}^T(\boldsymbol{r}_1)\boldsymbol{R}^{-1}\boldsymbol{L}(\boldsymbol{r}_1)$ is a 2×2 matrix. In such cases, it is easy to show that Eq. (6.44) changes to

$$\mathcal{Z}_V^{(\mathrm{I})}(\boldsymbol{r}_1) \approx 1 + \frac{1}{2}\alpha, \tag{6.46}$$

and

$$\Theta \approx \frac{1}{2}.$$

In this case, the output SNR of the spatial filter is half of the input SNR. This result is in accordance with that obtained by Vrba and Robinson[41] who assumed a special source-sensor configuration where a single source exists directly below the center of a rotationally-symmetric sensor array.

6.5 Influence of array mismatch

We next investigate the influence of array mismatch on the output SNR of the adaptive spatial filters. When computing a weight vector, the exact lead field is generally unknown and we must therefore use an approximate lead field. This situation always happens in bioelectromagnetic inverse modeling, because the lead field is usually estimated from some kind of model such as the spherical-homogeneous conductor model, as described in Section 13.7 of the Appendix. Even when we use a real-head shape model derived from each subject's anatomical information (such as subject's MRI), we cannot reduce the error between the true and approximated lead fields completely to zero. Let us define the true lead-field vector at $\boldsymbol{r}_1$ as $\boldsymbol{f}$ and the approximate lead-field vector as $\boldsymbol{f}_e$. Using $\boldsymbol{f}_e$ and Eq. (6.6), we have the following expression for the weight vector of the minimum-variance spatial filters such that,

$$\boldsymbol{w}(\boldsymbol{r}_1) = \xi_e \boldsymbol{R}^{-1} \boldsymbol{f}_e, \tag{6.47}$$

where $\xi_e = 1/[\boldsymbol{f}_e^T \boldsymbol{R}^{-1} \boldsymbol{f}_e]$, $\xi = \|\boldsymbol{f}_e\|/[\boldsymbol{f}_e^T \boldsymbol{R}^{-1} \boldsymbol{f}_e]$, and $\xi = 1/[\boldsymbol{f}_e^T \boldsymbol{R}^{-2} \boldsymbol{f}_e]^{-1/2}$ for the unit-gain, array-gain, and unit-noise-gain minimum-variance filters, respectively. Substituting the weight into Eq. (6.19), we derive

$$\Theta = \frac{[\boldsymbol{f}_e^T \boldsymbol{R}^{-1} \boldsymbol{f}]^2}{\|\boldsymbol{f}\|^2 (\boldsymbol{f}_e^T \boldsymbol{R}^{-2} \boldsymbol{f}_e)}. \tag{6.48}$$

Then, assuming the single-source scenario, and using again Eq. (13.47) and Eq. (13.49), we obtain

$$\Theta = \frac{\cos^2(\boldsymbol{f}_e, \boldsymbol{f})}{[1 + (2\alpha + \alpha^2)\sin^2(\boldsymbol{f}_e, \boldsymbol{f})]} \approx \frac{\cos^2(\boldsymbol{f}_e, \boldsymbol{f})}{[1 + \alpha^2 \sin^2(\boldsymbol{f}_e, \boldsymbol{f})]}, \tag{6.49}$$

where we use the fact that the value of α generally equals the order of the number of sensors M, and usually M is a large number in the range of 150–400. Thus, even if the difference between $\boldsymbol{f}_e$ and $\boldsymbol{f}$ is small, i.e., even if $\sin^2(\boldsymbol{f}_e, \boldsymbol{f})$ is small, the value of $\alpha^2 \sin^2(\boldsymbol{f}_e, \boldsymbol{f})$ can be much larger than 1, resulting in a significantly small Θ. That is, the array mismatch can cause a significant reduction of the output SNR. This is actually a well-known weak point of adaptive spatial filters. In the following sections, we describe methods to overcome this problem.

6.6 Diagonal loading

As suggested in Eq. (6.49), the SNR reduction caused by the array mismatch should become smaller if we use a smaller α. A smaller α value is attained by artificially adding noise, i.e., by adding $\epsilon \boldsymbol{I}$ to the measurement covariance matrix $\boldsymbol{R}$. Here, ϵ is a parameter that controls the amount of artificially-added noise equivalent. This parameter is called the loading factor. This technique has been called diagonal loading in the field of sensor array processing[44][45][3]. Using diagonal loading, the weight for the unit-gain minimum-variance spatial filter in Eq. (4.5) is expressed as

$$\boldsymbol{w}(\boldsymbol{r}) = \frac{(\boldsymbol{R} + \epsilon \boldsymbol{I})^{-1} \boldsymbol{l}(\boldsymbol{r})}{[\boldsymbol{l}^T(\boldsymbol{r})(\boldsymbol{R} + \epsilon \boldsymbol{I})^{-1} \boldsymbol{l}(\boldsymbol{r})]}. \tag{6.50}$$

Using Eq. (13.42), the inverse of $(\boldsymbol{R} + \epsilon \boldsymbol{I})^{-1}$ is given by:

$$(\boldsymbol{R} + \epsilon \boldsymbol{I})^{-1} = \frac{1}{(\sigma_0^2 + \epsilon)} (\boldsymbol{I} - \frac{\alpha'}{1 + \alpha'} \frac{\boldsymbol{f}\boldsymbol{f}^T}{\|\boldsymbol{f}\|^2}), \tag{6.51}$$

[3] Diagonal loading is referred to as Tikhonov regularization in numerical linear algebra, and it was already mentioned briefly in Section 3.2.

where

$$\alpha' = \frac{\sigma_1^2 \|\boldsymbol{f}\|^2}{(\sigma_0^2 + \epsilon)}. \tag{6.52}$$

Therefore, for the single-source scenario, Θ can be expressed in this case as

$$\Theta = \frac{\cos^2(\boldsymbol{f}_e, \boldsymbol{f})}{[1 + (2\alpha' + (\alpha')^2) \sin^2(\boldsymbol{f}_e, \boldsymbol{f})]}. \tag{6.53}$$

Here, since ϵ is a controllable parameter, α' can be made small by choosing a large ϵ (compared to σ_0^2), and we can recover the output SNR to some extent. However, the value of the input SNR (α' in this case) also determines the spatial resolution as shown in Eq. (5.30). If we make α' large, the spatial resolution is degraded accordingly. In other words, diagonal loading produces a trade-off between the output SNR and the spatial resolution, and it can improve output SNR by sacrificing the spatial resolution.

The diagonal loading spatial filter can also be formulated from the following quadratic constraint[45][28]:

$$\begin{aligned} \boldsymbol{w}(\boldsymbol{r}) = \underset{\boldsymbol{w}(\boldsymbol{r})}{\arg\min}\, \boldsymbol{w}(\boldsymbol{r}) \boldsymbol{R} \boldsymbol{w}^T(\boldsymbol{r}), \qquad &\text{subject to} \qquad \boldsymbol{w}^T(\boldsymbol{r}) \boldsymbol{l}(\boldsymbol{r}) = 1, \\ &\text{and} \qquad \|\boldsymbol{w}(\boldsymbol{r})\|^2 \le T_0. \end{aligned} \tag{6.54}$$

Because $\|\boldsymbol{w}(\boldsymbol{r})\|^2$ is the noise gain of the minimum-variance spatial filter, the weight vector obtained using Eq. (6.54) implements a spatial filter whose output noise power is less than T_0/σ_0^2. Namely, T_0 sets an upper limit on the power of output noise. To obtain an explicit form of the weight vector satisfying Eq. (6.54), we change the inequality constraint to an equality constraint, i.e.,

$$\begin{aligned} \boldsymbol{w}(\boldsymbol{r}) = \underset{\boldsymbol{w}(\boldsymbol{r})}{\arg\min}\, \boldsymbol{w}(\boldsymbol{r}) \boldsymbol{R} \boldsymbol{w}^T(\boldsymbol{r}), \qquad &\text{subject to} \qquad \boldsymbol{w}^T(\boldsymbol{r}) \boldsymbol{l}(\boldsymbol{r}) = 1, \\ &\text{and} \qquad \|\boldsymbol{w}(\boldsymbol{r})\|^2 = T_0. \end{aligned} \tag{6.55}$$

The solution satisfying Eq. (6.54) exists on the border of the inequality constraint $\|\boldsymbol{w}(\boldsymbol{r})\|^2 \le T_0$. Therefore, the solution satisfying Eq. (6.54) can be obtained as the solution for Eq. (6.55), which is obtained by using a derivation similar to the one used in Section 4.1.1. We first define two Lagrange multipliers, κ_1 and κ_2, and define the Lagrangian $\mathbb{L}$ as

$$\mathbb{L}(\boldsymbol{w}, \kappa_1, \kappa_2) = \boldsymbol{w}^T \boldsymbol{R} \boldsymbol{w} + \kappa_1(\boldsymbol{w}^T \boldsymbol{l}(\boldsymbol{r}) - 1) + \kappa_2(\boldsymbol{w}^T \boldsymbol{w} - T_0), \tag{6.56}$$

where we again omit the explicit notation of $(\boldsymbol{r})$ from the weight vector for simplicity. Then, calculating the derivative of $\boldsymbol{w}$ of the Lagrangian and setting it to zero gives

$$2\boldsymbol{R}\boldsymbol{w} + \kappa_1 \boldsymbol{l}(\boldsymbol{r}) + 2\kappa_2 \boldsymbol{w} = 0. \tag{6.57}$$

We can then obtain

$$\boldsymbol{w} = -\frac{\kappa_1}{2}(\boldsymbol{R} + \kappa_2 \boldsymbol{I})^{-1} \boldsymbol{l}(\boldsymbol{r}), \tag{6.58}$$

and by substituting the above weight expression back into the unit-gain constraint, we get

$$\kappa_1 = -2/[\boldsymbol{l}^T(\boldsymbol{R}+\kappa_2\boldsymbol{I})^{-1}\boldsymbol{l}],$$

and substituting the above expression into Eq. (6.58) gives the weight expression in Eq. (6.50).

Using the following constrained optimization:

$$\boldsymbol{w}(\boldsymbol{r}) = \arg\min_{\boldsymbol{w}(\boldsymbol{r})} \boldsymbol{w}(\boldsymbol{r})\boldsymbol{R}\boldsymbol{w}^T(\boldsymbol{r}), \quad \text{subject to} \quad \boldsymbol{w}^T(\boldsymbol{r})\boldsymbol{l}(\boldsymbol{r}) = \|\boldsymbol{l}(\boldsymbol{r})\|, \quad \text{and} \quad \|\boldsymbol{w}(\boldsymbol{r})\|^2 = T_0, \tag{6.59}$$

and exactly the same derivation, we can derive the array-gain-constraint version of the diagonal-loading spatial filter whose weight vector is expressed as

$$\boldsymbol{w}(\boldsymbol{r}) = \frac{(\boldsymbol{R}+\epsilon\boldsymbol{I})^{-1}\widetilde{\boldsymbol{l}}(\boldsymbol{r})}{[\widetilde{\boldsymbol{l}}^T(\boldsymbol{r})(\boldsymbol{R}+\epsilon\boldsymbol{I})^{-1}\widetilde{\boldsymbol{l}}(\boldsymbol{r})]}, \tag{6.60}$$

where $\widetilde{\boldsymbol{l}}(\boldsymbol{r}) = \boldsymbol{l}(\boldsymbol{r})/\|\boldsymbol{l}(\boldsymbol{r})\|$. As mentioned previously, diagonal loading degrades the spatial resolution of the reconstructed source spatial distribution. The following sections describe two methods that can produce a high output SNR without sacrificing the spatial resolution.

6.7 Asymmetric diagonal loading

In the preceding section, we describe the fact that diagonal loading produces a trade-off between the output SNR and the spatial resolution. A slight modification of diagonal loading, however, can avoid this trade-off and can overcome the array mismatch problem without sacrificing spatial resolution. This section describes this modification of the diagonal loading technique, called asymmetric diagonal loading, in which we give different values to the denominator and the numerator loading factors in Eq. (6.50). That is, the weight vector for the unit-gain minimum-variance filter with asymmetric diagonal loading is given by:

$$\boldsymbol{w}(\boldsymbol{r}) = \frac{(\boldsymbol{R}+\epsilon_n\boldsymbol{I})^{-1}\boldsymbol{l}}{\boldsymbol{l}^T(\boldsymbol{R}+\epsilon_d\boldsymbol{I})^{-1}\boldsymbol{l}}, \tag{6.61}$$

where ϵ_n and ϵ_d are the denominator and numerator loading factors, respectively.

When we do not know the true lead field, and we must use an approximated lead-field vector $\boldsymbol{f}_e$, the weight vector pointing at $\boldsymbol{r}_1$ with this asymmetric diagonal loading is expressed as

$$\boldsymbol{w}(\boldsymbol{r}_1) = \frac{(\boldsymbol{R}+\epsilon_n\boldsymbol{I})^{-1}\boldsymbol{f}_e}{[\boldsymbol{f}_e^T(\boldsymbol{R}+\epsilon_d\boldsymbol{I})^{-1}\boldsymbol{f}_e]}. \tag{6.62}$$

Then, substituting this equation into Eq. (6.19), the value of the SNR transfer factor, Θ, is expressed in this case as

$$\Theta = \frac{[\boldsymbol{f}_e^T(\boldsymbol{R}+\epsilon_n\boldsymbol{I})^{-1}\boldsymbol{f}]^2}{\|\boldsymbol{f}\|^2[\boldsymbol{f}_e^T(\boldsymbol{R}+\epsilon_n\boldsymbol{I})^{-2}\boldsymbol{f}_e]}. \tag{6.63}$$

Using the same derivation for Eq. (6.53), we finally obtain a value of Θ under the single-source scenario such that

$$\Theta = \frac{\cos^2(\boldsymbol{f}_e, \boldsymbol{f})}{[1+(2\alpha_n+\alpha_n^2)\sin^2(\boldsymbol{f}_e, \boldsymbol{f})]}, \tag{6.64}$$

where

$$\alpha_n = \frac{\sigma_1^2\|\boldsymbol{f}\|^2}{(\sigma_0^2+\epsilon_n)}.$$

The above equation shows that the Θ value (hence, the output SNR) is determined only by the numerator loading factor ϵ_n.

We next calculate the point-spread function using the weight vector with asymmetric diagonal loading. We use

$$(\boldsymbol{R}+\epsilon_n\boldsymbol{I})^{-1} = \frac{1}{\epsilon_n+\sigma_0^2}(\boldsymbol{I} - \frac{\alpha_n}{1+\alpha_n}\frac{\boldsymbol{f}\boldsymbol{f}^T}{\|\boldsymbol{f}\|^2}), \tag{6.65}$$

and

$$(\boldsymbol{R}+\epsilon_d\boldsymbol{I})^{-1} = \frac{1}{\epsilon_d+\sigma_0^2}(\boldsymbol{I} - \frac{\alpha_d}{1+\alpha_d}\frac{\boldsymbol{f}\boldsymbol{f}^T}{\|\boldsymbol{f}\|^2}), \tag{6.66}$$

where

$$\alpha_d = \frac{\sigma_1^2\|\boldsymbol{f}\|^2}{(\sigma_0^2+\epsilon_d)}.$$

We then derive the point-spread function for the array-gain minimum-variance filter

$$\digamma(\boldsymbol{r}) = \frac{(\epsilon_d+\sigma_0^2)(1+\alpha_d)}{(\epsilon_n+\sigma_0^2)(1+\alpha_n)}\|\boldsymbol{f}\|\frac{\cos(\boldsymbol{l}, \boldsymbol{f})}{1+\alpha_d[1-\cos^2(\boldsymbol{l}, \boldsymbol{f})]}. \tag{6.67}$$

The normalized point-spread function is obtained as

$$\digamma_N(\boldsymbol{r}) = \frac{\cos(\boldsymbol{l}, \boldsymbol{f})}{1+\alpha_d[1-\cos^2(\boldsymbol{l}, \boldsymbol{f})]}. \tag{6.68}$$

The above equation shows that the value of ϵ_d determines the shape of the point-spread function. (The value of ϵ_n is contained only in the multiplicative constant of the point-spread function.) Thus, we can see that only the denominator loading factor determines the spatial resolution, and the numerator loading factor has no effect on the resolution.

In summary, only the numerator loading factor determines the output SNR, whereas the denominator loading factor has no effect on it. Conversely, only the denominator loading factor determines the spatial resolution, whereas the numerator loading factor has no effect on it. Therefore, we can obtain a relatively high output SNR in the presence of array mismatch without sacrificing the spatial resolution, by controlling the numerator loading factor while keeping the denominator loading factor small.

So, what is the price we pay for the improvement of the output SNR when we apply asymmetric diagonal loading? To answer this question, let us consider the leakage problem. We consider a simple case where two sources exist at $\boldsymbol{r}_1$ and $\boldsymbol{r}_2$. The source distribution is expressed as

$$s(\boldsymbol{r},t) = s(\boldsymbol{r}_1,t)\delta(\boldsymbol{r}-\boldsymbol{r}_1) + s(\boldsymbol{r}_2,t)\delta(\boldsymbol{r}-\boldsymbol{r}_2). \tag{6.69}$$

Substituting the above equation into Eq. (2.82), we have

$$\widehat{s}(\boldsymbol{r},t) = s(\boldsymbol{r}_1,t)\mathbb{R}(\boldsymbol{r},\boldsymbol{r}_1) + s(\boldsymbol{r}_2,t)\mathbb{R}(\boldsymbol{r},\boldsymbol{r}_2), \tag{6.70}$$

where $\mathbb{R}(\boldsymbol{r},\boldsymbol{r}_1)$ and $\mathbb{R}(\boldsymbol{r},\boldsymbol{r}_2)$ are the resolution kernels in which a single source is located at $\boldsymbol{r}_1$ and $\boldsymbol{r}_2$, respectively. Thus, the spatial filter output at the first source location $\boldsymbol{r}_1$ is given by

$$\widehat{s}(\boldsymbol{r}_1,t) = s(\boldsymbol{r}_1,t)\mathbb{R}(\boldsymbol{r}_1,\boldsymbol{r}_1) + s(\boldsymbol{r}_2,t)\mathbb{R}(\boldsymbol{r}_1,\boldsymbol{r}_2). \tag{6.71}$$

Thus, the amount of leakage from the second-source time course into the estimated first-source time course is expressed by the ratio ψ, defined as

$$\psi = \frac{\mathbb{R}(\boldsymbol{r}_1,\boldsymbol{r}_2)}{\mathbb{R}(\boldsymbol{r}_1,\boldsymbol{r}_1)}. \tag{6.72}$$

Using again the weight expression in Eq. (6.61) and the matrix inverse relationships of Eqs. (6.65) and (6.66), we finally derive

$$\psi = \frac{\cos(\boldsymbol{f},\boldsymbol{g})}{[1+\alpha_n(1-\cos^2(\boldsymbol{f},\boldsymbol{g}))]}. \tag{6.73}$$

The above equation shows that the leakage ratio, ψ, is determined only by the numerator loading factor, and does not depend on the denominator loading factor. Therefore, if we use a smaller α_n to obtain a larger Θ, the amount of leakage will necessarily increase. In conclusion, the asymmetric regularization provides a trade-off between leakage and output SNR.

6.8 Eigenspace-projection spatial filter

6.8.1 Eigenspace projection

The SNR degradation due to array mismatch can be avoided by making use of the property of the low-rank signals described in Section 2.4. Using Eqs. (2.63) and (2.64), we can rewrite Eq. (6.48), such that

$$\Theta = \frac{[\boldsymbol{f}_e^T(\boldsymbol{E}_S\boldsymbol{\Lambda}_S^{-1}\boldsymbol{E}_S^T + \boldsymbol{E}_N\boldsymbol{\Lambda}_N^{-1}\boldsymbol{E}_N^T)\boldsymbol{f}]^2}{\|\boldsymbol{f}\|^2[\boldsymbol{f}_e^T(\boldsymbol{E}_S\boldsymbol{\Lambda}_S^{-2}\boldsymbol{E}_S^T + \boldsymbol{E}_N\boldsymbol{\Lambda}_N^{-2}\boldsymbol{E}_N^T)\boldsymbol{f}_e]}. \tag{6.74}$$

Then, using the property that the noise subspace is orthogonal to the lead-field vector at the true source locations, i.e.,

$$\boldsymbol{E}_N^T \boldsymbol{f} = 0, \tag{6.75}$$

Θ is given by:

$$\Theta = \frac{[\boldsymbol{f}_e^T(\boldsymbol{E}_S\boldsymbol{\Lambda}_S^{-1}\boldsymbol{E}_S^T)\boldsymbol{f}]^2}{\|\boldsymbol{f}\|^2[\boldsymbol{f}_e^T(\boldsymbol{E}_S\boldsymbol{\Lambda}_S^{-2}\boldsymbol{E}_S^T)\boldsymbol{f}_e + \boldsymbol{f}_e^T(\boldsymbol{E}_N\boldsymbol{\Lambda}_N^{-2}\boldsymbol{E}_N^T)\boldsymbol{f}_e]}. \tag{6.76}$$

Let us define the error contained in $\boldsymbol{f}_e$ as $\Delta\boldsymbol{f}$, i.e., $\boldsymbol{f}_e = \Delta\boldsymbol{f} + \boldsymbol{f}$. Then, Θ is expressed as[46][47]

$$\Theta = \frac{[\boldsymbol{f}_e^T(\boldsymbol{E}_S\boldsymbol{\Lambda}_S^{-1}\boldsymbol{E}_S^T)\boldsymbol{f}]^2}{\|\boldsymbol{f}\|^2[\boldsymbol{f}_e^T(\boldsymbol{E}_S\boldsymbol{\Lambda}_S^{-2}\boldsymbol{E}_S^T)\boldsymbol{f}_e + \Delta\boldsymbol{f}^T(\boldsymbol{E}_N\boldsymbol{\Lambda}_N^{-2}\boldsymbol{E}_N^T)\Delta\boldsymbol{f}]}. \tag{6.77}$$

We assume that $\Delta\boldsymbol{f}$ is small and the relationship $\boldsymbol{f}_e \approx \boldsymbol{f}$ holds, thus giving

$$\boldsymbol{f}_e^T(\boldsymbol{E}_S\boldsymbol{\Lambda}_S^{-1}\boldsymbol{E}_S^T)\boldsymbol{f} = \sum_{j=1}^{Q}\frac{\|\boldsymbol{f}_e^T\boldsymbol{e}_j\|\|\boldsymbol{f}^T\boldsymbol{e}_j\|}{\lambda_j} \approx \sum_{j=1}^{Q}\frac{\|\boldsymbol{f}^T\boldsymbol{e}_j\|^2}{\lambda_j}, \tag{6.78}$$

and

$$\boldsymbol{f}_e^T(\boldsymbol{E}_S\boldsymbol{\Lambda}_S^{-2}\boldsymbol{E}_S^T)\boldsymbol{f}_e = \sum_{j=1}^{Q}\frac{\|\boldsymbol{f}_e^T\boldsymbol{e}_j\|^2}{\lambda_j^2} \approx \sum_{j=1}^{Q}\frac{\|\boldsymbol{f}^T\boldsymbol{e}_j\|^2}{\lambda_j^2}. \tag{6.79}$$

Also, we have

$$\Delta\boldsymbol{f}^T(\boldsymbol{E}_N\boldsymbol{\Lambda}_N^{-2}\boldsymbol{E}_N^T)\Delta\boldsymbol{f} = \sum_{j=Q+1}^{M}\frac{\|\Delta\boldsymbol{f}^T\boldsymbol{e}_j\|^2}{\lambda_j^2}. \tag{6.80}$$

Therefore, Eq. (6.77) changes to

$$\Theta \approx \frac{\left[\sum_{j=1}^{Q}\|\boldsymbol{f}^T\boldsymbol{e}_j\|^2/\lambda_j\right]^2}{\|\boldsymbol{f}\|^2\left[\sum_{j=1}^{Q}\|\boldsymbol{f}^T\boldsymbol{e}_j\|^2/\lambda_j^2 + \sum_{j=Q+1}^{M}\|\Delta\boldsymbol{f}^T\boldsymbol{e}_j\|^2/\lambda_j^2\right]}. \tag{6.81}$$

The second term in the denominator causes the degradation of the output SNR, unless it is negligibly small compared to the first term. Equation (6.81) indicates that even when $\Delta\boldsymbol{f}$ is small, the second term in the denominator may not be negligibly small, because the noise-level eigenvalues are generally much smaller than the signal-level eigenvalues; i.e.,

$$\lambda_j \quad (j = 1, \ldots, Q) \quad \gg \quad \lambda_j \quad (j = Q+1, \ldots, M).$$

Therefore, a very small $\Delta\boldsymbol{f}$ can cause a significant reduction in the output SNR.

Equation (6.81) also shows that the second term in the denominator arises from the noise-subspace components of the weight vector, and therefore if the noise-subspace components are removed, this second term will be removed and no SNR reduction results. This consideration leads to a method called the eigenspace-projection spatial filter in which only the signal subspace components of a weight vector are used for spatial filtering. The weight vector of the minimum-variance spatial filters are expressed using the noise- and signal-subspace components,

$$\boldsymbol{w}(\boldsymbol{r}) = \xi \boldsymbol{R}^{-1}\boldsymbol{l}(\boldsymbol{r}) = \xi \boldsymbol{E}_S \boldsymbol{\Lambda}_S^{-1} \boldsymbol{E}_S^T \boldsymbol{l}(\boldsymbol{r}) + \xi \boldsymbol{E}_N \boldsymbol{\Lambda}_N^{-1} \boldsymbol{E}_N^T \boldsymbol{l}(\boldsymbol{r}). \tag{6.82}$$

At source locations, the weight vector can be expressed as

$$\boldsymbol{w}(\boldsymbol{r}) = \xi \boldsymbol{E}_S \boldsymbol{\Lambda}_S^{-1} \boldsymbol{E}_S^T \boldsymbol{f} + \xi \boldsymbol{E}_N \boldsymbol{\Lambda}_N^{-1} \boldsymbol{E}_N^T \boldsymbol{f}. \tag{6.83}$$

The second term on the right-hand side of the equation above should be zero due to the orthogonality between $\boldsymbol{E}_N^T$ and $\boldsymbol{f}$. However, because the lead-field vector $\boldsymbol{f}$ used for deriving the weight vector is not exactly equal to the true $\boldsymbol{f}$, the product $\boldsymbol{E}_N^T \boldsymbol{f}$ should have some non-zero value and this non-zero $\boldsymbol{E}_N^T \boldsymbol{f}$ results in the second term in the denominator in Eq. (6.81). Therefore, the eigenspace-projection spatial filter only uses the signal-subspace components of the minimum-variance weight vector. That is, defining the weight vector of the eigenspace-projection spatial filter as $\boldsymbol{w}_e(\boldsymbol{r})$, this weight vector $\boldsymbol{w}_e(\boldsymbol{r})$ is obtained using[48].

$$\boldsymbol{w}_e(\boldsymbol{r}) = \boldsymbol{E}_S \boldsymbol{E}_S^T \boldsymbol{w}(\boldsymbol{r}) = \xi \boldsymbol{E}_S \boldsymbol{E}_S^T (\boldsymbol{E}_S \boldsymbol{\Lambda}_S^{-1} \boldsymbol{E}_S^T + \boldsymbol{E}_N \boldsymbol{\Lambda}_N^{-1} \boldsymbol{E}_N^T) \boldsymbol{l}(\boldsymbol{r}) = \xi \boldsymbol{E}_S \boldsymbol{\Lambda}_S^{-1} \boldsymbol{E}_S^T \boldsymbol{l}(\boldsymbol{r}). \tag{6.84}$$

Using $\boldsymbol{w}_e(\boldsymbol{r})$, Θ in Eq. (6.19) is expressed as

$$\Theta \approx \frac{\left[\sum_{j=1}^{Q} \|\boldsymbol{f}^T \boldsymbol{e}_j\|^2/\lambda_j\right]^2}{\|\boldsymbol{f}\|^2 \left[\sum_{j=1}^{Q} \|\boldsymbol{f}^T \boldsymbol{e}_j\|^2/\lambda_j^2\right]}, \tag{6.85}$$

where we again use the approximation that $\boldsymbol{f}_e \approx \boldsymbol{f}$. The above Θ does not have the second term in the denominator, and thus the eigenspace-projection spatial filter can prevent the degradation of the output SNR.

However, to implement the eigenspace spatial filter, the information regarding the separation of the noise and signal subspaces, i.e., the value of the signal-subspace dimension Q in Eq. (2.53) should be known. The signal subspace dimension Q is in principle determined by separating distinctly-large eigenvalues from the small eigenvalues in the eigenvalue spectrum of the measurement covariance matrix. This separation, however, may not be easy if there is no clear threshold in the eigenvalue spectrum.

The well-known work-around for this problem concerns the overestimation of Q. In general, the overestimation of the signal subspace dimensionality gives intermediate results between those of the (non-eigenspace-projected) minimum-variance spatial filter and those of the eigenspace-projection spatial filter obtained with the correct signal-subspace dimension. Let us consider the case where the signal subspace dimension is overestimated at $Q + \mathit{\Delta} Q$. Then, the SNR transfer

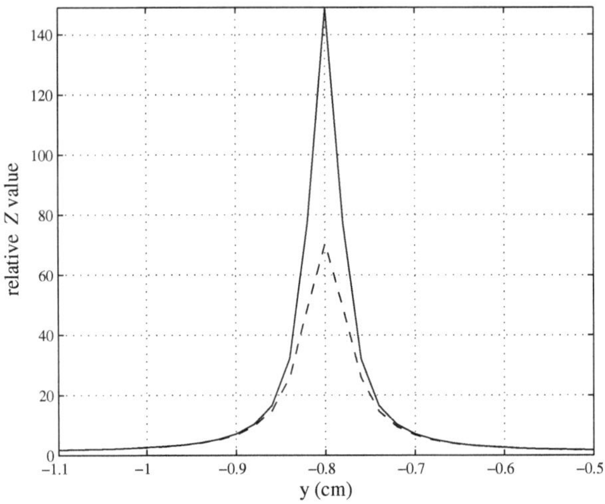

Figure 6.1: The plot of the reconstructed $\mathcal{Z}$ value. The ordinate shows the relative $\mathcal{Z}$ value because σ_0^2 was set to one in this numerical experiments. A cross-sectional view of the source at $(0, -0.8, -6)$ along the line $(x = 0; z = -6)$ cm. The solid line shows the results of $\mathcal{Z}_V^{(\mathrm{II})}$ and the broken line shows the results of $\mathcal{Z}_V^{(\mathrm{I})}$.

factor of such an eigenspace spatial filter, $\Theta_{(\Delta Q)}$, is expressed as

$$\Theta_{(\Delta Q)} = \frac{[\sum_{j=1}^{Q} \|\boldsymbol{f}^T \boldsymbol{e}_j\|^2/\lambda_j]^2}{\|\boldsymbol{f}\|^2[\sum_{j=1}^{Q} \|\boldsymbol{f}^T \boldsymbol{e}_j\|^2/\lambda_j^2 + \sum_{j=Q+1}^{Q+\Delta Q} \|\Delta \boldsymbol{f}^T \boldsymbol{e}_j\|^2/\lambda_j^2]}. \tag{6.86}$$

Here, the relationship

$$0 \leq \sum_{j=Q+1}^{Q+\Delta Q} \|\Delta \boldsymbol{f}^T \boldsymbol{e}_j\|^2/\lambda_j^2 \leq \sum_{j=Q+1}^{M} \|\Delta \boldsymbol{f}^T \boldsymbol{e}_j\|^2/\lambda_j^2$$

holds. Namely, $\Theta_{(\Delta Q)}$ has an intermediate value between Θ from the (non-eigenspace-projected) minimum variance spatial filter and from the eigenspace-projection spatial filter with the correct Q. If the relationship

$$\sum_{j=1}^{Q} \|\boldsymbol{f}^T \boldsymbol{e}_j\|^2/\lambda_j^2 \gg \sum_{j=Q+1}^{Q+\Delta Q} \|\Delta \boldsymbol{f}^T \boldsymbol{e}_j\|^2/\lambda_j^2$$

holds, the eigenspace-projection spatial filter with an overestimated signal-subspace dimensionality gives nearly the same SNR as that of the eigenspace-projection spatial filter with a correct signal-subspace dimension.

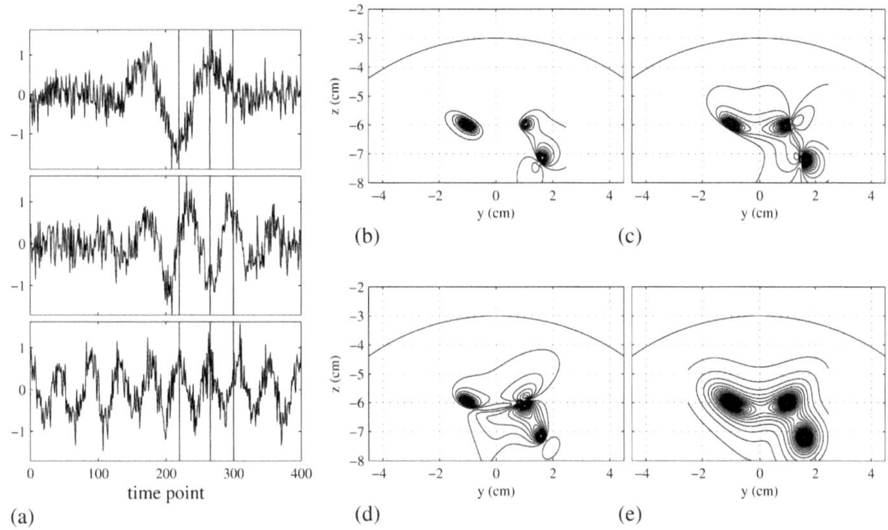

Figure 6.2: (a) The outputs of the minimum-variance scalar spatial filter at the three source locations, $\widehat{s}(\boldsymbol{r}_1, t)$ (top), $\widehat{s}(\boldsymbol{r}_2, t)$ (middle), and $\widehat{s}(\boldsymbol{r}_3, t)$ (bottom). The three vertical lines show the time corresponding to the 220th, 268th, and 300th time points. (b)–(e) The results of source reconstruction on the plane $x = 0$ cm. Results of the instantaneous reconstruction $|\widehat{s}(\boldsymbol{r}, t)|$ at (b) the 220th, (c) the 268th, and (d) the 300th time points. (e) The power reconstruction $\sqrt{\langle \widehat{s}(\boldsymbol{r}, t)^2 \rangle}$ where $\langle \widehat{s}(\boldsymbol{r}, t)^2 \rangle = \widehat{P}_s(\boldsymbol{r})$.

6.8.2 Extension to vector spatial-filter formulation

The eigenspace projection can also be applied to the vector formulation with no modification. However, since applying the signal-subspace projector generally invalidates the null constraints for the LCMV spatial filter[49], it may seem a bit puzzling that the eigenspace-projected vector spatial filter can reconstruct the three orthogonal source components, in spite of the fact that the null constrains are not preserved. We explain this fact in the following manner[40].

The eigenspace-projected weight vectors of the vector unit-gain minimum-variance spatial filter are denoted $\boldsymbol{w}_x^e, \boldsymbol{w}_y^e, \boldsymbol{w}_z^e$, where we omit the notation of $(\boldsymbol{r})$ for simplicity. Let us consider, for example, the case of $\boldsymbol{w}_x^e$, which is obtained from

$$\boldsymbol{w}_x^e = \boldsymbol{E}_S \boldsymbol{E}_S^T \boldsymbol{w}_x.$$

The null constraints in this case should satisfy

$$(\boldsymbol{w}_x^e)^T \boldsymbol{l}_y(\boldsymbol{r}) = (\boldsymbol{E}_S \boldsymbol{E}_S^T \boldsymbol{w}_x)^T \boldsymbol{l}_y(\boldsymbol{r}) = \boldsymbol{w}_x^T \boldsymbol{E}_S \boldsymbol{E}_S^T \boldsymbol{l}_y(\boldsymbol{r}) = 0,$$

$$\text{and} \quad (\boldsymbol{w}_x^e)^T \boldsymbol{l}_z(\boldsymbol{r}) = (\boldsymbol{E}_S \boldsymbol{E}_S^T \boldsymbol{w}_x)^T \boldsymbol{l}_z(\boldsymbol{r}) = \boldsymbol{w}_x^T \boldsymbol{E}_S \boldsymbol{E}_S^T \boldsymbol{l}_z(\boldsymbol{r}) = 0.$$

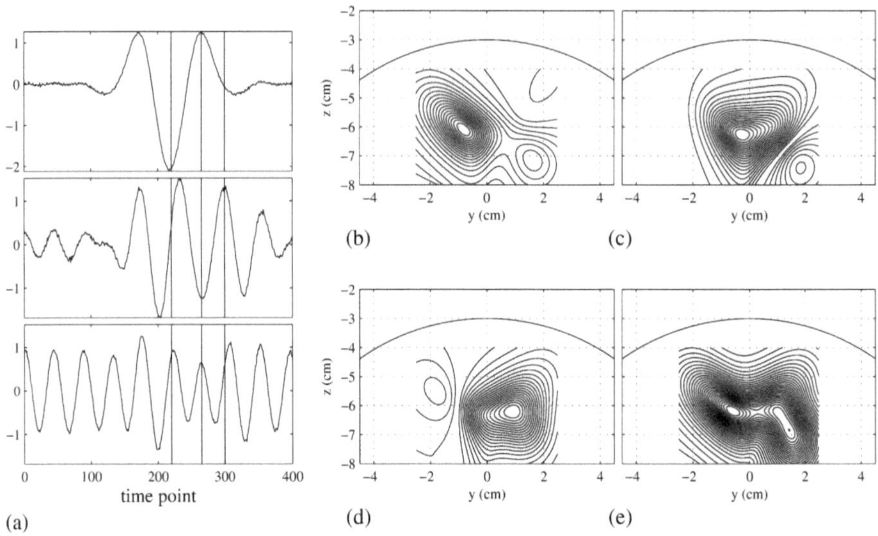

Figure 6.3: (a) The outputs of the diagonal-loading minimum-variance spatial filter at the three source locations, $\widehat{s}(\boldsymbol{r}_1,t)$ (top), $\widehat{s}(\boldsymbol{r}_2,t)$ (middle), and $\widehat{s}(\boldsymbol{r}_3,t)$ (bottom). The three vertical lines show the time corresponding to the 220th, 268th, and 300th time points. The loading factor ϵ was set to $0.01\lambda_{max}\{\boldsymbol{R}\}$. (b)–(e) The results of source reconstruction on the plane $x = 0$ cm. Results of instantaneous reconstruction $|\widehat{s}(\boldsymbol{r},t)|$ at (b) the 220th, (c) the 268th, and (d) the 300th time points. (e) The power reconstruction $\sqrt{\langle \widehat{s}(\boldsymbol{r},t)^2\rangle}$ where $\langle \widehat{s}(\boldsymbol{r},t)^2\rangle = \widehat{P}_s(\boldsymbol{r})$.

However, since $\boldsymbol{l}_y(\boldsymbol{r})$ and $\boldsymbol{l}_z(\boldsymbol{r})$ are not necessarily in the signal subspace, we generally have

$$\boldsymbol{E}_S\boldsymbol{E}_S^T\boldsymbol{l}_y(\boldsymbol{r}) \neq \boldsymbol{l}_y(\boldsymbol{r}), \quad \text{and} \quad \boldsymbol{E}_S\boldsymbol{E}_S^T\boldsymbol{l}_z(\boldsymbol{r}) \neq \boldsymbol{l}_z(\boldsymbol{r}),$$

and therefore

$$\boldsymbol{w}_x^T\boldsymbol{E}_S\boldsymbol{E}_S^T\boldsymbol{l}_y(\boldsymbol{r}) \neq 0, \quad \text{and} \quad \boldsymbol{w}_x^T\boldsymbol{E}_S\boldsymbol{E}_S^T\boldsymbol{l}_z(\boldsymbol{r}) \neq 0,$$

leading to the relationships

$$(\boldsymbol{w}_x^e)^T\boldsymbol{l}_y(\boldsymbol{r}) \neq 0, \quad \text{and} \quad (\boldsymbol{w}_x^e)^T\boldsymbol{l}_z(\boldsymbol{r}) \neq 0. \tag{6.87}$$

Consequently, we should conclude that the signal subspace projector $\boldsymbol{E}_S\boldsymbol{E}_S^T$ does not preserve the null constraints.

However, we can show that the eigenspace-projected weight vectors can reconstruct the three orthogonal components of the source moment, even though the projection operator does not preserve the null constraints. Let us assume the single-source scenario, i.e., a single source exists at $\boldsymbol{r}$ with its orientation equal to

$[\eta_x, \eta_y, \eta_z]$. Denoting its magnitude as $s(t)$, the sensor output generated by this source is expressed as

$$\boldsymbol{b}(t) = [\eta_x \boldsymbol{l}_x(\boldsymbol{r}) + \eta_x \boldsymbol{l}_y(\boldsymbol{r}) + \eta_z \boldsymbol{l}_z(\boldsymbol{r})]s(t).$$

The x component of the source, $\widehat{s}_x(t)$, estimated using the eigenspace-projected vector spatial filter is given by:

$$\widehat{s}_x(t) = (\boldsymbol{w}_x^{\mathrm{e}})^T \boldsymbol{b}(t) = \boldsymbol{w}_x^T \boldsymbol{E}_S \boldsymbol{E}_S^T [\eta_x \boldsymbol{l}_x(\boldsymbol{r}) + \eta_x \boldsymbol{l}_y(\boldsymbol{r}) + \eta_z \boldsymbol{l}_z(\boldsymbol{r})]s(t). \tag{6.88}$$

Here, the vector $(\eta_x \boldsymbol{l}_x(\boldsymbol{r}) + \eta_x \boldsymbol{l}_y(\boldsymbol{r}) + \eta_z \boldsymbol{l}_z(\boldsymbol{r}))$ lies in the signal subspace, and therefore the relationship

$$\boldsymbol{E}_S \boldsymbol{E}_S^T (\eta_x \boldsymbol{l}_x(\boldsymbol{r}) + \eta_x \boldsymbol{l}_y(\boldsymbol{r}) + \eta_z \boldsymbol{l}_z(\boldsymbol{r})) = (\eta_x \boldsymbol{l}_x(\boldsymbol{r}) + \eta_x \boldsymbol{l}_y(\boldsymbol{r}) + \eta_z \boldsymbol{l}_z(\boldsymbol{r})) \tag{6.89}$$

holds. Consequently, we get

$$\begin{aligned} \widehat{s}_x(t) &= (\boldsymbol{w}_x^{\mathrm{e}})^T \boldsymbol{b}(t) \\ &= \boldsymbol{w}_x^T (\eta_x \boldsymbol{l}_x(\boldsymbol{r}) + \eta_x \boldsymbol{l}_y(\boldsymbol{r}) + \eta_z \boldsymbol{l}_z(\boldsymbol{r}))s(t) = \eta_x s(t)(\boldsymbol{w}_x^T \boldsymbol{l}_x(\boldsymbol{r})) = \eta_x s(t). \end{aligned} \tag{6.90}$$

Using exactly the same argument, we can also show

$$\widehat{s}_y(t) = (\boldsymbol{w}_y^{\mathrm{e}})^T \boldsymbol{b}(t) = \eta_y s(t), \tag{6.91}$$

$$\text{and} \quad \widehat{s}_z(t) = (\boldsymbol{w}_z^{\mathrm{e}})^T \boldsymbol{b}(t) = \eta_z s(t). \tag{6.92}$$

Therefore, the vector-type eigenspace-projection spatial filter can detect the three orthogonal components of the source moment, even though the null constraints are not preserved. Here we also give a numerical example of the results obtained from the eigenspace-projected vector minimum-variance spatial filter.

6.9 Numerical examples

A series of numerical examples are presented to illustrate the results of the arguments in this chapter. We first present a numerical example for comparing $\mathcal{Z}_V^{(\mathrm{I})}$ and $\mathcal{Z}_V^{(\mathrm{II})}$, which are the two-types of the output SNRs of the vector minimum-variance spatial filter. For these numerical experiments, the same computer simulation scheme shown in Fig. 5.2 is used, except that a single point source exists at $\boldsymbol{r}_1 = (0, -0.8, -6)$. The time course shown in the top panel in Fig. 4.3(a) is assigned to this source. Since the lead-field vector $\boldsymbol{f}$ is calculated using the spherically-homogeneous conductor model, we use two tangential components to express the source orientation. The cross sections of the reconstructed relative $\mathcal{Z}$ values along the line $(x = 0; z = -6)$ cm are shown in Fig. 6.1. Here, $\mathcal{Z}_V^{(\mathrm{II})}$ is shown using the solid line, and $\mathcal{Z}_V^{(\mathrm{I})}$ is shown using the broken line. The peak value of $\mathcal{Z}_V^{(\mathrm{II})}$ is equal to 147.5, and that of $\mathcal{Z}_V^{(\mathrm{I})}$ is equal to 70.5. These results show that the peak of the orientation-optimized $\mathcal{Z}_V^{(\mathrm{II})}$ is nearly twice as high as the peak of $\mathcal{Z}_V^{(\mathrm{I})}$. This is in accordance with the results of the discussion in Section 6.4.

We next perform a series of numerical experiments that demonstrate what occurs in the presence of array mismatch. We here use the same computer simulation as in Section 4.7. The input SNRs for the three sources, α_1, α_2, and α_3, are equal to $8M$. In the numerical experiments presented here, instead of using the theoretical covariance matrix, we use a sample covariance matrix calculated from

$$\widehat{\boldsymbol{R}} = \frac{1}{400} \sum_{k=1}^{400} \boldsymbol{b}(t_k)\boldsymbol{b}^T(t_k), \tag{6.93}$$

where we have a total of 400 time points, and the sample covariance matrix calculated from such a relatively few number of samples should contain some error.

We first apply the array-gain constraint scalar minimum-variance spatial filter, and the results of the spatio-temporal reconstruction are shown in Fig. 6.2. Fig. 6.2(a) shows the estimated time courses as spatial filter outputs at the three source locations, $\widehat{s}(\boldsymbol{r}_1, t)$, $\widehat{s}(\boldsymbol{r}_2, t)$, and $\widehat{s}(\boldsymbol{r}_3, t)$. Fig. 6.2(b)–(e) show the results of source reconstruction on the plane $x = 0$ cm. The instantaneous reconstruction results at the 220th, 268th, and 300th time points are shown, respectively, in Fig. 6.2(b), (c), and (d). The square root of source power reconstruction $\sqrt{\langle \widehat{s}(\boldsymbol{r}, t)^2 \rangle}$ is shown in Fig. 6.2(e). The reconstructed source time courses contain a considerable amount of noise, and accordingly, results of the instantaneous reconstruction show some influence of this low output SNR. This low output SNR is in this case due to the fact that the noise subspace obtained from the sample covariance matrix is not exactly equal to the true noise subspace and the error term in the denominator in Eq. (6.81) remains non-zero, resulting in the degradation of the output SNR.

Next, we show the results from the diagonal-loading array-gain minimum-variance spatial filter in Fig. 6.3. Here, the loading factor ϵ is set to $0.01\lambda_{max}\{\boldsymbol{R}\}$. Compared to the non-diagonal loading minimum-variance results in Fig. 6.2, the SNR of the spatial filter output is greatly improved, but a considerable amount of blur is introduced. The results here confirm the trade-off relationship between the spatial resolution and the output SNR.

We show the results of the asymmetric diagonal-loading spatial filter in Fig. 6.4. The loading factor for the denominator, ϵ_d, is set to 0, and the loading factor for the numerator, ϵ_n, is set to $0.1\lambda_{max}\{\boldsymbol{R}\}$. Compared to the results in Figs. 6.2 and 6.3, it can be seen that the asymmetric diagonal loading can avoid the blur, while the output SNR is recovered. The spatial resolution is almost the same as that of the eigenspace-projection spatial filter shown in Fig. 6.5. However, the source time courses contain considerable amounts of leakage from other sources. The results here demonstrate the trade-off relationship between the leakage and the output SNR improvements with the asymmetric diagonal loading, even while the spatial resolution is maintained.

We then show results for the eigenspace-projection minimum-variance spatial filter. The signal subspace dimension Q was set to three. The reconstructed results

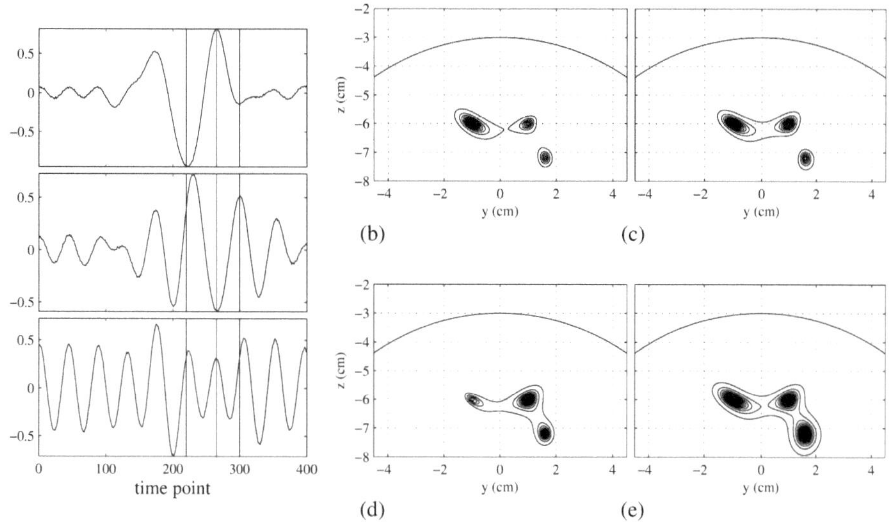

Figure 6.4: (a) The outputs of the asymmetric diagonal-loading minimum-variance spatial filter at the three source locations, $\widehat{s}(\boldsymbol{r}_1, t)$ (top), $\widehat{s}(\boldsymbol{r}_2, t)$ (middle), and $\widehat{s}(\boldsymbol{r}_3, t)$ (bottom). The three vertical lines show the time corresponding to the 220th, 268th, and 300th time points. The numerator loading factor ϵ_n was set to $0.1\lambda_{max}\{\boldsymbol{R}\}$ and the denominator loading factor ϵ_d was set to 0. (b)–(e) The results of the source reconstruction on the plane $x = 0$ cm. Results of instantaneous reconstruction, $|\widehat{s}(\boldsymbol{r}, t)|$, at (b) the 220th, (c) the 268th, and (d) the 300th time points. (e) The time-averaged reconstruction $\sqrt{\langle \widehat{s}(\boldsymbol{r}, t)^2 \rangle}$ where $\langle \widehat{s}(\boldsymbol{r}, t)^2 \rangle = \widehat{P}_s(\boldsymbol{r})$.

are shown in Fig. 6.5. This figure shows that the eigenspace-projection considerably improves the output SNR with almost no sacrifice of spatial resolution and leakage. Finally, we show results for the vector eigenspace-projection minimum-variance spatial filter. The signal subspace dimension Q was again set to three. The reconstructed results are shown in Fig. 6.6. This figure shows that the results almost identical to those from the scalar eigenspace-projection minimum-variance spatial filter are obtained, showing that the eigenspace projection operation works for vector formulation, even though the eigenspace projection does not preserve the null constraint, as discussed in Section 6.8.2.

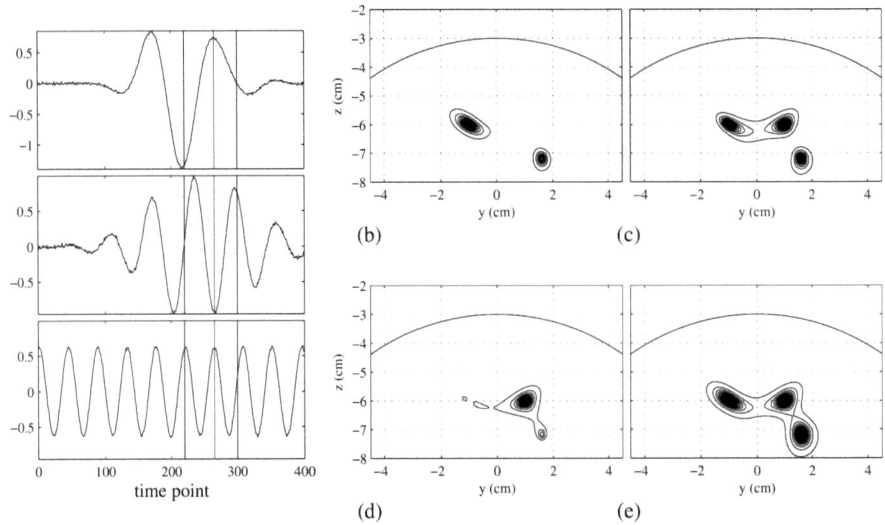

Figure 6.5: (a) The outputs of the eigenspace-projected minimum-variance spatial filter at the three source locations, $\widehat{s}(\boldsymbol{r}_1,t)$ (top), $\widehat{s}(\boldsymbol{r}_2,t)$ (middle), and $\widehat{s}(\boldsymbol{r}_3,t)$ (bottom). The three vertical lines show the time corresponding to the 220th, 268th, and 300th time points. The signal-subspace dimension was set to three. (b)–(e) The results of the source reconstruction on $x = 0$ cm. Results of instantaneous reconstruction, $|\widehat{s}(\boldsymbol{r},t)|$, at (b) the 220th, (c) the 268th, and (d) the 300th time points. (e) The power reconstruction $\sqrt{\langle \widehat{s}(\boldsymbol{r},t)^2 \rangle}$ where $\langle \widehat{s}(\boldsymbol{r},t)^2 \rangle = \boldsymbol{w}_e^T(\boldsymbol{r})\boldsymbol{R}\boldsymbol{w}_e(\boldsymbol{r})$. The weight vector $\boldsymbol{w}_e(\boldsymbol{r})$ is defined in Eq. (6.84).

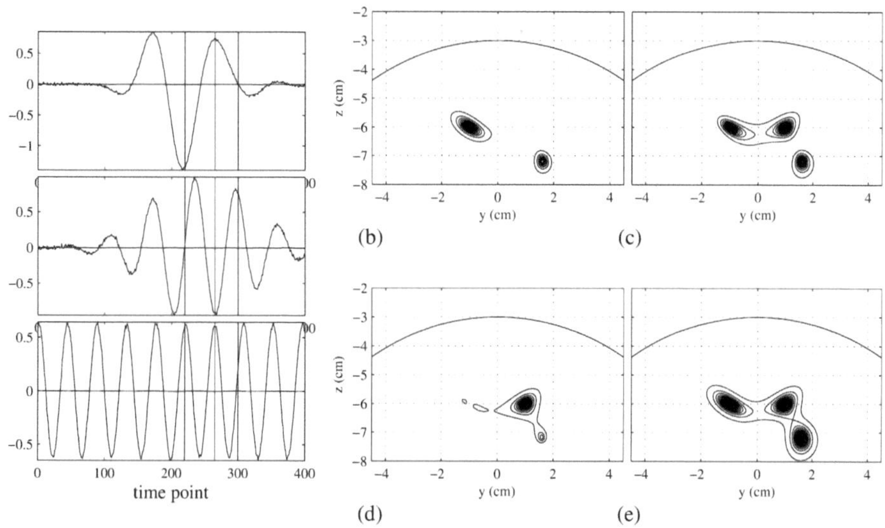

Figure 6.6: (a) The outputs of the eigenspace-projected vector minimum-variance spatial filter at the first source (top), the second source (middle), and the third source (bottom) locations. The two time courses correspond to $\widehat{s}_{\parallel}(\boldsymbol{r},t)$ and $\widehat{s}_{\perp}(\boldsymbol{r},t)$. The signal-subspace dimension was set to three. The three vertical lines show the time corresponding to the 220th, 268th, and 300th time points. (b)–(e) The results of the source reconstruction on $x = 0$ cm. Results of instantaneous reconstruction, $|\widehat{s}(\boldsymbol{r},t)|$, at (b) the 220th, (c) the 268th, and (d) the 300th time points. (e) The power reconstruction $\sqrt{\widehat{P}_V^{(\mathrm{I})}(\boldsymbol{r})}$.

Chapter 7

Effects of low-rank interference

Bioelectromagnetic measurements are often contaminated by various types of overlapping external interference even when the measurements are performed in a shielded room. This chapter discusses the influence of external interference of non-biological origin on the adaptive spatial filter output. Typical examples of such interference include magnetic noise from power lines or electrical appliances such as elevators, automobiles or the subway. We assume that the interference has the following two properties: First, it is additive and uncorrelated with brain activity. Second, the interference can be modeled as a low-rank signal. Under these assumptions, our analysis shows that the adaptive spatial filters are insensitive to interference when its spatial singular vectors are very different from the lead-field vector of any brain sources. Since this condition is approximately met for many types of non-biological interferences, we conclude that the adaptive spatial filters are robust to the superimposition of such interference.

7.1 Influence of low-rank interference

7.1.1 Low-rank interference

We denote the interference carried by the mth sensor channel at time t as $d_m(t)$. The column vector $\boldsymbol{d}(t) = [d_1(t), \ldots, d_M(t)]^T$ represents the interference contained in the measurements of the whole sensor array. Assuming that $\boldsymbol{d}(t)$ is additive, the measured data $\boldsymbol{b}(t)$ is expressed as

$$\boldsymbol{b}(t) = \sum_{q=1}^{Q} \boldsymbol{l}(\boldsymbol{r}_q) s(\boldsymbol{r}_q, t) + \boldsymbol{d}(t) + \boldsymbol{n}(t) = \boldsymbol{b}_{s+n}(t) + \boldsymbol{d}(t), \tag{7.1}$$

where $\boldsymbol{n}(t)$ is the sensor noise, and $\boldsymbol{b}_{s+n}(t)$ is the signal plus sensor noise defined as

$$\boldsymbol{b}_{s+n}(t) = \sum_{q=1}^{Q} \boldsymbol{l}(\boldsymbol{r}_q) s(\boldsymbol{r}_q, t) + \boldsymbol{n}(t). \tag{7.2}$$

We define the spatio-temporal matrix of the measurement $\boldsymbol{b}(t)$ as $\boldsymbol{B}$:

$$\boldsymbol{B} = [\boldsymbol{b}(t_1), \ldots, \boldsymbol{b}(t_K)].$$

The spatio-temporal matrix of the signal plus sensor noise, $\boldsymbol{b}_{s+n}(t)$, is defined as $\boldsymbol{B}_{s+n}$, where

$$\boldsymbol{B}_{s+n} = [\boldsymbol{b}_{s+n}(t_1), \ldots, \boldsymbol{b}_{s+n}(t_K)],$$

and the spatio-temporal matrix of the interference, $\boldsymbol{B}_d$, is defined as

$$\boldsymbol{B}_d = [\boldsymbol{d}(t_1), \ldots, \boldsymbol{d}(t_K)].$$

Thus, we have the relationship,

$$\boldsymbol{B} = \boldsymbol{B}_{s+n} + \boldsymbol{B}_d. \tag{7.3}$$

We next define the signal plus sensor noise covariance matrix obtained from $\boldsymbol{b}_{s+n}(t)$, as $\boldsymbol{R}_{s+n}$, i.e.,

$$\boldsymbol{R}_{s+n} = \langle \boldsymbol{b}_{s+n}(t) \boldsymbol{b}_{s+n}^T(t) \rangle.$$

We also define the covariance matrix of the interference as $\boldsymbol{R}_d$, i.e.,

$$\boldsymbol{R}_d = \langle \boldsymbol{d}(t) \boldsymbol{d}^T(t) \rangle.$$

Assuming that the interference $\boldsymbol{d}(t)$ is uncorrelated with $\boldsymbol{b}_{s+n}(t)$, we have the relationship,

$$\boldsymbol{R} = \boldsymbol{R}_{s+n} + \boldsymbol{R}_d. \tag{7.4}$$

The output of the minimum-variance spatial filter is given by

$$[\widehat{s}(\boldsymbol{r}, t_1), \ldots, \widehat{s}(\boldsymbol{r}, t_K)] = \xi \boldsymbol{l}^T(\boldsymbol{r}) \boldsymbol{R}^{-1} \boldsymbol{B}, \tag{7.5}$$

where

$$\begin{aligned} \xi &= 1/[\boldsymbol{l}^T(\boldsymbol{r}) \boldsymbol{R}^{-1} \boldsymbol{l}(\boldsymbol{r})], \\ \xi &= \|\boldsymbol{l}(\boldsymbol{r})\| / [\boldsymbol{l}^T(\boldsymbol{r}) \boldsymbol{R}^{-1} \boldsymbol{l}(\boldsymbol{r})], \\ \text{and} \quad \xi &= 1/\sqrt{\boldsymbol{l}^T(\boldsymbol{r}) \boldsymbol{R}^{-2} \boldsymbol{l}(\boldsymbol{r})}, \end{aligned}$$

corresponding to the unit-gain, array-gain, and unit-noise-gain constraints, respectively. In the following discussion, we assume the unit-gain minimum-variance spatial filter. The discussion here can be applied to the array-gain minimum-variance spatial filter with a very minor modification. The case of the unit-noise-gain minimum-variance filter is discussed in Section 7.2.

The key assumption for $\boldsymbol{R}_d$ in our analysis is that $\boldsymbol{R}_d$ has only a few (relatively) large eigenvalues[50]. That is, $\boldsymbol{R}_d$ can be expressed as

$$\boldsymbol{R}_d = \sum_{j=1}^{Q_D} \chi_j \boldsymbol{u}_j \boldsymbol{u}_j^T, \tag{7.6}$$

where χ_j and $\boldsymbol{u}_j$ are the jth eigenvalue of $\boldsymbol{R}_d$ and its corresponding eigenvector, respectively. Here, Q_D is the number of non-zero eigenvalues, and we assume that $Q_D \ll M$. Using the singular value decomposition, the spatio-temporal matrix $\boldsymbol{B}_d$ is expressed as

$$\boldsymbol{B}_d = \sum_{j=1}^{Q_D} \sqrt{\chi_j} \boldsymbol{u}_j \boldsymbol{v}_j^T, \tag{7.7}$$

where $\boldsymbol{v}_j$ is the jth temporal singular vector of $\boldsymbol{B}_d$.

7.1.2 Analysis when $\boldsymbol{R}_d$ is a rank-one matrix

We first analyze the simplest case where $\boldsymbol{R}_d$ is a rank-one matrix. In such a case, omitting the subscript for eigenvalue numbering, $\boldsymbol{R}_d$ is expressed as $\boldsymbol{R}_d = \chi \boldsymbol{u}\boldsymbol{u}^T$. Then, we derive

$$\boldsymbol{R}^{-1} = (\boldsymbol{R}_{s+n} + \chi \boldsymbol{u}\boldsymbol{u}^T)^{-1} = \boldsymbol{R}_{s+n}^{-1} - \boldsymbol{R}_{s+n}^{-1} \frac{\boldsymbol{u}\boldsymbol{u}^T}{1/\chi + \boldsymbol{u}^T \boldsymbol{R}_{s+n}^{-1} \boldsymbol{u}} \boldsymbol{R}_{s+n}^{-1}. \tag{7.8}$$

We can show that the relationship $\boldsymbol{u}^T \boldsymbol{R}_{s+n}^{-1} \boldsymbol{u} \approx 1/\sigma_0^2$ holds[1]. The power of the interference is equal to χ, and we assume in this chapter that the power of the interference is much larger than the power of the sensor noise. Thus, we have the relationship, $1/\chi \ll \boldsymbol{u}^T \boldsymbol{R}_{s+n}^{-1} \boldsymbol{u}$. Then, Eq. (7.8) changes to

$$\boldsymbol{R}^{-1} \approx \boldsymbol{R}_{s+n}^{-1} - \boldsymbol{R}_{s+n}^{-1} \frac{\boldsymbol{u}\boldsymbol{u}^T}{\boldsymbol{u}^T \boldsymbol{R}_{s+n}^{-1} \boldsymbol{u}} \boldsymbol{R}_{s+n}^{-1}. \tag{7.9}$$

[1] It is easy to show this relationship for the single source scenario. According to Eq. (13.43), $\boldsymbol{u}^T \boldsymbol{R}_{s+n}^{-1} \boldsymbol{u}$ is expressed as

$$\boldsymbol{u}^T \boldsymbol{R}_{s+n}^{-1} \boldsymbol{u} = \frac{1}{\sigma_0^2} \left(1 - \frac{\alpha}{1+\alpha} \cos^2(\boldsymbol{u}, \boldsymbol{f}) \right) \approx \frac{1}{\sigma_0^2},$$

where we assume that $\cos^2(\boldsymbol{u}, \boldsymbol{f}) \approx 0$. When two sources exist, by substituting $\boldsymbol{u}$ for $\boldsymbol{l}$ in Eqs. (13.57) and assuming $\cos^2(\boldsymbol{u}, \boldsymbol{f}) \approx 0$ and $\cos^2(\boldsymbol{u}, \boldsymbol{g}) \approx 0$, we can derive

$$\boldsymbol{u}^T \boldsymbol{R}_{s+n}^{-1} \boldsymbol{u} \approx \frac{1}{\Gamma} \left[1 + \alpha_1 [1 - \frac{\alpha_2}{1+\alpha_2} \cos^2(\boldsymbol{f}, \boldsymbol{g})] \right] = \frac{1}{\sigma_0^2}.$$

We can show that this relationship holds for cases where the number of sources are more than two.

Substituting Eq. (7.9) and $\boldsymbol{B}_d = \sqrt{\chi}\boldsymbol{u}\boldsymbol{v}^T$ into Eq. (7.5), we obtain

$$
\begin{aligned}
[\widehat{s}(\boldsymbol{r},t_1),\ldots,\widehat{s}(\boldsymbol{r},t_K)] &\approx \xi \boldsymbol{l}^T(\boldsymbol{r})\boldsymbol{R}_{s+n}^{-1}(\boldsymbol{B}_{s+n} + \sqrt{\chi}\boldsymbol{u}\boldsymbol{v}^T) \\
&\quad - \xi \frac{[\boldsymbol{l}^T(\boldsymbol{r})\boldsymbol{R}_{s+n}^{-1}\boldsymbol{u}][\boldsymbol{u}^T\boldsymbol{R}_{s+n}^{-1}(\boldsymbol{B}_{s+n} + \sqrt{\chi}\boldsymbol{u}\boldsymbol{v}^T)]}{\boldsymbol{u}^T\boldsymbol{R}_{s+n}^{-1}\boldsymbol{u}} \\
&= \xi \boldsymbol{l}^T(\boldsymbol{r})\boldsymbol{R}_{s+n}^{-1}\boldsymbol{B}_{s+n} - \xi \left[\frac{\boldsymbol{l}^T(\boldsymbol{r})\boldsymbol{R}_{s+n}^{-1}\boldsymbol{u}}{\boldsymbol{u}^T\boldsymbol{R}_{s+n}^{-1}\boldsymbol{u}}\right] \boldsymbol{u}^T\boldsymbol{R}_{s+n}^{-1}\boldsymbol{B}_{s+n} \\
&= \xi \boldsymbol{l}_d^T(\boldsymbol{r})\boldsymbol{R}_{s+n}^{-1}\boldsymbol{B}_{s+n},
\end{aligned}
\tag{7.10}
$$

where

$$
\boldsymbol{l}_d(\boldsymbol{r}) = \boldsymbol{l}(\boldsymbol{r}) - \left[\frac{\boldsymbol{l}^T(\boldsymbol{r})\boldsymbol{R}_{s+n}^{-1}\boldsymbol{u}}{\boldsymbol{u}^T\boldsymbol{R}_{s+n}^{-1}\boldsymbol{u}}\right]\boldsymbol{u}. \tag{7.11}
$$

Using Eq. (7.9), the value of ξ is found to be

$$
1/\xi = \boldsymbol{l}^T(\boldsymbol{r})\boldsymbol{R}^{-1}\boldsymbol{l}(\boldsymbol{r}) \approx \boldsymbol{l}^T(\boldsymbol{r})\boldsymbol{R}_{s+n}^{-1}\,\boldsymbol{l}(\boldsymbol{r}) - \frac{[\boldsymbol{l}^T(\boldsymbol{r})\boldsymbol{R}_{s+n}^{-1}\boldsymbol{u}]^2}{\boldsymbol{u}^T\boldsymbol{R}_{s+n}^{-1}\boldsymbol{u}} = \boldsymbol{l}_d^T(\boldsymbol{r})\boldsymbol{R}_{s+n}^{-1}\boldsymbol{l}_d(\boldsymbol{r}). \tag{7.12}
$$

Therefore, substituting Eq. (7.12) into Eq. (7.10), we finally derive

$$
\widehat{s}(\boldsymbol{r},t) = \frac{\boldsymbol{l}^T(\boldsymbol{r})\boldsymbol{R}^{-1}\boldsymbol{b}(t)}{\boldsymbol{l}^T(\boldsymbol{r})\boldsymbol{R}^{-1}\boldsymbol{l}(\boldsymbol{r})} \approx \frac{\boldsymbol{l}_d^T(\boldsymbol{r})\boldsymbol{R}_{s+n}^{-1}\boldsymbol{b}_{s+n}(t)}{\boldsymbol{l}_d^T(\boldsymbol{r})\boldsymbol{R}_{s+n}^{-1}\boldsymbol{l}_d(\boldsymbol{r})}. \tag{7.13}
$$

The above equations indicate that the temporal behavior of the interference represented by the temporal singular vector $\boldsymbol{v}$ does not affect the output time courses of the minimum-variance filter. These equations also indicate that the interference affects the outputs through its spatial singular vector $\boldsymbol{u}$ by modifying the lead-field vector from $\boldsymbol{l}(\boldsymbol{r})$ to $\boldsymbol{l}_d(\boldsymbol{r})$ according to Eq. (7.11).

The minimum-variance filter output is derived from Eq. (7.13). That is, using Eq. (7.2) with ignoring the sensor noise term, and denoting $\boldsymbol{l}(\boldsymbol{r})$ as $\boldsymbol{l}$ and $\boldsymbol{l}_d(\boldsymbol{r})$ as $\boldsymbol{l}_d$, the time course output at $\boldsymbol{r}$ is expressed as

$$
\widehat{s}(\boldsymbol{r},t) = \frac{\boldsymbol{l}_d^T\boldsymbol{R}_{s+n}^{-1}\boldsymbol{b}_{s+n}(t)}{\boldsymbol{l}_d^T\boldsymbol{R}_{s+n}^{-1}\boldsymbol{l}_d} = \sum_{q=1}^{Q} s(\boldsymbol{r}_q,t)\frac{\boldsymbol{l}_d^T\boldsymbol{R}_{s+n}^{-1}\boldsymbol{l}(\boldsymbol{r}_q)}{\boldsymbol{l}_d^T\boldsymbol{R}_{s+n}^{-1}\boldsymbol{l}_d}, \tag{7.14}
$$

where

$$
\begin{aligned}
\boldsymbol{l}_d^T\boldsymbol{R}_{s+n}^{-1}\boldsymbol{l}(\boldsymbol{r}_q) &= (\boldsymbol{l} - \frac{\boldsymbol{l}^T\boldsymbol{R}_{s+n}^{-1}\boldsymbol{u}}{\boldsymbol{u}^T\boldsymbol{R}_{s+n}^{-1}\boldsymbol{u}}\boldsymbol{u})^T\boldsymbol{R}_{s+n}^{-1}\boldsymbol{l}(\boldsymbol{r}_q) \\
&= \boldsymbol{l}^T\boldsymbol{R}_{s+n}^{-1}\boldsymbol{l}(\boldsymbol{r}_q)\left[1 - \frac{\cos(\boldsymbol{l},\boldsymbol{u}|\boldsymbol{R}_{s+n}^{-1})\cos(\boldsymbol{u},\boldsymbol{l}(\boldsymbol{r}_q)|\boldsymbol{R}_{s+n}^{-1})}{\cos(\boldsymbol{l},\boldsymbol{l}(\boldsymbol{r}_q)|\boldsymbol{R}_{s+n}^{-1})}\right]
\end{aligned}
\tag{7.15}
$$

and

$$\boldsymbol{l}_d^T \boldsymbol{R}_{s+n}^{-1} \boldsymbol{l}_d = (\boldsymbol{l} - \frac{\boldsymbol{l}^T \boldsymbol{R}_{s+n}^{-1} \boldsymbol{u}}{\boldsymbol{u}^T \boldsymbol{R}_{s+n}^{-1} \boldsymbol{u}} \boldsymbol{u})^T \boldsymbol{R}_{s+n}^{-1} (\boldsymbol{l} - \frac{\boldsymbol{l}^T \boldsymbol{R}_{s+n}^{-1} \boldsymbol{u}}{\boldsymbol{u}^T \boldsymbol{R}_{s+n}^{-1} \boldsymbol{u}} \boldsymbol{u})$$
$$= \boldsymbol{l}^T \boldsymbol{R}_{s+n}^{-1} \boldsymbol{l} [1 - \cos^2(\boldsymbol{l}, \boldsymbol{u} | \boldsymbol{R}_{s+n}^{-1})]. \quad (7.16)$$

Therefore, Eq. (7.14) can be rewritten as

$$\widehat{s}(\boldsymbol{r}, t) = \sum_{q=1}^{Q} s(\boldsymbol{r}_q, t) \frac{\boldsymbol{l}^T \boldsymbol{R}_{s+n}^{-1} \boldsymbol{l}(\boldsymbol{r}_q)}{\boldsymbol{l}^T \boldsymbol{R}_{s+n}^{-1} \boldsymbol{l}} \frac{\left[1 - \frac{\cos(\boldsymbol{l}, \boldsymbol{u} | \boldsymbol{R}_{s+n}^{-1}) \cos(\boldsymbol{u}, \boldsymbol{l}(\boldsymbol{r}_q) | \boldsymbol{R}_{s+n}^{-1})}{\cos(\boldsymbol{l}, \boldsymbol{l}(\boldsymbol{r}_q) | \boldsymbol{R}_{s+n}^{-1})}\right]}{[1 - \cos^2(\boldsymbol{l}, \boldsymbol{u} | \boldsymbol{R}_{s+n}^{-1})]}. \quad (7.17)$$

We can also show that the output power is given by

$$\langle \widehat{s}(\boldsymbol{r}, t)^2 \rangle = \frac{1}{\boldsymbol{l}_d^T \boldsymbol{R}_{s+n}^{-1} \boldsymbol{l}_d} = \frac{1}{\boldsymbol{l}^T \boldsymbol{R}_{s+n}^{-1} \boldsymbol{l} [1 - \cos^2(\boldsymbol{l}, \boldsymbol{u} | \boldsymbol{R}_{s+n}^{-1})]}. \quad (7.18)$$

Because the generalized cosine quantifies the similarity or the difference of the two vectors, when any lead-field vector in the source space is very different from the spatial eigenvector of the interference $\boldsymbol{u}$, the relationships $\cos(\boldsymbol{l}, \boldsymbol{u} | \boldsymbol{R}_{s+n}^{-1}) \ll 1$ and $\cos(\boldsymbol{u}, \boldsymbol{l}(\boldsymbol{r}_q) | \boldsymbol{R}_{s+n}^{-1}) \ll 1$ hold. In this case, ignoring the sensor noise, Eq. (7.17) is rewritten as

$$\widehat{s}(\boldsymbol{r}, t) \approx \sum_{q=1}^{Q} s(\boldsymbol{r}_q, t) \frac{\boldsymbol{l}^T \boldsymbol{R}_{s+n}^{-1} \boldsymbol{l}(\boldsymbol{r}_q)}{\boldsymbol{l}^T \boldsymbol{R}_{s+n}^{-1} \boldsymbol{l}} = \frac{\boldsymbol{l}^T \boldsymbol{R}_{s+n}^{-1} \boldsymbol{b}_{s+n}(t)}{\boldsymbol{l}^T \boldsymbol{R}_{s+n}^{-1} \boldsymbol{l}}, \quad (7.19)$$

and Eq. (7.18) can be rewritten as

$$\langle \widehat{s}(\boldsymbol{r}, t)^2 \rangle \approx \frac{1}{\boldsymbol{l}^T(\boldsymbol{r}) \boldsymbol{R}_{s+n}^{-1} \boldsymbol{l}(\boldsymbol{r})}. \quad (7.20)$$

We can conclude that the influence of the interference is negligible in the output of the minimum-variance spatial filter.

7.1.3 Analysis when $\boldsymbol{R}_d$ is a rank-two matrix

The analysis can be extended to the case where $\boldsymbol{R}_d$ is a rank-two matrix. In this case, $\boldsymbol{R}_d$ is expressed as $\boldsymbol{R}_d = \chi_1 \boldsymbol{u}_1 \boldsymbol{u}_1^T + \chi_2 \boldsymbol{u}_2 \boldsymbol{u}_2^T$. We assume the relationships hold:

$$(\boldsymbol{u}_1^T \boldsymbol{R}_{s+n}^{-1} \boldsymbol{u}_2) / (\boldsymbol{u}_1^T \boldsymbol{R}_{s+n}^{-1} \boldsymbol{u}_1) \approx 0 \quad \text{and} \quad (\boldsymbol{u}_1^T \boldsymbol{R}_{s+n}^{-1} \boldsymbol{u}_2) / (\boldsymbol{u}_2^T \boldsymbol{R}_{s+n}^{-1} \boldsymbol{u}_2) \approx 0. \quad (7.21)$$

Then, based on Eq. (7.9), we can derive

$$\boldsymbol{R}^{-1} = (\boldsymbol{R}_{s+n} + \chi_1 \boldsymbol{u}_1 \boldsymbol{u}_1^T + \chi_2 \boldsymbol{u}_2 \boldsymbol{u}_2^T)^{-1}$$
$$\approx \boldsymbol{R}_{s+n}^{-1} - \boldsymbol{R}_{s+n}^{-1} \frac{\boldsymbol{u}_1 \boldsymbol{u}_1^T}{\boldsymbol{u}_1^T \boldsymbol{R}_{s+n}^{-1} \boldsymbol{u}_1} \boldsymbol{R}_{s+n}^{-1} - \boldsymbol{R}_{s+n}^{-1} \frac{\boldsymbol{u}_2 \boldsymbol{u}_2^T}{\boldsymbol{u}_2^T \boldsymbol{R}_{s+n}^{-1} \boldsymbol{u}_2} \boldsymbol{R}_{s+n}^{-1}. \quad (7.22)$$

Substituting the equation above and $\boldsymbol{B}_d = \sqrt{\chi_1}\boldsymbol{u}_1\boldsymbol{v}_1^T + \sqrt{\chi_2}\boldsymbol{u}_2\boldsymbol{v}_2^T$ into Eq. (7.5), after lengthy calculations, we finally get

$$[\widehat{s}(\boldsymbol{r},t_1),\ldots,\widehat{s}(\boldsymbol{r},t_K)] = \xi\boldsymbol{l}_d^T(\boldsymbol{r})\boldsymbol{R}_{s+n}^{-1}\boldsymbol{B}_{s+n} \tag{7.23}$$

where

$$\boldsymbol{l}_d = \boldsymbol{l} - \left[\frac{\boldsymbol{l}^T\boldsymbol{R}_{s+n}^{-1}\boldsymbol{u}_1}{\boldsymbol{u}_1^T\boldsymbol{R}_{s+n}^{-1}\boldsymbol{u}_1}\right]\boldsymbol{u}_1 - \left[\frac{\boldsymbol{l}^T\boldsymbol{R}_{s+n}^{-1}\boldsymbol{u}_2}{\boldsymbol{u}_2^T\boldsymbol{R}_{s+n}^{-1}\boldsymbol{u}_2}\right]\boldsymbol{u}_2. \tag{7.24}$$

We can then derive

$$\widehat{s}(\boldsymbol{r},t) = \sum_{q=1}^{Q} s(\boldsymbol{r}_q,t)\frac{\boldsymbol{l}^T\boldsymbol{R}_{s+n}^{-1}\boldsymbol{l}(\boldsymbol{r}_q)}{\boldsymbol{l}^T\boldsymbol{R}_{s+n}^{-1}\boldsymbol{l}}\cdot$$
$$\frac{\left[1 - \frac{\cos(\boldsymbol{l},\boldsymbol{u}_1|\boldsymbol{R}_{s+n}^{-1})\cos(\boldsymbol{l}(\boldsymbol{r}_q),\boldsymbol{u}_1|\boldsymbol{R}_{s+n}^{-1})}{\cos(\boldsymbol{l},\boldsymbol{l}(\boldsymbol{r}_q)|\boldsymbol{R}_{s+n}^{-1})} - \frac{\cos(\boldsymbol{l},\boldsymbol{u}_2|\boldsymbol{R}_{s+n}^{-1})\cos(\boldsymbol{l}(\boldsymbol{r}_q),\boldsymbol{u}_2|\boldsymbol{R}_{s+n}^{-1})}{\cos(\boldsymbol{l},\boldsymbol{l}(\boldsymbol{r}_q)|\boldsymbol{R}_{s+n}^{-1})}\right]}{[1-\cos^2(\boldsymbol{l},\boldsymbol{u}_1|\boldsymbol{R}_{s+n}^{-1}) - \cos^2(\boldsymbol{l},\boldsymbol{u}_2|\boldsymbol{R}_{s+n}^{-1})]}. \tag{7.25}$$

The output power in this case is expressed as

$$\langle\widehat{s}(\boldsymbol{r},t)^2\rangle = \frac{1}{\boldsymbol{l}^T\boldsymbol{R}_{s+n}^{-1}\boldsymbol{l}}\frac{1}{[1-\cos^2(\boldsymbol{l},\boldsymbol{u}_1|\boldsymbol{R}_{s+n}^{-1}) - \cos^2(\boldsymbol{l},\boldsymbol{u}_2|\boldsymbol{R}_{s+n}^{-1})]}. \tag{7.26}$$

That is, these equations above show that the output of the minimum-variance spatial filter is not affected by the low-rank interference also in this case, assuming that any lead-field vector in the source space is very different from the spatial eigenvectors of the interference. The above equations also show that the second eigenvector $\boldsymbol{u}_2$ affects the filter output in an additive manner. The analysis can be further extended to the general case where the rank of $\boldsymbol{R}_d$ is greater than 2, and it can be shown that each eigenvector affects the spatial filter output in exactly the same additive manner.

7.2 Influence on output of the unit-noise-gain minimum-variance filter

Let us consider the influence of the low-rank external interference on the output of the unit-noise-gain minimum-variance filter. We first note that equation (7.10) holds, regardless of the value of ξ. The value of ξ for the unit-noise-gain minimum-variance filter is found to be

$$(1/\xi)^2 = \boldsymbol{l}^T(\boldsymbol{r})\boldsymbol{R}^{-2}\boldsymbol{l}(\boldsymbol{r}) = \boldsymbol{l}^T(\boldsymbol{r})\boldsymbol{R}_{s+n}^{-2}\boldsymbol{l}(\boldsymbol{r})(1-\nu), \tag{7.27}$$

where, denoting $\boldsymbol{l}(\boldsymbol{r})$ as $\boldsymbol{l}$, ν is expressed as

$$\nu = \frac{\boldsymbol{l}^T\boldsymbol{R}_{s+n}^{-1}\boldsymbol{u}}{\boldsymbol{u}^T\boldsymbol{R}_{s+n}^{-1}\boldsymbol{u}}\left[2\frac{\boldsymbol{l}^T\boldsymbol{R}_{s+n}^{-2}\boldsymbol{u}}{\boldsymbol{l}^T\boldsymbol{R}_{s+n}^{-2}\boldsymbol{l}} - (\frac{\boldsymbol{l}^T\boldsymbol{R}_{s+n}^{-1}\boldsymbol{u}}{\boldsymbol{u}^T\boldsymbol{R}_{s+n}^{-1}\boldsymbol{u}})(\frac{\boldsymbol{u}^T\boldsymbol{R}_{s+n}^{-2}\boldsymbol{u}}{\boldsymbol{l}^T\boldsymbol{R}_{s+n}^{-2}\boldsymbol{l}})\right]. \tag{7.28}$$

When deriving these relationships, we use,

$$\boldsymbol{R}^{-2} \approx \boldsymbol{R}_{s+n}^{-2} - \boldsymbol{R}_{s+n}^{-2} \frac{\boldsymbol{u}\boldsymbol{u}^T}{\boldsymbol{u}^T \boldsymbol{R}_{s+n}^{-1} \boldsymbol{u}} \boldsymbol{R}_{s+n}^{-1} - \boldsymbol{R}_{s+n}^{-1} \frac{\boldsymbol{u}\boldsymbol{u}^T}{\boldsymbol{u}^T \boldsymbol{R}_{s+n}^{-1} \boldsymbol{u}} \boldsymbol{R}_{s+n}^{-2} + \frac{\boldsymbol{u}^T \boldsymbol{R}_{s+n}^{-2} \boldsymbol{u}}{\boldsymbol{u}^T \boldsymbol{R}_{s+n}^{-1} \boldsymbol{u}} \boldsymbol{R}_{s+n}^{-1} \frac{\boldsymbol{u}\boldsymbol{u}^T}{\boldsymbol{u}^T \boldsymbol{R}_{s+n}^{-1} \boldsymbol{u}} \boldsymbol{R}_{s+n}^{-1}. \quad (7.29)$$

Then, we can derive,

$$\widehat{s}(\boldsymbol{r},t) = \sum_{q=1}^{Q} s(\boldsymbol{r}_q,t) \frac{\boldsymbol{l}_d^T \boldsymbol{R}_{s+n}^{-1} \boldsymbol{l}(\boldsymbol{r}_q)}{\sqrt{\boldsymbol{l}^T \boldsymbol{R}^{-2} \boldsymbol{l}}} = \frac{\boldsymbol{l}^T \boldsymbol{R}_{s+n}^{-1}\, \boldsymbol{l}(\boldsymbol{r}_q)}{\sqrt{\boldsymbol{l}^T \boldsymbol{R}_{s+n}^{-2}\, \boldsymbol{l}}} \frac{\left[1 - \frac{\cos(\boldsymbol{l},\boldsymbol{u}|\boldsymbol{R}_{s+n}^{-1}) \cos(\boldsymbol{u},\boldsymbol{l}(\boldsymbol{r}_q)|\boldsymbol{R}_{s+n}^{-1})}{\cos(\boldsymbol{l},\boldsymbol{l}(\boldsymbol{r}_q)|\boldsymbol{R}_{s+n}^{-1})}\right]}{\sqrt{(1-\nu)}}, \quad (7.30)$$

and

$$\langle \widehat{s}(\boldsymbol{r},t)^2 \rangle = \frac{\boldsymbol{l}_d^T \boldsymbol{R}_{s+n}^{-1} \boldsymbol{l}_d}{\boldsymbol{l}^T \boldsymbol{R}^{-2} \boldsymbol{l}} = \frac{\boldsymbol{l}^T \boldsymbol{R}_{s+n}^{-1}\, \boldsymbol{l}\, [1 - \cos^2(\boldsymbol{l},\boldsymbol{u}|\boldsymbol{R}_{s+n}^{-1})]}{\boldsymbol{l}^T \boldsymbol{R}_{s+n}^{-2}\, \boldsymbol{l}\, (1-\nu)}, \quad (7.31)$$

Therefore, when the assumptions that $\cos(\boldsymbol{l},\boldsymbol{u}|\boldsymbol{R}_{s+n}^{-1}) \ll 1$, $\cos(\boldsymbol{u},\boldsymbol{l}(\boldsymbol{r}_q)|\boldsymbol{R}_{s+n}^{-1}) \ll 1$, and $|\nu| \ll 1$ hold, the influence of the interference on the output of the unit-noise-gain minimum-variance filter is also negligible. The value ν is small when $\boldsymbol{l}$ and $\boldsymbol{u}$ are very different because only the numerators in Eq. (7.28) contain the cross products of $\boldsymbol{l}$ and $\boldsymbol{u}$. In the numerical experiments in Section 7.4, ν is always less than 10^{-6}.

7.3 Effects on the output of the eigenspace-projected spatial filter

Using Eq. (7.5), the output of the eigenspace-projection spatial filter is given by

$$[\widehat{s}(\boldsymbol{r},t_1),\ldots,\widehat{s}(\boldsymbol{r},t_K)] = [\boldsymbol{E}_S \boldsymbol{E}_S^T \boldsymbol{w}(\boldsymbol{r})]^T \boldsymbol{B} = \xi \boldsymbol{l}^T \boldsymbol{R}^{-1} \boldsymbol{E}_S \boldsymbol{E}_S^T (\boldsymbol{B}_{s+n} + \sum_{j=1}^{Q_D} \sqrt{\chi_j} \boldsymbol{u}_j \boldsymbol{v}_j^T). \quad (7.32)$$

Here, the columns of $\boldsymbol{E}_S$ span the signal subspace. As mentioned in Section 6.8, the signal subspace is determined by separating distinctly-large eigenvalues from the small eigenvalues in the eigenvalue spectrum of the measurement covariance matrix. When external interference is significantly greater than the sensor noise, the signal subspace determined in the above-mentioned manner generally contains the interference subspace spanned by $\boldsymbol{u}_j$ ($j = 1,\ldots,Q_D$), and the columns of $\boldsymbol{E}_S$ span the signal-plus-interference subspace. Since the source lead-field vectors $\boldsymbol{l}(\boldsymbol{r}_q)$ ($q = 1,\ldots,Q$) and the interference eigenvectors $\boldsymbol{u}_j$ ($j = 1,\ldots,Q_D$) both exist in the signal-plus-interference subspace, $\boldsymbol{B}_{s+n}$ and $\boldsymbol{u}_j$ are unaffected by the application of the signal-plus-interference-subspace projector $\boldsymbol{E}_S\boldsymbol{E}_S^T$; i.e.,

$$\boldsymbol{E}_S\boldsymbol{E}_S^T(\boldsymbol{B}_{s+n}+\sum_{j=1}^{Q_D}\sqrt{\chi_j}\boldsymbol{u}_j\boldsymbol{v}_j^T)=\boldsymbol{B}_{s+n}+\sum_{j=1}^{Q_D}\sqrt{\chi_j}\boldsymbol{u}_j\boldsymbol{v}_j^T. \tag{7.33}$$

Therefore, the argument in this chapter also holds for the eigenspace-projected spatial filter and the output of the eigenspace-projected spatial filter is not affected by the external interference, $\boldsymbol{d}(t)$.

7.4 Numerical examples

A series of numerical experiments are conducted to illustrate the results of the arguments in this chapter. The same source-sensor configuration and the coordinate system as illustrated in Fig. 4.2 are used, and the time courses shown in Fig. 4.3(a) are assigned to the three sources. The simulated MEG recordings are then calculated at 400 time points. We assume that the sampling interval is 2 ms. Gaussian noise is added to the generated magnetic field so that the total input SNR, defined as $\alpha_T = \alpha_1 + \alpha_2 + \alpha_3$, is equal to 8. The generated signal plus sensor noise, $\boldsymbol{b}_{s+n}(t)$, is shown in Fig. 7.1(a). The (square-root of the) power reconstruction obtained from $\boldsymbol{b}_{s+n}(t)$ are shown in Fig. 7.1(b), and the reconstructed source time courses, obtained as the spatial filter outputs at the source locations, are shown in Fig. 7.1(c). Here, the eigenspace-projected array-gain minimum-variance spatial filter is used for the reconstruction.

In our numerical experiments, four types of interference $\boldsymbol{d}(t)$ are simulated. The first is the periodic interference, where the interference for the mth sensor recording, $d_m(t)$, was calculated using $d_m(t) = \sin(2\pi f_d t + \phi)$ for the 60 sensors over the right hemisphere, and $d_m(t) = 0$ for the other sensors. The frequency of this periodic interference, f_d, was set to 14 Hz, which is fairly close to the 10 Hz-frequency of the third source. The interference $\boldsymbol{d}(t)$ was then added to $\boldsymbol{b}_{s+n}(t)$ to generate simulated recordings $\boldsymbol{b}(t)$: $\boldsymbol{b}(t) = \boldsymbol{b}_{s+n}(t) + \boldsymbol{d}(t)$. The resultant simulated recordings are shown in Fig. 7.2(a).

The second type of interference is the same periodic interference except that the phase offset varies from sensor to sensor. That is, the interference $d_m(t)$ for the 60 sensors over the right hemisphere was calculated using $d_m(t) = \sin(2\pi f_d t + \phi_m)$, where each sensor had a different ϕ_m that was determined by using a uniformly-distributed random number between 0 and 2π. The simulated recordings containing this interference are shown in Fig. 7.2(b).

The third type of interference is a linear trend whose inclination varies from channel to channel. That is, $d_m(t)$ was calculated using $d_m(t) = \Delta_m t$ where each sensor had a different value of Δ_m, which was determined by using a Gaussian-distributed random number. Simulated recordings containing such a linear trend are shown in Fig. 7.2(c). The fourth type of interference is a combination of the linear trend just described with a low-frequency noise. In this case, $d_m(t)$ was calculated by using $d_m(t) = \Delta_m t + \sin[2\pi f_d t + \phi]$ where f_d was set at 1.1 Hz and ϕ was the same for all sensor recordings. The simulated recordings containing this interference are shown in Fig. 7.2(d).

The covariance matrix of the interference, $\boldsymbol{R}_d$, was calculated numerically using $\boldsymbol{R}_d = \langle \boldsymbol{d}(t)\boldsymbol{d}^T(t) \rangle$. The eigenvalue spectra of $\boldsymbol{R}_d$ for these four types of interference are shown in Fig. 7.3. The spectra show that the first and third interferences have a single large eigenvalue, indicating that these interferences are rank-one interferences. The second and the fourth interferences have two relatively large eigenvalues, indicating that they are rank-two interferences. The contour plots of the first spatial eigenvector (the one associated with the largest eigenvalue) of each of the four kinds of interference are shown in Fig. 7.4(b)–(e). The lead-field vector associated with the second source is also shown in Fig. 7.4(a). This contour plot is typical of all (focal) brain sources.

It can be seen that the spatial eigenvectors for all four cases are very different from the lead fields associated with brain sources. The values of $\cos^2(\boldsymbol{l}, \boldsymbol{u}|\boldsymbol{R}_{s+n}^{-1})$ between the second-source lead field and the first eigenvector are less than 3×10^{-3} for all four cases. These very small values confirm our visually-obtained interpretation that the eigenvectors in Fig. 7.4 are very different from the lead-field vector associated with a brain source. The value of $(\boldsymbol{u}_1^T \boldsymbol{R}_{s+n}^{-1} \boldsymbol{u}_2)/(\boldsymbol{u}_1^T \boldsymbol{R}_{s+n}^{-1} \boldsymbol{u}_1)$ is also calculated for the two cases of rank-two interference, and this value is on the order of 10^{-6}. These numerical evaluations suggest that the influence of the interferences should be very small. The eigenspace-projection array-gain minimum-variance spatial filter was applied to the four cases of the simulated recordings in Fig. 7.2. The results of (the square-root of) the source power reconstruction are shown in Fig. 7.5. The reconstructed source time courses are shown in Fig. 7.6. There is no observable influence in any of the results in Figs. 7.5 and 7.6. These results illustrate the conclusions of the arguments in this chapter.

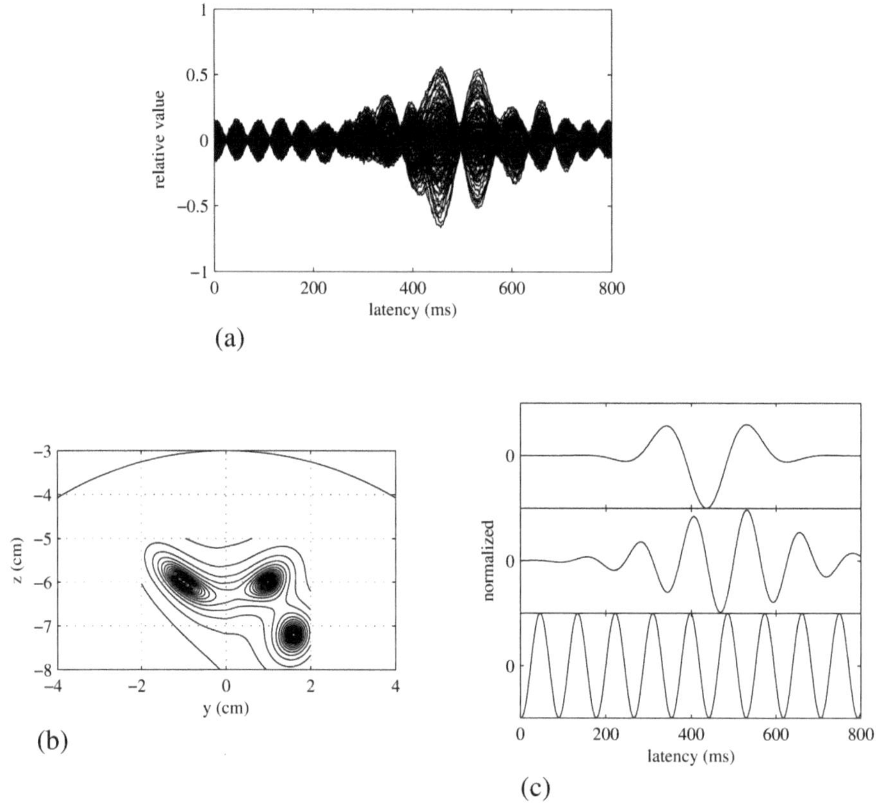

Figure 7.1: (a) The simulated MEG recordings with no interference, $\boldsymbol{b}_{s+n}(t)$. (b) The square root of the power reconstruction, $\sqrt{\langle \widehat{s}(\boldsymbol{r},t)^2 \rangle}$, obtained using $\boldsymbol{b}_{s+n}(t)$. (c) Time course outputs at the three source locations, $\widehat{s}(\boldsymbol{r}_1,t)$, $\widehat{s}(\boldsymbol{r}_2,t)$, and $\widehat{s}(\boldsymbol{r}_3,t)$ (from top to bottom), obtained using $\boldsymbol{b}_{s+n}(t)$. The eigenspace-projected array-gain minimum-variance filter was used.

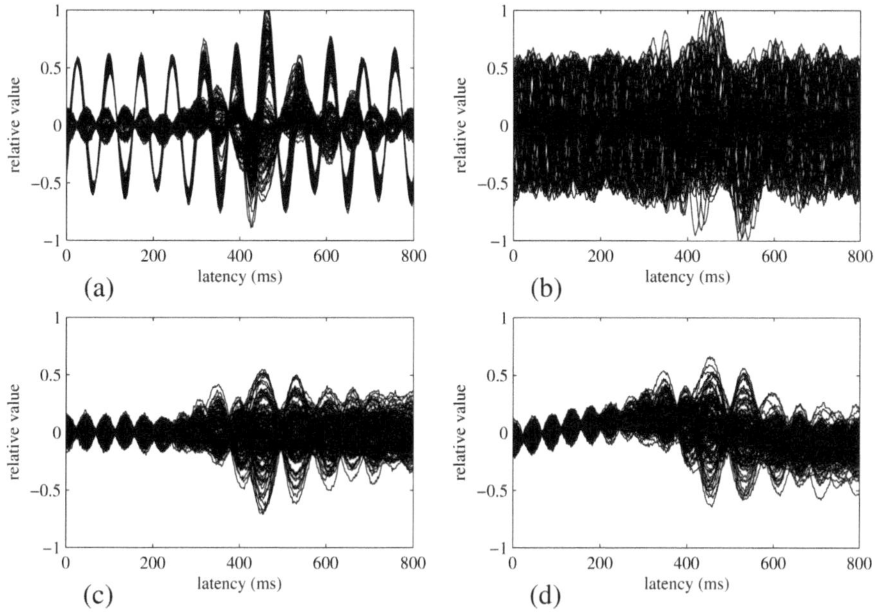

Figure 7.2: Simulated MEG recordings containing low-rank interferences. (a) A periodic interference with a frequency of 13.7 Hz overlapped the recordings of the 60 sensors located over the right hemisphere. (b) The same periodic interference except that a phase offset varies from sensor to sensor. (c) A linear trend, with its inclination varying from sensor to sensor, overlapped all sensor recordings. (d) A combination of the linear trend and low-frequency (1.1 Hz) interference overlapped all sensor recordings.

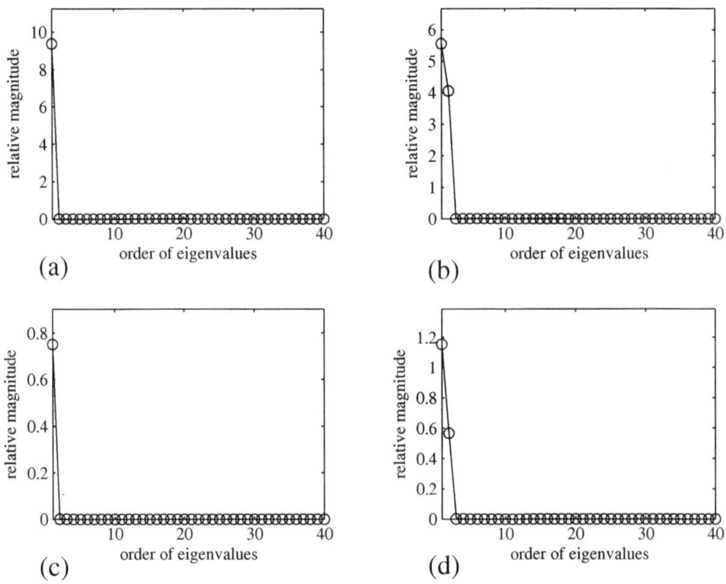

Figure 7.3: Eigenvalue spectrum of $\boldsymbol{R}_d$ shown up to the 40th eigenvalue. The eigenvalues are sorted from largest to smallest. (a) The spectrum for the periodic interference in Fig. 7.2(a). (b) The spectrum for the periodic interference with a random phase offset in Fig. 7.2(b). (c) The spectrum for the linear trend in Fig. 7.2(c). (d) The spectrum for the linear trend and low-frequency (1.1 Hz) interference in Fig. 7.2(d).

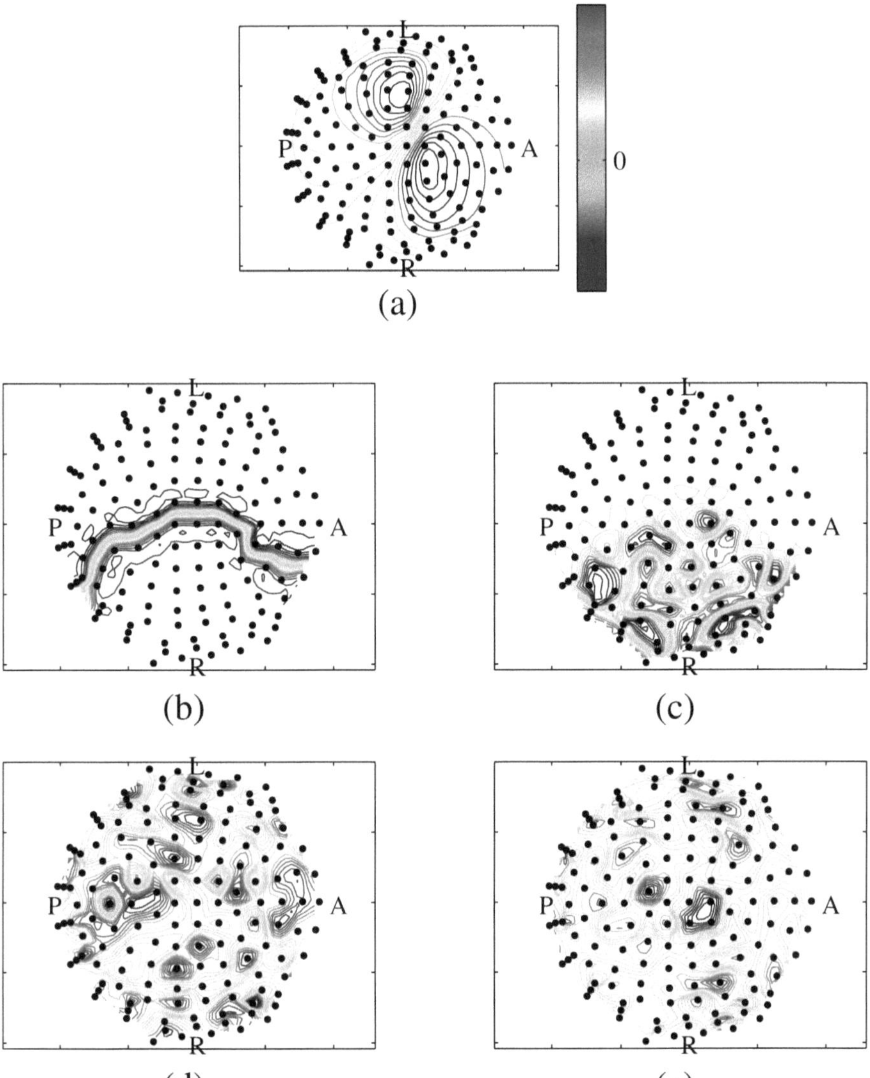

Figure 7.4: (a) The contour plot of the lead-field vector for the second source. (b)–(e) The contour plots of the eigenvector of $\boldsymbol{R}_d$ associated with the largest eigenvalue. (b) Eigenvector contour for the periodic interference in Fig. 7.2(a). (c) Eigenvector contour for the periodic interference with a random phase offset shown in Fig. 7.2(b). (d) Eigenvector contour for the linear trend shown in Fig. 7.2(c). (e) Eigenvector contour for the linear trend with low-frequency interference shown in Fig. 7.2(d). The dots represent the locations of the sensors (mapped from three dimensions to two dimensions). The anterior, posterior, left and right directions are indicated by A, P, L, and R, respectively. The contours represent the relative intensity, as indicated by the color bar.

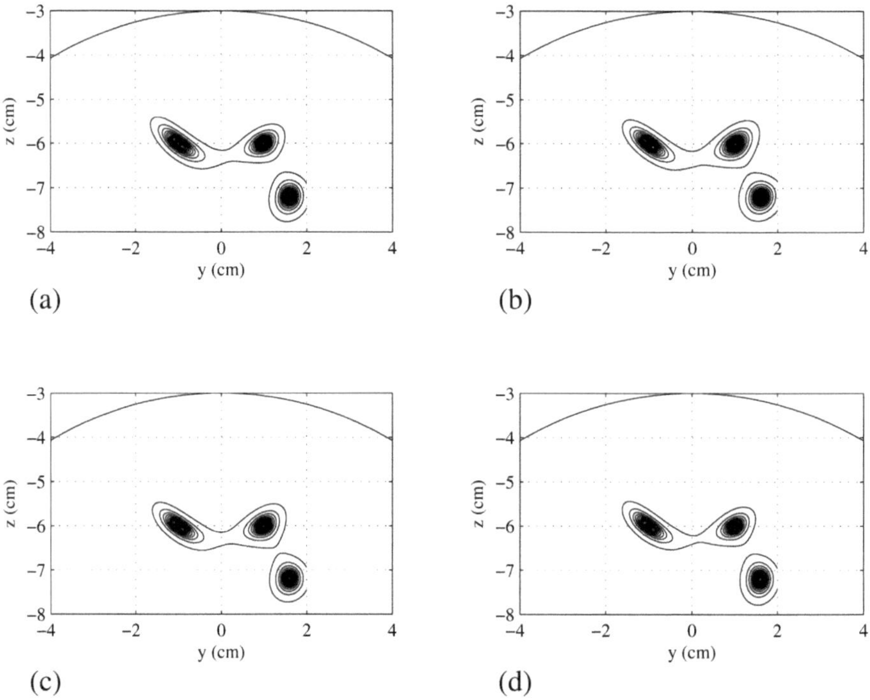

Figure 7.5: Results of square-root of the power reconstruction, $\sqrt{\langle \widehat{s}(\boldsymbol{r},t)^2 \rangle}$, obtained using the eigenspace-projected array-gain minimum-variance spatial filter. (a) The simulated recordings in Fig. 7.2(a) were used. (b) The simulated recordings in Fig. 7.2(b) were used. (c) The simulated recordings in Fig. 7.2(c) were used. (d) The simulated recordings in Fig. 7.2(d) were used.

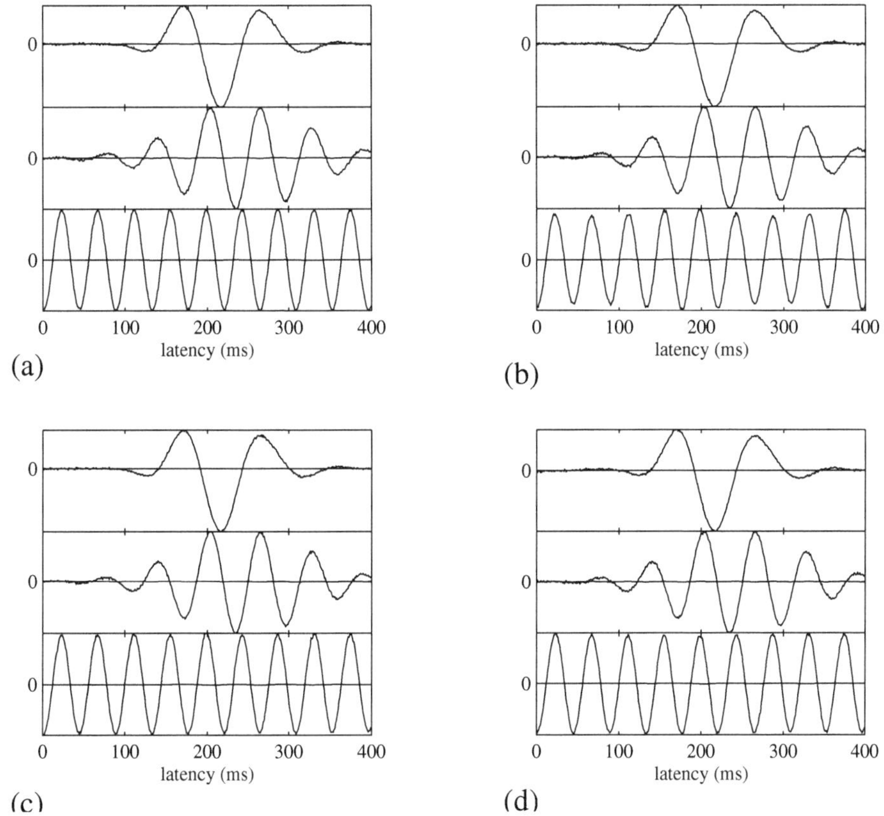

Figure 7.6: The reconstructed source time-courses, $\widehat{s}(\boldsymbol{r}_q, t)$ $q = 1, 2, 3$, obtained as outputs of the eigenspace-projected array-gain minimum-variance spatial filter. (a) The simulated recordings in Fig. 7.2(a) were used. (b) The simulated recordings in Fig. 7.2(b) were used. (c) The simulated recordings in Fig. 7.2(c) were used. (d) The simulated recordings in Fig. 7.2(d) were used.

Chapter 8

Effects of high-rank interference

The major problem with brain electromagnetic measurements is that the measured signal generally contains a large amount of interfering magnetic fields. The preceding chapter dealt with the influence of external interference of non-biological origin. This chapter deals with interference of a physiological origin. In particular, this chapter focuses on the background brain activity, which is sometimes referred to as brain noise or physiological noise. The prominent characteristic of the background interference is that it is generated by a large number of sources[51][52]. Hence, the underlying low-rank signal assumption, which is necessary for formulating adaptive spatial filters, is invalidated. This chapter first presents a theoretical analysis showing that such high-rank interference can cause a severe spatial blur in the reconstruction results. We then describe a prewhitening eigenspace-projection spatial filter, which can achieve the source reconstruction free from the influence of background interference even when the power of the interference is significantly large.

8.1 Influence of background brain activity

8.1.1 Point-spread function under background interference

Here, we derive the point-spread function when background interference exists[53]. For this derivation, we use a model for the measurements $\boldsymbol{b}(t)$ expressed as

$$\boldsymbol{b}(t) = \boldsymbol{b}_s(t) + \boldsymbol{b}_I(t) + \boldsymbol{n}(t), \tag{8.1}$$

where $\boldsymbol{b}_s(t)$ is the signal part, $\boldsymbol{b}_I(t)$ is the interference generated by the background activity, and $\boldsymbol{n}(t)$ is the sensor noise. The spatial-filter output is then expressed as

$$\widehat{s}(\boldsymbol{r},t) = \boldsymbol{w}^T(\boldsymbol{r})\boldsymbol{b}(t) = \boldsymbol{w}^T(\boldsymbol{r})\boldsymbol{b}_s(t) + \boldsymbol{w}^T(\boldsymbol{r})\boldsymbol{b}_I(t) + \boldsymbol{w}^T(\boldsymbol{r})\boldsymbol{n}(t). \tag{8.2}$$

For non-adaptive spatial filters, the influence of the interference $\boldsymbol{b}_I(t)$ is simply additive. More specifically, the influence is the overlap of $\boldsymbol{w}^T(\boldsymbol{r})\boldsymbol{b}_I(t)$ onto the signal part, $\boldsymbol{w}^T(\boldsymbol{r})\boldsymbol{b}_s(t)$. For adaptive spatial filters, however, the interference $\boldsymbol{b}_I(t)$ affects the source reconstruction results in a more complex manner because $\boldsymbol{b}_I(t)$ also affects the filter weight $\boldsymbol{w}(\boldsymbol{r})$ through the covariance matrix $\boldsymbol{R}$. We derive the point-spread function of the minimum-variance spatial filter that takes the background interference into account. Numerical examples of the point-spread function are presented in Section 8.1.2.

The magnitude and the orientation of an interference source are denoted $\mu(\boldsymbol{r},t)$ and $\boldsymbol{\eta}(\boldsymbol{r})$, respectively. We assume that the background sources are continuously distributed. Then, we have

$$\boldsymbol{b}_I(t) = \int_\Omega \mu(\boldsymbol{r},t)\boldsymbol{L}(\boldsymbol{r})\boldsymbol{\eta}(\boldsymbol{r})\, d^3r = \int_\Omega \mu(\boldsymbol{r},t)\boldsymbol{l}(\boldsymbol{r})\, d^3r, \tag{8.3}$$

where the integral is computed over the whole source space Ω, and the lead-field vector $\boldsymbol{l}(\boldsymbol{r})$ is defined such that $\boldsymbol{l}(\boldsymbol{r}) = \boldsymbol{L}(\boldsymbol{r})\boldsymbol{\eta}(\boldsymbol{r})$. We also assume that only a single target source exists at $\boldsymbol{r}_1$ with an orientation equal to $\boldsymbol{\eta}_1$. The magnitude of the target source is denoted $s(\boldsymbol{r}_1,t)$. Then, defining $\boldsymbol{f}$ such that $\boldsymbol{f} = \boldsymbol{L}(\boldsymbol{r}_1)\boldsymbol{\eta}_1$, the measurement $\boldsymbol{b}(t)$ is expressed as

$$\boldsymbol{b}(t) = s(\boldsymbol{r}_1,t)\boldsymbol{f} + \int_\Omega \mu(\boldsymbol{r},t)\boldsymbol{l}(\boldsymbol{r})\, d^3r + \boldsymbol{n}(t). \tag{8.4}$$

Therefore, assuming that the relationship $\langle \boldsymbol{n}(t)\boldsymbol{n}^T(t)\rangle = \sigma_0^2\boldsymbol{I}$ holds, the covariance matrix of the measurements is given by

$$\boldsymbol{R} = \langle \boldsymbol{b}(t)\boldsymbol{b}^T(t)\rangle = \sigma_1^2\boldsymbol{f}\boldsymbol{f}^T + \iint_\Omega \langle \mu(\boldsymbol{r},t)\mu(\boldsymbol{r}',t)\rangle \boldsymbol{l}(\boldsymbol{r})\boldsymbol{l}^T(\boldsymbol{r}')\, d^3r\, d^3r' + \sigma_0^2\boldsymbol{I}, \tag{8.5}$$

where the signal power σ_1^2 is defined such that $\sigma_1^2 = \langle s(\boldsymbol{r}_1,t)^2\rangle$. We assume that the background source activity is spatially uniform and incoherent, i.e.,

$$\langle \mu(\boldsymbol{r},t)\mu(\boldsymbol{r}',t)\rangle = \sigma_c^2\delta(\boldsymbol{r}-\boldsymbol{r}'), \tag{8.6}$$

where σ_c^2 is the power of the background source activity. Substituting Eq. (8.6) into (8.5), we obtain

$$\boldsymbol{R} = \sigma_1^2\boldsymbol{f}\boldsymbol{f}^T + \sigma_c^2\iint_\Omega \boldsymbol{l}(\boldsymbol{r})\boldsymbol{l}^T(\boldsymbol{r})\, d^3r + \sigma_0^2\boldsymbol{I}. \tag{8.7}$$

In the right-hand side of the above equation, the integral in the second term can be approximated using the gram matrix $\boldsymbol{G}$. Thus, we finally have

$$\boldsymbol{R} = \sigma_1^2\boldsymbol{f}\boldsymbol{f}^T + \sigma_c^2\boldsymbol{G} + \sigma_0^2\boldsymbol{I}. \tag{8.8}$$

According to the discussion in Section 5.1, omitting the explicit notation of $(\boldsymbol{r})$ from $\boldsymbol{l}(\boldsymbol{r})$, the point-spread function of the array-gain-constraint minimum-variance spatial filter is expressed as

$$F(\boldsymbol{r}) = \boldsymbol{w}^T(\boldsymbol{r})\boldsymbol{f} = \frac{\widetilde{\boldsymbol{l}}^T\boldsymbol{R}^{-1}\boldsymbol{f}}{\widetilde{\boldsymbol{l}}^T\boldsymbol{R}^{-1}\widetilde{\boldsymbol{l}}}. \tag{8.9}$$

Substituting Eq. (8.8) into (8.9), we can derive the explicit form of the point-spread function, which is given by

$$F(\boldsymbol{r}) = \|\boldsymbol{l}\|\sqrt{\frac{\boldsymbol{f}^T\boldsymbol{D}^{-1}\boldsymbol{f}}{\boldsymbol{l}^T\boldsymbol{D}^{-1}\boldsymbol{l}}}\left[\frac{\cos(\widetilde{\boldsymbol{l}, \boldsymbol{f}}|\boldsymbol{D}^{-1})}{[1+\alpha_D(1-\cos^2(\widetilde{\boldsymbol{l}, \boldsymbol{f}}|\boldsymbol{D}^{-1}))]}\right], \tag{8.10}$$

where $\boldsymbol{D} = \boldsymbol{I} + (\sigma_c^2/\sigma_0^2)\boldsymbol{G}$, and $\alpha_D = (\sigma_1^2/\sigma_0^2)\boldsymbol{f}^T\boldsymbol{D}^{-1}\boldsymbol{f}$. When the background interference is negligibly small, substituting $\sigma_c = 0$ into (8.10), (namely, substituting $\boldsymbol{I}$ for $\boldsymbol{D}$ in (8.10),) the point-spread function in this case becomes

$$F(\boldsymbol{r}) = \|\boldsymbol{f}\|\frac{\cos(\widetilde{\boldsymbol{l}, \boldsymbol{f}})}{[1+\alpha(1-\cos^2(\widetilde{\boldsymbol{l}, \boldsymbol{f}}))]}, \tag{8.11}$$

where α is equal to $\alpha = (\sigma_1^2/\sigma_0^2)\|\boldsymbol{f}\|^2$. The above equation is exactly equal to Eq. (5.14). When the power of the background interference is much larger than the power of the sensor noise, substituting $\boldsymbol{D} \approx (\sigma_c^2/\sigma_0^2)\boldsymbol{G}$ into (8.10), we derive

$$F(\boldsymbol{r}) = \|\boldsymbol{l}\|\sqrt{\frac{\boldsymbol{f}^T\boldsymbol{G}^{-1}\boldsymbol{f}}{\boldsymbol{l}^T\boldsymbol{G}^{-1}\boldsymbol{l}}}\frac{\cos(\widetilde{\boldsymbol{l}, \boldsymbol{f}}|\boldsymbol{G}^{-1})}{[1+\alpha_G(1-\cos^2(\widetilde{\boldsymbol{l}, \boldsymbol{f}}|\boldsymbol{G}^{-1}))]}, \tag{8.12}$$

where $\alpha_G = (\sigma_1^2/\sigma_c^2)(\boldsymbol{f}^T\boldsymbol{G}^{-1}\boldsymbol{f})$, which represents the input SNR when only the background brain noise is present. We present numerical examples of these point-spread functions in the following section. These examples show that, when large background activities exist, a significant degradation of the spatial resolution is caused in the output of the minimum-variance filters.

8.1.2 Numerical examples

To calculate the point-spread functions, the same source-sensor configuration and coordinate system illustrated in Fig. 5.2 are used except that the single source is located at $(0, -1, -6)$ cm. We calculated the point-spread function using Eq. (8.10) for five values of σ_c/σ_1. Here, the input SNR, which is defined as $\alpha = (\sigma_1^2/\sigma_0^2)\|\boldsymbol{f}\|^2$, is set equal to $4M$. Also, in this calculation, the gram matrix $\boldsymbol{G}$ was obtained by numerically integrating over a volume defined as $-8 \le x \le 8$, $-8 \le y \le 8$, and $-11 \le z \le -3$. The results plotted in Fig. 8.1(a) show that when the power of the background activity, σ_c^2, is increased, the point-spread function is blurred and the full-width-at-half-maximum (FWHM) of the point-spread function becomes greater. The FWHM of the point-spread functions is plotted with respect to σ_c/σ_1 for three different values of input SNR, α, in Fig. 8.1(b). The results show that the degree of the spatial blur is approximately linearly increased, depending on the value of σ_c/σ_1, and the FWHM reaches nearly 1.5 cm when $\sigma_c \approx 0.5\sigma_1$.

As discussed in Section 5.3, sensor noise is known to cause degradation of the spatial resolution. Let us compare the degree of spatial blur caused by background interference and sensor noise. The results of plotting the point-spread functions in Eqs. (8.11) and (8.12) are shown in Figs. 8.2(a) and (b). Here, α in Eq. (8.11)

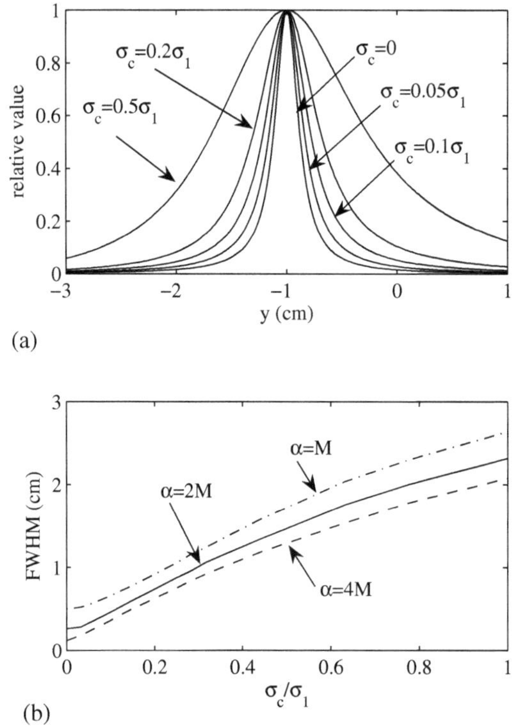

Figure 8.1: (a) Point-spread functions of a source located at $(0, -1, -6)$ cm. Five values of σ_c were used: $\sigma_c = 0$, $\sigma_c = 0.05\sigma_1$, $\sigma_c = 0.1\sigma_1$, $\sigma_c = 0.2\sigma_1$, and $\sigma_c = 0.5\sigma_1$. The input SNR, defined as $\alpha = (\sigma_1^2/\sigma_0^2)\|\boldsymbol{f}\|^2$, was set equal to $4M$. (b) Plot of the full-width-at-half-maximum (FWHM) of the point-spread function with respect to σ_c/σ_1 for three input SNR values: $\alpha = M$, $\alpha = 2M$, and $\alpha = 4M$.

and α_G in Eq. (8.12) were set equal to each other; these values were set to $2M$ for the results in (a) and to M for the results in (b). Note that, by setting α and α_G equal, the powers of these two kinds of noises were set equal when comparing their effects on the point-spread function. These calculated results show that the FWHM of the point-spread function in Eq. (8.11) is approximately two-fold greater than that for the point-spread function in Eq. (8.12). That is, the blur caused by the sensor noise is two-fold greater than that caused by the interference. The difference between the two point-spread functions is caused by the difference between $\cos(\boldsymbol{l}, \boldsymbol{f})$ and $\cos(\boldsymbol{l}, \boldsymbol{f}|\boldsymbol{G}^{-1})$. We can see that $\cos(\boldsymbol{l}, \boldsymbol{f}|\boldsymbol{G}^{-1})$ decays twice as fast as $\cos(\boldsymbol{l}, \boldsymbol{f})$, thus explaining the difference in these two types of point-spread functions. The amount of sensor noise in a modern MEG scanner is around 5 fT/$\sqrt{\text{Hz}}$, and therefore we expect a noise amplitude to approximately equal 50 fT (0.05 pT) in the measurement if the bandwidth is set to 100 Hz. On

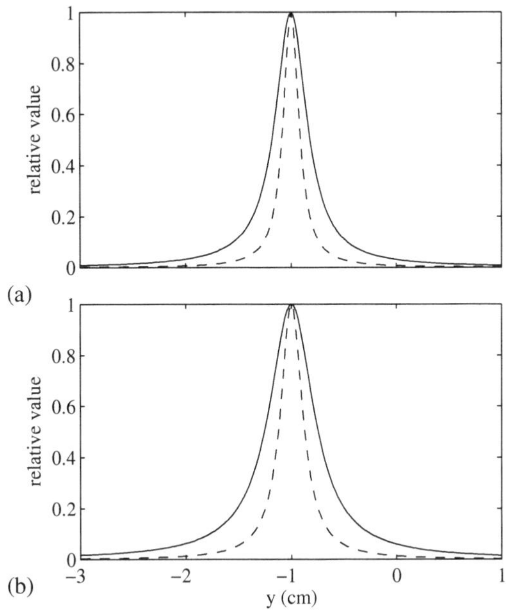

Figure 8.2: Comparison of the point spread functions in Eqs. (8.11) and (8.12). The solid lines indicate the point-spread function in Eq. (8.11) and the broken lines indicate the point-spread function in Eq. (8.12). The value of α in Eq. (8.11) and the value of α_G in Eq. (8.12) were set equal to each other. (a) The point-spread functions when $\alpha = \alpha_G = 2M$ and (b) The point-spread functions when $\alpha = \alpha_G = M$.

the other hand, the amplitude of the background interference is, on average, equal to $0.5 - 2$ pT. Hence, the background interference is $10 - 40$ times greater than the sensor noise. Therefore, the background interference is the major cause of the spatial blur in electromagnetic source imaging. This is true despite the fact that the sensor noise has a two-fold greater effect on the spatial resolution.

8.2 Prewhitening eigenspace-projection spatial filter

8.2.1 Prewhitening signal covariance estimation

We next describe the prewhitening eigenspace-projection spatial filter, which can significantly reduce the influence of background interference[53]. To describe the method, we first make several definitions. We define the covariance matrix of the signal data, $\boldsymbol{b}_s(t)$, as $\boldsymbol{R}_s$:

$$\boldsymbol{R}_s = \langle \boldsymbol{b}_s(t)\boldsymbol{b}_s^T(t)\rangle. \tag{8.13}$$

We also define the signal-plus-sensor-noise covariance matrix, $\boldsymbol{R}_{s+n}$:

$$\boldsymbol{R}_{s+n} = \langle [\boldsymbol{b}_s(t) + \boldsymbol{n}(t)][\boldsymbol{b}_s(t) + \boldsymbol{n}(t)]^T\rangle, \tag{8.14}$$

and the interference-plus-sensor-noise covariance matrix, $\boldsymbol{R}_{i+n}$:

$$\boldsymbol{R}_{i+n} = \langle [\boldsymbol{b}_I(t) + \boldsymbol{n}(t)][\boldsymbol{b}_I(t) + \boldsymbol{n}(t)]^T\rangle. \tag{8.15}$$

We further assume that the background interference is uncorrelated with the signal source activity. Under this assumption, the relationship

$$\boldsymbol{R} = \boldsymbol{R}_s + \boldsymbol{R}_{i+n} \tag{8.16}$$

holds.

The fundamental assumption for the prewhitening eigenspace-projection spatial filter is that the control-state measurements, $\boldsymbol{b}_c(t)$, which contain only the contributions from the background interference and sensor noise, are available[1],

$$\boldsymbol{b}_c(t) = \boldsymbol{b}_I(t) + \boldsymbol{n}(t). \tag{8.17}$$

The interference-plus-noise covariance matrix, $\boldsymbol{R}_{i+n}$, can be obtained from such control-state measurements by

$$\boldsymbol{R}_{i+n} = \langle \boldsymbol{b}_c(t)\boldsymbol{b}_c^T(t)\rangle. \tag{8.18}$$

To obtain an estimate of the signal covariance matrix, $\boldsymbol{R}_s$, we first compute the prewhitened measurement covariance matrix, $\widetilde{\boldsymbol{R}}$, which is defined as

$$\widetilde{\boldsymbol{R}} = \boldsymbol{R}_{i+n}^{-1/2}\boldsymbol{R}\boldsymbol{R}_{i+n}^{-1/2}. \tag{8.19}$$

Thus, from Eq. (8.16), we have the relationship

$$\widetilde{\boldsymbol{R}} = \boldsymbol{R}_{i+n}^{-1/2}\boldsymbol{R}_s\boldsymbol{R}_{i+n}^{-1/2} + \boldsymbol{I} = \widetilde{\boldsymbol{R}}_s + \boldsymbol{I}, \tag{8.20}$$

where

$$\widetilde{\boldsymbol{R}}_s = \boldsymbol{R}_{i+n}^{-1/2}\boldsymbol{R}_s\boldsymbol{R}_{i+n}^{-1/2}. \tag{8.21}$$

We define the eigenvalues and eigenvectors of an $M \times M$ matrix $\widetilde{\boldsymbol{R}}_s$ as $\gamma_1, \ldots, \gamma_M$ and $\boldsymbol{h}_1, \ldots, \boldsymbol{h}_M$, respectively. Since $\boldsymbol{R}_s$ is a positive semi-definite matrix with rank Q, and $\boldsymbol{R}_{i+n}^{-1/2}$ is a non-singular matrix, $\widetilde{\boldsymbol{R}}_s$ is also a positive semi-definite

[1] Equation (8.17) may not always hold in real-life task-and-control-type measurements. There may be a situation in which there are some sources that appear only in the control state and do not appear in the task state. Such sources are called control-only sources, and the prewhitening method is also significantly robust to the existence of such control-only sources, as discussed in Section 13.5 in the Appendix.

matrix with rank Q. Thus, the eigenvalues $\gamma_1, \ldots, \gamma_Q$ are positive and the other eigenvalues $\gamma_{Q+1}, \ldots, \gamma_M$ are zero. Namely, we have

$$\widetilde{\boldsymbol{R}}_s = \sum_{j=1}^{Q} \gamma_j \boldsymbol{h}_j \boldsymbol{h}_j^T. \tag{8.22}$$

Therefore, the eigendecomposition of $\widetilde{\boldsymbol{R}}$ can be expressed as

$$\widetilde{\boldsymbol{R}} = \sum_{j=1}^{Q} \gamma_j \boldsymbol{h}_j \boldsymbol{h}_j^T + \boldsymbol{I} = \sum_{j=1}^{Q} (\gamma_j + 1) \boldsymbol{h}_j \boldsymbol{h}_j^T + \sum_{j=Q+1}^{M} \boldsymbol{h}_j \boldsymbol{h}_j^T. \tag{8.23}$$

The equation above indicates that the Q largest eigenvalues of $\widetilde{\boldsymbol{R}}$ are greater than 1 and the corresponding eigenvectors are the same as those associated with the non-zero eigenvalues of $\widetilde{\boldsymbol{R}}_s$. The eigenvalues greater than 1 are referred to as the signal-level eigenvalues of $\widetilde{\boldsymbol{R}}$ and their corresponding eigenvectors are referred to as the signal-level eigenvectors of $\widetilde{\boldsymbol{R}}$.

Equation (8.23) indicates that it is possible to retrieve $\widetilde{\boldsymbol{R}}_s$ from the signal-level eigenvectors of $\widetilde{\boldsymbol{R}}$. That is, defining a matrix $\boldsymbol{H}_S$ as $\boldsymbol{H}_S = [\boldsymbol{h}_1, \ldots, \boldsymbol{h}_Q]$, the signal covariance matrix can be obtained using

$$\boldsymbol{R}_{i+n}^{1/2} \boldsymbol{H}_S \boldsymbol{H}_S^T (\widetilde{\boldsymbol{R}} - \boldsymbol{I}) \boldsymbol{R}_{i+n}^{1/2} = \boldsymbol{R}_{i+n}^{1/2} [\sum_{j=1}^{Q} \gamma_j \boldsymbol{h}_j \boldsymbol{h}_j^T] \boldsymbol{R}_{i+n}^{1/2} = \boldsymbol{R}_{i+n}^{1/2} \widetilde{\boldsymbol{R}}_s \boldsymbol{R}_{i+n}^{1/2} = \boldsymbol{R}_s. \tag{8.24}$$

In actual cases, the theoretical covariance matrices $\boldsymbol{R}$ and $\boldsymbol{R}_{i+n}$ are unknown, and we should use the sample covariance matrices, which are obtained using

$$\widehat{\boldsymbol{R}} = \frac{1}{K} \sum_{k=1}^{K} \boldsymbol{b}(t_k) \boldsymbol{b}^T(t_k), \quad \text{and} \quad \widehat{\boldsymbol{R}}_{i+n} = \frac{1}{K_C} \sum_{k=1}^{K_C} \boldsymbol{b}_c(t_k) \boldsymbol{b}_c^T(t_k), \tag{8.25}$$

assuming that we have K and K_C time points for the task and the control periods, respectively. We define $\widehat{\widetilde{\boldsymbol{R}}}$ such that $\widehat{\widetilde{\boldsymbol{R}}} = \widehat{\boldsymbol{R}}_{i+n}^{-1/2} \widehat{\boldsymbol{R}} \widehat{\boldsymbol{R}}_{i+n}^{-1/2}$. Using Eq. (8.24), the estimate of the signal covariance matrix $\widehat{\boldsymbol{R}}_s$ can be obtained using

$$\widehat{\boldsymbol{R}}_s = \widehat{\boldsymbol{R}}_{i+n}^{1/2} \widehat{\boldsymbol{H}}_S \widehat{\boldsymbol{H}}_S^T (\widehat{\widetilde{\boldsymbol{R}}} - \boldsymbol{I}) \widehat{\boldsymbol{R}}_{i+n}^{1/2}, \tag{8.26}$$

where $\widehat{\boldsymbol{H}}_S = [\widehat{\boldsymbol{h}}_1, \ldots, \widehat{\boldsymbol{h}}_Q]$, and $\widehat{\boldsymbol{h}}_1, \ldots, \widehat{\boldsymbol{h}}_Q$ are the signal-level eigenvectors of $\widehat{\widetilde{\boldsymbol{R}}}$. Ideally, since the noise-level eigenvalues of $\widetilde{\boldsymbol{R}}$ are equal to 1, multiplying $\boldsymbol{H}_S \boldsymbol{H}_S^T$ by $(\widetilde{\boldsymbol{R}} - \boldsymbol{I})$ is not needed to extract the signal subspace. However, in actual cases, the noise-level eigenvalues obtained from $\widehat{\widetilde{\boldsymbol{R}}}$ may not be equal to 1. Thus, this multiplication is still needed in the prewhitening estimation of the signal covariance matrix.

8.2.2 Prewhitening eigenspace-projection spatial filter

Given the estimate of the signal covariance matrix, $\widehat{\boldsymbol{R}}_s$, a reasonable estimate of the signal-plus-sensor-noise covariance matrix, $\widehat{\boldsymbol{R}}_{s+n}$, can be obtained using

$$\widehat{\boldsymbol{R}}_{s+n} = \widehat{\boldsymbol{R}}_s + \epsilon \boldsymbol{I}, \tag{8.27}$$

where ϵ is the diagonal loading factor that should be set close to the variance of the sensor noise, σ_0^2, in this case. Consequently, using the minimum-variance spatial filter, source power reconstruction free from the influence of the background activity can be obtained using

$$\langle \widehat{s}(\boldsymbol{r},t)^2 \rangle = \frac{1}{\boldsymbol{l}^T(\boldsymbol{r})\widehat{\boldsymbol{R}}_{s+n}^{-1}\boldsymbol{l}(\boldsymbol{r})} = \frac{1}{\boldsymbol{l}^T(\boldsymbol{r})(\widehat{\boldsymbol{R}}_s + \epsilon \boldsymbol{I})^{-1}\boldsymbol{l}(\boldsymbol{r})}, \tag{8.28}$$

We then define the spatio-temporal matrix of the measurement, $\boldsymbol{b}(t)$, as $\boldsymbol{B}$:

$$\boldsymbol{B} = [\boldsymbol{b}(t_1), \boldsymbol{b}(t_2), \ldots, \boldsymbol{b}(t_K)].$$

The spatio-temporal matrices for $\boldsymbol{b}_s(t)$ and $\boldsymbol{b}_I(t)$ are defined in exactly the same manner, i.e.,

$$\boldsymbol{B}_s = [\boldsymbol{b}_s(t_1), \boldsymbol{b}_s(t_2), \ldots, \boldsymbol{b}_s(t_K)],$$

and

$$\boldsymbol{B}_I = [\boldsymbol{b}_I(t_1), \boldsymbol{b}_I(t_2), \ldots, \boldsymbol{b}_I(t_K)].$$

Then, in accordance with Eq. (8.1), the relationship

$$\boldsymbol{B} = \boldsymbol{B}_s + \boldsymbol{B}_I + \boldsymbol{N} \tag{8.29}$$

holds where $\boldsymbol{N}$ is the noise matrix defined as $\boldsymbol{N} = [\boldsymbol{n}(t_1), \ldots, \boldsymbol{n}(t_K)]$. A reasonable estimate of the spatio-temporal matrix of the signal, $\boldsymbol{B}_s$, can also be obtained in a similar manner as described in Section 8.2.1.

The prewhitened version of $\boldsymbol{B}$ is defined as $\widetilde{\boldsymbol{B}}$ such that

$$\widetilde{\boldsymbol{B}} = \boldsymbol{R}_{i+n}^{-1/2}\boldsymbol{B} = \boldsymbol{R}_{i+n}^{-1/2}\boldsymbol{B}_s + \boldsymbol{R}_{i+n}^{-1/2}(\boldsymbol{B}_I + \boldsymbol{N}). \tag{8.30}$$

The singular value decomposition (SVD) of $\widetilde{\boldsymbol{B}}$ is obtained as

$$\widetilde{\boldsymbol{B}} = \sum_{j=1}^{Q} \sqrt{\gamma_j + 1}\,\boldsymbol{h}_j \boldsymbol{m}_j^T + \sum_{j=Q+1}^{M} \boldsymbol{h}_j \boldsymbol{m}_j^T, \tag{8.31}$$

where $\boldsymbol{m}_j^T$ indicates the jth temporal singular vector. Once again, the first Q singular values of $\widetilde{\boldsymbol{B}}$ are greater than 1 and they are associated with the signal portion of the prewhitened data, $\boldsymbol{R}_{i+n}^{-1/2}\boldsymbol{B}_s$. The other singular values are equal to

1 and they are associated with the interference and noise portion $\boldsymbol{R}_{i+n}^{-1/2}(\boldsymbol{B}_I + \boldsymbol{N})$. Therefore, by applying the projector $\boldsymbol{H}_S\boldsymbol{H}_S^T$ to $\widetilde{\boldsymbol{B}}$, we have

$$\boldsymbol{H}_S\boldsymbol{H}_S^T\widetilde{\boldsymbol{B}} = \sum_{j=1}^{Q}(\sqrt{\gamma_j + 1})\boldsymbol{h}_j\boldsymbol{m}_j^T \approx \boldsymbol{R}_{i+n}^{-1/2}\boldsymbol{B}_s, \tag{8.32}$$

and the spatio-temporal matrix of the signal magnetic field $\boldsymbol{B}_s$ can be extracted by

$$\boldsymbol{R}_{i+n}^{1/2}[\boldsymbol{H}_S\boldsymbol{H}_S^T\widetilde{\boldsymbol{B}}] = \boldsymbol{R}_{i+n}^{1/2}\boldsymbol{R}_{i+n}^{-1/2}\boldsymbol{B}_s = \boldsymbol{B}_s. \tag{8.33}$$

Defining $\boldsymbol{\Phi}_S$ as $\boldsymbol{\Phi}_S = \widehat{\boldsymbol{R}}_{i+n}^{1/2}[\widehat{\boldsymbol{H}}_S\widehat{\boldsymbol{H}}_S^T]\widehat{\boldsymbol{R}}_{i+n}^{-1/2}$, we have

$$\boldsymbol{\Phi}_S\boldsymbol{B} = \boldsymbol{B}_s, \tag{8.34}$$

and the weight vector of the prewhitening eigenspace-projection spatial filter is expressed as

$$\boldsymbol{w}_p(\boldsymbol{r}) = \frac{\boldsymbol{\Phi}_S^T(\widehat{\boldsymbol{R}}_s + \epsilon\boldsymbol{I})^{-1}\boldsymbol{l}(\boldsymbol{r})}{\boldsymbol{l}^T(\boldsymbol{r})(\widehat{\boldsymbol{R}}_s + \epsilon\boldsymbol{I})^{-1}\boldsymbol{l}(\boldsymbol{r})}. \tag{8.35}$$

Interference-free spatio-temporal source reconstruction is obtained using this $\boldsymbol{w}_p(\boldsymbol{r})$, i.e.,

$$[\widehat{s}(\boldsymbol{r}, t_1), \ldots, \widehat{s}(\boldsymbol{r}, t_K)] = \boldsymbol{w}_p^T(\boldsymbol{r})\boldsymbol{B} = \frac{\boldsymbol{l}^T(\boldsymbol{r})(\widehat{\boldsymbol{R}}_s + \epsilon\boldsymbol{I})^{-1}\boldsymbol{\Phi}_S\boldsymbol{B}}{\boldsymbol{l}^T(\boldsymbol{r})(\widehat{\boldsymbol{R}}_s + \epsilon\boldsymbol{I})^{-1}\boldsymbol{l}(\boldsymbol{r})} = \frac{\boldsymbol{l}^T(\boldsymbol{r})(\widehat{\boldsymbol{R}}_s + \epsilon\boldsymbol{I})^{-1}\boldsymbol{B}_s}{\boldsymbol{l}^T(\boldsymbol{r})(\widehat{\boldsymbol{R}}_s + \epsilon\boldsymbol{I})^{-1}\boldsymbol{l}(\boldsymbol{r})}, \tag{8.36}$$

where $\widehat{\boldsymbol{R}}_s$ is obtained using Eq. (8.26).

8.3 Overestimation of signal-subspace dimensionality

One problem of the prewhitening spatial filter is that it requires the determination of the prewhitened signal subspace dimension from the eigenvalue spectrum of $\widetilde{\boldsymbol{R}}$. In actual bioelectromagnetic measurements, it is often problematic to accurately determine this dimension because the eigenvalue spectrum often does not have a clear separation between the two subspaces. In the following, we show that the method is quite insensitive to the overestimation of the signal subspace dimensionality . Therefore, accurate determination of the prewhitened signal subspace dimension is not needed, and we can use an intentionally large Q to implement the prewhitening spatial filter.

Let us assume that the signal subspace dimension is overestimated as $Q + Q_\epsilon$ and define $\boldsymbol{H}_\varepsilon$ as $\boldsymbol{H}_\varepsilon = [\boldsymbol{h}_{Q+1}, \ldots, \boldsymbol{h}_{Q+Q_\epsilon}]$. Ideally, the prewhitened data covariance matrix, $\widetilde{\boldsymbol{R}}$, has Q signal-level eigenvalues greater than 1 and $M - Q$ eigenvalues equal to 1. According to Eq. (8.24), the relationship $\boldsymbol{H}_\varepsilon\boldsymbol{H}_\varepsilon^T(\widetilde{\boldsymbol{R}} - \boldsymbol{I}) = 0$ holds; hence overestimation of Q does not affect the signal covariance estimate $\widehat{\boldsymbol{R}}_s$.

However, the data covariance matrix is usually estimated from a finite number of time samples, and in such cases, the noise-level eigenvalues are generally not equal to 1. We denote the noise-level eigenvalues of $\widehat{\widehat{\boldsymbol{R}}}$ as δ_j+1 and assume $\delta_j \geq 0$ for $j=Q+1,\ldots,Q_\epsilon$. Then, the estimated signal covariance matrix, $\widehat{\boldsymbol{R}}_s$, is expressed as

$$\begin{aligned}\widehat{\boldsymbol{R}}_s &= \widehat{\boldsymbol{R}}_{i+n}^{1/2}[\widehat{\boldsymbol{H}}_S\widehat{\boldsymbol{H}}_S^T + \widehat{\boldsymbol{H}}_\varepsilon\widehat{\boldsymbol{H}}_\varepsilon^T](\widehat{\widehat{\boldsymbol{R}}}-\boldsymbol{I})\widehat{\boldsymbol{R}}_{i+n}^{1/2} \\ &= \widehat{\boldsymbol{R}}_{i+n}^{1/2}[\sum_{j=1}^{Q}\gamma_j\widehat{\boldsymbol{h}}_j\widehat{\boldsymbol{h}}_j^T]\widehat{\boldsymbol{R}}_{i+n}^{1/2} + \widehat{\boldsymbol{R}}_{i+n}^{1/2}[\sum_{j=Q+1}^{Q+Q_\epsilon}\delta_j\widehat{\boldsymbol{h}}_j\widehat{\boldsymbol{h}}_j^T]\widehat{\boldsymbol{R}}_{i+n}^{1/2} = \boldsymbol{R}_s + \boldsymbol{R}_\varepsilon. \qquad (8.37)\end{aligned}$$

The second term on the right-hand side of the equation above indicates the error caused by the overestimation. This error term is expressed as

$$\boldsymbol{R}_\varepsilon = \widehat{\boldsymbol{R}}_{i+n}^{1/2}[\sum_{j=Q+1}^{Q+Q_\epsilon}\delta_j\widehat{\boldsymbol{h}}_j\widehat{\boldsymbol{h}}_j^T]\widehat{\boldsymbol{R}}_{i+n}^{1/2} = \sum_{j=Q+1}^{Q+Q_\epsilon}\delta_j\bar{\boldsymbol{h}}_j\bar{\boldsymbol{h}}_j^T, \qquad (8.38)$$

where $\bar{\boldsymbol{h}}_j = \widehat{\boldsymbol{R}}_{i+n}^{1/2}\widehat{\boldsymbol{h}}_j$.

We then decompose $\bar{\boldsymbol{h}}_j$ using $\boldsymbol{e}_i$ ($i=1,\ldots,M$), which are the eigenvectors of the measurement covariance matrix $\boldsymbol{R}$, such that

$$\bar{\boldsymbol{h}}_j = \sum_{i=1}^{M}\tau_{j,i}\boldsymbol{e}_i, \qquad (8.39)$$

where $\tau_{j,i}$ is the expansion coefficient. We can obtain

$$\boldsymbol{R}_\varepsilon = \sum_{i=1}^{Q}\Delta\lambda_i\boldsymbol{e}_i\boldsymbol{e}_i^T + \sum_{i=Q+1}^{M}\Delta\lambda_i\boldsymbol{e}_i\boldsymbol{e}_i^T, \qquad (8.40)$$

where

$$\Delta\lambda_i = \sum_{j=Q+1}^{Q+Q_\epsilon}\delta_j\tau_{j,i}^2 \quad (i=1,\ldots,M). \qquad (8.41)$$

As we discussed in Section 2.4.2, the spans of $\{\boldsymbol{e}_1,\ldots,\boldsymbol{e}_Q\}$ and $\{\boldsymbol{e}_{Q+1},\ldots,\boldsymbol{e}_M\}$ are respectively the signal and noise subspaces of the measurement. Therefore, on the right-hand side of Eq. (8.40), the first term is the signal-subspace component and the second term is the noise-subspace component. To obtain the source reconstruction in Eq. (8.28), we need to calculate the signal-plus-sensor-noise covariance matrix. The estimate of the signal-plus-sensor-noise covariance matrix, $\widehat{\boldsymbol{R}}_{s+n}$, in this case is given by

$$\widehat{\boldsymbol{R}}_{s+n} = \boldsymbol{R}_s + \boldsymbol{R}_\varepsilon + \epsilon\boldsymbol{I} = \sum_{i=1}^{Q}(\lambda_i' + \Delta\lambda_i + \epsilon)\boldsymbol{e}_i\boldsymbol{e}_i^T + \sum_{i=Q+1}^{M}(\Delta\lambda_i + \epsilon)\boldsymbol{e}_i\boldsymbol{e}_i^T, \qquad (8.42)$$

where λ'_i is the ith eigenvalue of $\boldsymbol{R}_s$. The equation above indicates that the influence of the overestimation is mainly an increase of the diagonal-loading factor. As discussed in Section 6, a large diagonal-loading factor introduces some blur in the reconstruction results, so the overestimation of the signal-subspace dimension may cause some amount of spatial blur. The blur should not be significant as long as $\Delta\lambda_i$ is small.

8.4 Reconstruction of induced activity

8.4.1 General background

A growing interest has developed in studying so-called induced responses, which are the intensity modulation of brain activities caused by a stimulus[2]. Since the induced activity is generally frequency specific, it is appropriate to use the frequency-domain minimum-variance spatial filter described in Section 4.6 for imaging induced activities. Also, since the induced activity is not time-locked to the stimulus, the traditional waveform-based averaging averages out the activity of interest. Therefore, the sample cross-spectrum matrix should be computed from non-averaged raw epochs, such that

$$\widehat{\boldsymbol{\Gamma}}(F_w) = \frac{1}{K_E}\sum_{k=1}^{K_E}\sum_{f\in F_w} \boldsymbol{g}_k(f)(\boldsymbol{g}_k(f))^H, \tag{8.43}$$

where $\boldsymbol{g}_k(f)$ is the Fourier transform of the kth raw epoch, K_E is the total number of measured epochs, and $\sum_{f\in F_w}$ indicates the summation over a target frequency window F_w.

Since we use the non-averaged epoch data to calculate the sample covariance matrix, it naturally contains a significant amount of influence from background interference. We assume that task and control data sets are available, and, using Eq. (8.43), we calculate the frequency-specific covariance matrix for the task period, $\widehat{\boldsymbol{\Gamma}}(F_w)$, and for the control period, $\widehat{\boldsymbol{\Gamma}}_c(F_w)$. To reconstruct the induced activity, Robinson *et al.* [34] proposed one method, which is called the F-image method and is based on the subtraction between the task and control images. That is, using $\widehat{\boldsymbol{\Gamma}}(F_w)$ and $\widehat{\boldsymbol{\Gamma}}_c(F_w)$, this method first calculates the spatial filter weight such that

$$\boldsymbol{w}(\boldsymbol{r}, F_w) = \frac{\widehat{\boldsymbol{\Gamma}}_{total}^{-1}(F_w)\boldsymbol{l}(\boldsymbol{r})}{\boldsymbol{l}^T(\boldsymbol{r})\widehat{\boldsymbol{\Gamma}}_{total}^{-1}(F_w)\boldsymbol{l}(\boldsymbol{r})}, \tag{8.44}$$

where $\widehat{\boldsymbol{\Gamma}}_{total}(F_w) = \widehat{\boldsymbol{\Gamma}}(F_w) + \widehat{\boldsymbol{\Gamma}}_c(F_w)$. Then, the so-called pseudo F-image is calculated as

$$F(\boldsymbol{r}, F_w) = \frac{\langle \widehat{s}(\boldsymbol{r}, F_w)^2\rangle - \langle \widehat{s}_C(\boldsymbol{r}, F_w)^2\rangle}{\langle \widehat{s}_C(\boldsymbol{r}, F_w)^2\rangle}, \tag{8.45}$$

[2]This modulation is referred to as the event-related spectral power change. When the power change is negative, it is customarily termed an event-related desynchronization (ERD), and when it is positive, it is termed an event-related synchronization (ERS)[54].

where $\langle \widehat{s}(\boldsymbol{r}, F_w)^2 \rangle$ and $\langle \widehat{s}_C(\boldsymbol{r}, F_w)^2 \rangle$ are the power reconstruction results, which are obtained using

$$\langle \widehat{s}(\boldsymbol{r}, F_w)^2 \rangle = \boldsymbol{w}^T(\boldsymbol{r}, F_w)\widehat{\boldsymbol{\Gamma}}(F_w)\boldsymbol{w}(\boldsymbol{r}, F_w), \tag{8.46}$$

and

$$\langle \widehat{s}_C(\boldsymbol{r}, F_w)^2 \rangle = \boldsymbol{w}^T(\boldsymbol{r}, F_w)\widehat{\boldsymbol{\Gamma}}_c(F_w)\boldsymbol{w}(\boldsymbol{r}, F_w). \tag{8.47}$$

The F-image method attempts to remove the influence of the background activity based on the subtraction between the task and control images. However, the effectiveness for this removal is limited because, as described in Section 8.1.1, the influence of the background activity is not simply additive. The influence contains spatial blur, which cannot be removed by using the F-image method, as we demonstrate with numerical examples in Fig. 8.8.

8.4.2 Prewhitening method

We can use the prewhitening method to image the induced source activity[55]. However, as we pointed out, the induced activity is a stimulus-elicited intensity modulation of the source activity, and the target source activity exists both in the task and control periods, although its intensity differs between the two periods. Since the control covariance matrix contains the signal activity, it is not obvious whether the prewhitening method can be applied. To clarify this point, we momentarily use the notations for the covariance matrices defined in Chapter 8.2, and discuss the property of the prewhitening method. In this discussion, it is assumed that the control state contains the signal activity, and the covariance matrix of the signal activity in the control state is defined as $\boldsymbol{R}_s'$. Thus, the covariance matrix relationship is expressed as

$$\begin{aligned} \text{Control:} \quad \boldsymbol{R}_c &= \boldsymbol{R}_s' + \boldsymbol{R}_{i+n}, \\ \text{Task:} \quad \boldsymbol{R} &= \boldsymbol{R}_s + \boldsymbol{R}_{i+n}, \end{aligned} \tag{8.48}$$

and therefore,

$$\boldsymbol{R} = \boldsymbol{R}_s - \boldsymbol{R}_s' + \boldsymbol{R}_c. \tag{8.49}$$

We consider a general case where some signal sources have greater intensities in the control state than in the task state, but others have smaller intensities in the control state than in the task state. The power of the jth signal source in the task and the control conditions are denoted, respectively, σ_j^2 and $(\sigma_j')^2$. We assume that $\sigma_j^2 > (\sigma_j')^2$ for $j = 1, \ldots, Q_p$ and that $(\sigma_j')^2 > \sigma_j^2$ for $j = Q_p + 1, \ldots, Q$. Therefore, defining $(\Delta\sigma_j)^2 = |\sigma_j^2 - (\sigma_j')^2|$, we obtain[3]

$$\boldsymbol{R}_s - \boldsymbol{R}_s' = \sum_{j=1}^{Q_p} (\Delta\sigma_j)^2 \boldsymbol{l}(\boldsymbol{r}_j)\boldsymbol{l}^T(\boldsymbol{r}_j) - \sum_{j=Q_p+1}^{Q} (\Delta\sigma_j)^2 \boldsymbol{l}(\boldsymbol{r}_j)\boldsymbol{l}^T(\boldsymbol{r}_j) = \boldsymbol{D}_p - \boldsymbol{D}_n, \tag{8.50}$$

[3]The implicit assumption here is that sources are uncorrelated.

where

$$\boldsymbol{D}_p = \sum_{j=1}^{Q_p} (\Delta\sigma_j)^2 \boldsymbol{l}(\boldsymbol{r}_j)\boldsymbol{l}^T(\boldsymbol{r}_j), \tag{8.51}$$

and

$$\boldsymbol{D}_n = \sum_{j=Q_p+1}^{Q} (\Delta\sigma_j)^2 \boldsymbol{l}(\boldsymbol{r}_j)\boldsymbol{l}^T(\boldsymbol{r}_j). \tag{8.52}$$

This gives

$$\boldsymbol{R} = \boldsymbol{D}_p - \boldsymbol{D}_n + \boldsymbol{R}_c, \tag{8.53}$$

and thus,

$$\widetilde{\boldsymbol{R}} = \widetilde{\boldsymbol{D}}_p + \boldsymbol{I} - \widetilde{\boldsymbol{D}}_n, \tag{8.54}$$

where

$$\widetilde{\boldsymbol{D}}_p = \boldsymbol{R}_c^{-1/2}\boldsymbol{D}_p\boldsymbol{R}_c^{-1/2}, \quad \text{and} \quad \widetilde{\boldsymbol{D}}_n = \boldsymbol{R}_c^{-1/2}\boldsymbol{D}_n\boldsymbol{R}_c^{-1/2}.$$

Because both $\widetilde{\boldsymbol{D}}_p$ and $\widetilde{\boldsymbol{D}}_n$ are positive semi-definite matrices, Equation (8.54) is in principle the same as Eq. (13.70), and exactly the same arguments hold as those in Section 13.5 of the Appendix. Accordingly, the following relationship holds:

$$\boldsymbol{R}_c^{1/2}\boldsymbol{H}_S\boldsymbol{H}_S^T(\widetilde{\boldsymbol{R}} - \boldsymbol{I})\boldsymbol{R}_c^{1/2} = \boldsymbol{D}_p. \tag{8.55}$$

We now switch to the notations using the frequency-domain covariance matrices $\widehat{\boldsymbol{\Gamma}}$ and $\widehat{\boldsymbol{\Gamma}}_c$, where the explicit notation of (F_w) is omitted for simplicity. Using $\widehat{\boldsymbol{\Gamma}}_c$ as $\boldsymbol{R}_c$, and $\widehat{\widetilde{\boldsymbol{\Gamma}}}$ (obtained such that $\widehat{\widetilde{\boldsymbol{\Gamma}}} = \widehat{\boldsymbol{\Gamma}}_c^{-1/2}\widehat{\boldsymbol{\Gamma}}\widehat{\boldsymbol{\Gamma}}_c^{-1/2}$) as $\widetilde{\boldsymbol{R}}$ in Eq. (8.55), we can derive the formula to estimate $\boldsymbol{D}_p$ such that

$$\widehat{\boldsymbol{D}}_p = \widehat{\boldsymbol{\Gamma}}_c^{1/2}\widehat{\boldsymbol{H}}_S\widehat{\boldsymbol{H}}_S^H(\widehat{\widetilde{\boldsymbol{\Gamma}}} - \boldsymbol{I})\widehat{\boldsymbol{\Gamma}}_c^{1/2}, \tag{8.56}$$

where $\widehat{\boldsymbol{H}}_S = [\widehat{\boldsymbol{h}}_1, \ldots, \widehat{\boldsymbol{h}}_{Q_p}]$ is a matrix containing the Q_p signal-level eigenvectors of $\widehat{\widetilde{\boldsymbol{\Gamma}}}$. We can estimate $\boldsymbol{D}_n$ by changing the role of $\widehat{\boldsymbol{\Gamma}}$ and $\widehat{\boldsymbol{\Gamma}}_c$. That is, we first calculate $\widehat{\widetilde{\boldsymbol{\Gamma}}}_c$ such that $\widehat{\widetilde{\boldsymbol{\Gamma}}}_c = \widehat{\boldsymbol{\Gamma}}^{-1/2}\widehat{\boldsymbol{\Gamma}}_c\widehat{\boldsymbol{\Gamma}}^{-1/2}$, and then obtain an estimate of $\boldsymbol{D}_n$ using

$$\widehat{\boldsymbol{D}}_n = \widehat{\boldsymbol{\Gamma}}^{1/2}\widehat{\boldsymbol{H}}_S^c(\widehat{\boldsymbol{H}}_S^c)^H(\widehat{\widetilde{\boldsymbol{\Gamma}}}_c - \boldsymbol{I})\widehat{\boldsymbol{\Gamma}}^{1/2}. \tag{8.57}$$

In the equation above, $\widehat{\boldsymbol{H}}_S^c$ is defined as $\widehat{\boldsymbol{H}}_S^c = [\widehat{\boldsymbol{h}}_1^c, \ldots, \widehat{\boldsymbol{h}}_{Q_n}^c]$, which is a matrix containing the Q_n (where $Q_n = Q - Q_p$) signal-level eigenvectors of $\widehat{\widetilde{\boldsymbol{\Gamma}}}_c$. Here, the prewhitening method, in which the roles of $\widehat{\boldsymbol{\Gamma}}$ and $\widehat{\boldsymbol{\Gamma}}_c$ are reversed, is referred to as the "flipped" prewhitening method. Therefore, the sources that are stronger in the task state than in the control state can be reconstructed using

$$\langle \widehat{s}_p(\boldsymbol{r}, F_w)^2 \rangle = \frac{1}{\boldsymbol{l}^T(\boldsymbol{r})(\widehat{\boldsymbol{D}}_p(F_w) + \epsilon\boldsymbol{I})^{-1}\boldsymbol{l}(\boldsymbol{r})}. \tag{8.58}$$

The sources that are stronger in the control state than in the task state can be reconstructed using the flipped prewhitening method, i.e.,

$$\langle \widehat{s}_n(\boldsymbol{r}, F_w)^2 \rangle = \frac{1}{\boldsymbol{l}^T(\boldsymbol{r})(\widehat{\boldsymbol{D}}_n(F_w) + \epsilon \boldsymbol{I})^{-1}\boldsymbol{l}(\boldsymbol{r})}. \tag{8.59}$$

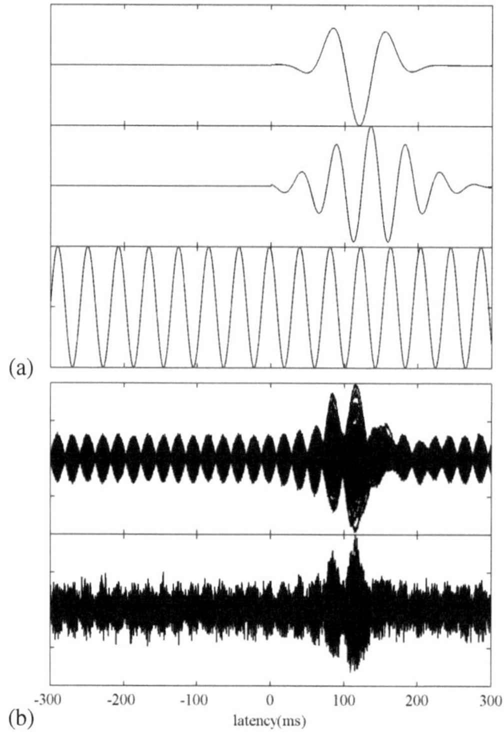

Figure 8.3: (a) Time courses of the three sources assumed for the numerical experiments. The time courses in the panels from top to bottom were assigned to the first to third sources, respectively. (b) Generated magnetic recordings obtained when no background sources exist (upper panel), and when one-hundred background sources exist (lower panel).

8.5 Numerical examples

Numerical experiments were performed to illustrate the results of our arguments on the prewhitening spatial filter. The same source-sensor configuration and the coordinate system as illustrated in Fig. 4.2 are again used except the third source location was set to $(0, 1.2, -6.4)$ cm. The time courses assigned to the three sources are shown in Fig. 8.3(a). They are basically the same as those in Fig. 4.3(a), but they have a prestimulus period, and the third source is active throughout both the pre- and post-stimulus periods. The simulated sensor recordings are calculated from -300 ms to 300 ms with the sampling frequency assumed to equal 16 kHz. This results in a total 2400 time points in the whole recordings between -300 ms to 300 ms. The data portion between -300 ms and 0 ms is considered the prestimulus (control) period, and that between 0 ms and 300 ms the post-stimulus (task) period. A small amount of white Gaussian noise to simulate the sensor noise is added to the generated recordings, resulting in the total input SNR ($\alpha_T = \alpha_1 + \alpha_2 + \alpha_3$) equal to 16. The generated sensor recordings, $\boldsymbol{b}_s(t) + \boldsymbol{n}(t)$, are shown in the upper panel of Fig. 8.3(b). We then generated the simulated background interference, $\boldsymbol{b}_I(t)$. We used one hundred background sources with random locations and random orientations. Each of these isotropic interference sources has a random incoherent time course. The background interference, $\boldsymbol{b}_I(t)$, is added to $\boldsymbol{b}_s(t) + \boldsymbol{n}(t)$ and the resultant simulated sensor data, $\boldsymbol{b}(t)$, are shown in the lower panel of Fig. 8.3(b). Here the resultant interference-to-signal ratio $\langle \boldsymbol{b}_s(t)^2 \rangle / \langle \boldsymbol{b}_I(t)^2 \rangle$ was set to 2.

The conventional eigenspace spatial filter was first applied to these two sets of simulated recordings in Fig. 8.3(b). The sample covariance matrix was calculated using the post-stimulus period with 1200 time points, and then the source-power reconstruction $\langle \widehat{s}(\boldsymbol{r}, t)^2 \rangle$ was obtained. Figure 8.4(a) shows the results for the case with no background interference, and Figure 8.4(b) shows the results for the case with background interference present. Here, the comparison between these two sets of results again confirms that background interference causes a severe blur in the reconstruction results.

We then applied the prewhitening spatial filter in Eq. (8.28) to the data with the background interference. The results are shown in Fig. 8.4(c). In this application, $\boldsymbol{R}_{i+n}$ is obtained using the whole pre-stimulus period, and the signal subspace dimension Q was set to 2. In the results in Fig. 8.4(c), the blur due to the background activity is significantly reduced relative to the results shown in Fig. 8.4(b). Also, the third source is not reconstructed because it was also active during the prestimulus period. These results clearly demonstrate the effectiveness of the proposed spatial filter. Figure 8.5 shows the reconstructed time courses of the three sources. In the conventional reconstruction in (b), the reconstructed time course of the second source shows some influence from the third source. In other words, there is leakage between the second and the third source time course estimates. In the prewhitening reconstruction in (c), the influence of the third source on the second source time course is removed.

Next, we tested the robustness of the prewhitening method when the signal-subspace dimension, Q, is overestimated. The power reconstruction obtained with Q set to 5 is shown in Fig. 8.4(d). The time-course estimate obtained with Q set to 5 is shown in Fig. 8.5(d). In these results, despite the fact that Q is significantly overestimated, the first and second sources are successfully reconstructed. Here, the overestimation causes only a slight increase of the spatial blur and a slight decrease of the output SNR.

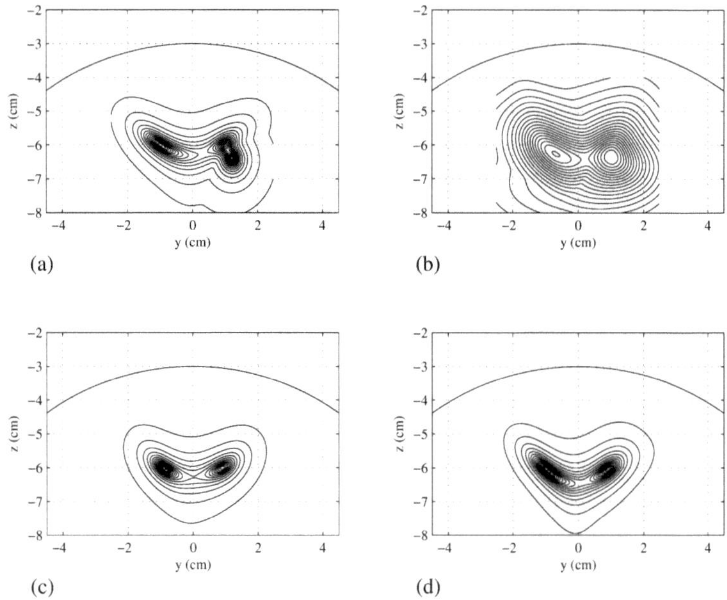

Figure 8.4: Results of the power reconstruction, $\langle \widehat{s}(\boldsymbol{r},t)^2 \rangle$. (a) Conventional eigenspace spatial filter was applied to the recordings with no background interference. (b) Conventional eigenspace spatial filter was applied to the recordings with background interference. (c) Prewhitening spatial filter (Eq. (8.28)) was applied to the recordings with background interference. The signal subspace dimension Q was set to 2. (d) Prewhitening spatial filter (Eq. (8.28)) was applied to the recordings with background interference. The signal subspace dimension Q was set to 5.

Numerical examples are presented for imaging induced activities. In this experiment, we used the sensor alignment of the 275-sensor array from the Omega^{TM} (VMS Medtech, Coquitlam, Canada) neuromagnetometer. This was because this experiment called for real, spontaneous MEG data from this system. Three sources were assumed to exist on a single plane ($x = 0$ cm), and their (y, z) coordinates were $(-2.1, 9.5)$ cm, $(2.6, 10.5)$ cm, and $(1.4, 7.5)$ cm, respectively. The sphere origin was set to $(0, 0, 4)$ cm. The sources and the coordinate system used in this experiment are shown schematically in Fig. 8.6. The powers of the three sources were set equal in the sensor-domain, i.e., $\sigma_1^2\|\boldsymbol{l}(\boldsymbol{r}_1)\|^2 = \sigma_2^2\|\boldsymbol{l}(\boldsymbol{r}_2)\|^2 = \sigma_3^2\|\boldsymbol{l}(\boldsymbol{r}_3)\|^2$, where $\boldsymbol{r}_j$, σ_j^2, and $\boldsymbol{l}(\boldsymbol{r}_j)$ are the location, power, and lead-field vector of the jth source, respectively.

Multi-epoch measurements were simulated. Each epoch consists of two sets of data: the task and the control data sets. Uncorrelated sinusoidal time courses for the three sources were assumed, and the frequency was set to 10 Hz for the first

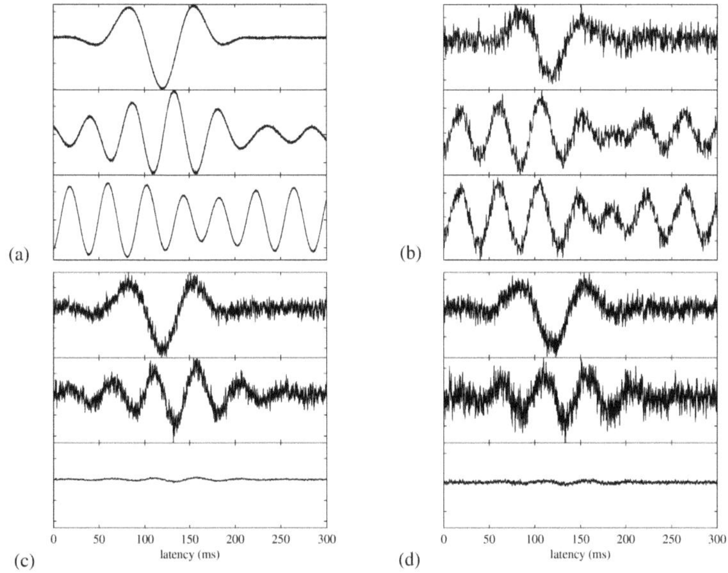

Figure 8.5: Results of the reconstructed time courses at the three source locations. (a) Conventional eigenspace spatial filter was applied to the recordings with no background interference. (b) Conventional eigenspace spatial filter was applied to the recordings with background interference. (c) Prewhitening spatial filter (Eq. (8.36)) was applied to the recordings with background interference. The signal subspace dimension was set to 2. (d) Prewhitening spatial filter (Eq. (8.36)) was applied. The signal subspace dimension Q was set to 5.

source, 16 Hz for the second source, and 28 Hz for the third source. The power spectra of these three source time courses are shown in Fig. 8.7. These sinusoidal time courses have phase offsets that were generated using a uniformly-distributed random number between 0 and 2π and different random numbers were used across sources, epochs, and the two conditions. We therefore simulate induced source activities, which are elicited by the stimulus but not phase-locked to it. Real, spontaneous MEG was used as interference, and the signal-to-interference ratio (SIR) was set to 0.3. The intensity of the first source was decreased by 30% from the control to the task conditions, while the intensity of the second source was increased by 30% from the control to the task conditions. The intensity of the third source remained the same between the two conditions.

The results of the source reconstruction experiments are shown in Fig. 8.8. Here we selected the frequency window between 8 and 18 Hz (shown by the vertical broken lines in Fig. 8.7), which does not contain the activity of the third source. The sample cross-spectral matrices are computed using the frequency components in this frequency window. In the pseudo-F image in Fig. 8.8 (a), the second source

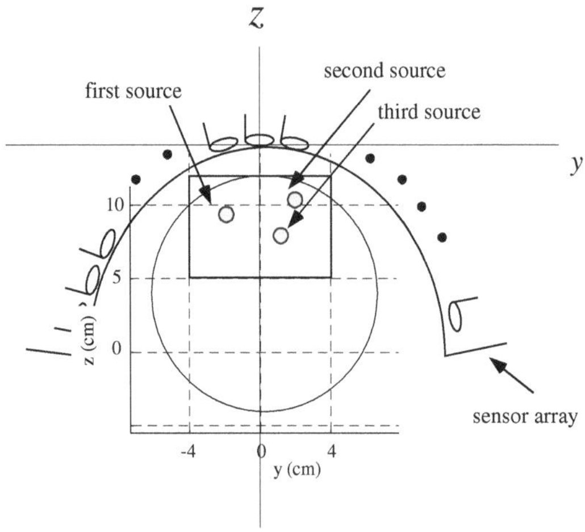

Figure 8.6: The geometry of the computer simulation used for imaging induced activities. Note that, because the subject-specific head coordinate system is used, the z values of the sources are significantly different from the z values used in the numerical experiments in the previous chapters.

forms a positive peak, and the first source forms a negative peak. Although these peaks are slightly blurred, the pseudo-F image detects these two sources. Next, the methods of prewhitening and flipped prewhitening source reconstruction were applied, and the results are shown in Fig. 8.8 (b) and (c), respectively. In these results, the first and the second sources form clear peaks at the correct locations of these sources.

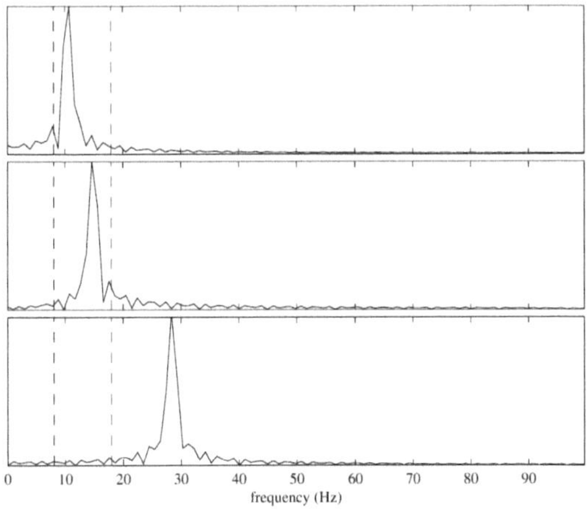

Figure 8.7: Power spectra of the time courses of the first source(top), the second source(middle), and the third source(bottom). The vertical broken lines indicate the frequency window between 8 and 18 Hz.

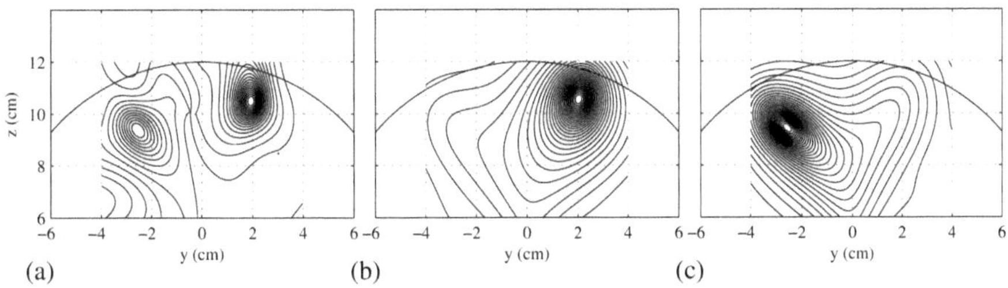

Figure 8.8: Results of the source reconstruction experiments. (a) Results of the F-image method. (b) Results of the prewhitening spatial filter. (c) Results of the flipped prewhitening spatial filter.

Chapter 9

Effects of source correlation

As discussed in Section 4.2, one of the fundamental assumptions of the adaptive spatial filter formulation is that source activities are uncorrelated. However, brain sources are inevitably correlated to some degree, and in a strict sense, the above assumption cannot be satisfied. This chapter investigates the influence of temporally correlated sources on adaptive-spatial-filter source imaging. We discuss how the source correlation affects the output of the spatial filter, and how robust the adaptive spatial filter is to source correlation. We also present the LCMV spatial filter, which can reduce the influence of the source correlation by imposing null sensitivity over an extended region in which correlated interference exists. Finally, we briefly describe the source-coherence imaging, which can provide mapping of the functional connectivity between different brain regions.

9.1 Performance of adaptive spatial filters in the presence of correlated sources

Let us assume that a total of Q sources exist at $\boldsymbol{r}_1, \boldsymbol{r}_2, \ldots, \boldsymbol{r}_Q$, and denote their time courses as $s(\boldsymbol{r}_1, t), s(\boldsymbol{r}_2, t), \ldots, s(\boldsymbol{r}_Q, t)$. If these Q sources are uncorrelated, the blocking capability of the minimum-variance spatial filters is expressed as

$$\boldsymbol{w}^T(\boldsymbol{r}_p)\boldsymbol{l}(\boldsymbol{r}_q) = \tau\delta_{p,q}, \tag{9.1}$$

where τ is a constant that is determined by the constraint, as discussed in Eq. (4.27). In general cases where partially correlated sources exist, the blocking capability of the spatial filter weight is expressed as

$$\boldsymbol{w}^T(\boldsymbol{r}_p)\boldsymbol{l}(\boldsymbol{r}_q) = \tau\frac{[\boldsymbol{R}_\nu^{-1}]_{p,q}}{[\boldsymbol{R}_\nu^{-1}]_{p,p}}, \tag{9.2}$$

where $\boldsymbol{R}_\nu$ indicates the source covariance matrix defined in Eq. (2.35) and $[\boldsymbol{R}_\nu^{-1}]_{p,q}$ indicates the (p, q) element of $\boldsymbol{R}_\nu^{-1}$. Equation (9.2), derived in Section 4.2.2, is the basis of the analysis in this chapter. In the following discussion, we assume for simplicity that the unit-gain minimum-variance filter is used and τ is set to 1. It is easy to extend the discussion to cases where τ has other values.

We assume that a target source exists at $\boldsymbol{r}_p$ and that the Q_I sources are correlated with the target source. (A total of Q_I+1 sources are mutually correlated.) The spatial filter output $\widehat{s}(\boldsymbol{r}_p,t)$, is expressed as[56]

$$\widehat{s}(\boldsymbol{r}_p,t)=s(\boldsymbol{r}_p,t)+\sum_{q=1}^{Q_I}\frac{[\boldsymbol{R}_\nu^{-1}]_{p,q}}{[\boldsymbol{R}_\nu^{-1}]_{p,p}}s(\boldsymbol{r}_q,t), \tag{9.3}$$

where the locations of the correlated interferences are denoted $\boldsymbol{r}_q$ ($q=1,\ldots,Q_I$). This equation shows that the spatial filter outputs for the correlated sources contain leakage from the other correlated sources, and such leakage causes errors in the time-course estimates of the source activities.

We consider a simple case where $Q_I=1$. That is, among Q sources, the first and the second sources are correlated and the other sources have no correlation with any other sources. We define the correlation coefficient between the first and second sources as μ, and the average power of the first and the second sources as σ_1^2 and σ_2^2, respectively. The source covariance matrix can then be expressed as

$$\boldsymbol{R}_\nu=\begin{bmatrix}\begin{bmatrix}\sigma_1^2 & \mu\sigma_1\sigma_2\\ \mu\sigma_1\sigma_2 & \sigma_2^2\end{bmatrix} & 0 & \cdots & 0\\ 0 & \sigma_3^2 & \cdot & \vdots\\ \vdots & \cdot & \ddots & 0\\ 0 & \cdots & 0 & \sigma_Q^2\end{bmatrix}, \tag{9.4}$$

and its inverse $\boldsymbol{R}_\nu^{-1}$ is expressed as

$$\boldsymbol{R}_\nu^{-1}=\begin{bmatrix}\frac{1}{\sigma_1^2\sigma_2^2(1-\mu^2)}\begin{bmatrix}\sigma_2^2 & -\mu\sigma_1\sigma_2\\ -\mu\sigma_1\sigma_2 & \sigma_1^2\end{bmatrix} & 0 & \cdots & 0\\ 0 & 1/\sigma_3^2 & \cdot & \vdots\\ \vdots & \cdot & \ddots & 0\\ 0 & \cdots & 0 & 1/\sigma_Q^2\end{bmatrix}. \tag{9.5}$$

Using Eqs. (9.2) and (9.5), we can derive

$$\begin{aligned}&\boldsymbol{w}(\boldsymbol{r}_1)\boldsymbol{l}(\boldsymbol{r}_1)=1, && \boldsymbol{w}(\boldsymbol{r}_1)\boldsymbol{l}(\boldsymbol{r}_2)=-\frac{\sigma_1}{\sigma_2}\mu\\ &\boldsymbol{w}(\boldsymbol{r}_2)\boldsymbol{l}(\boldsymbol{r}_1)=-\frac{\sigma_2}{\sigma_1}\mu, && \boldsymbol{w}(\boldsymbol{r}_2)\boldsymbol{l}(\boldsymbol{r}_2)=1,\end{aligned} \tag{9.6}$$

and

$$\boldsymbol{w}(\boldsymbol{r}_1)\boldsymbol{l}(\boldsymbol{r}_q)=0,\quad \boldsymbol{w}(\boldsymbol{r}_2)\boldsymbol{l}(\boldsymbol{r}_q)=0,\quad \text{for}\quad q=3,\ldots,Q. \tag{9.7}$$

Therefore, when the spatial filter is pointing at the first source location $\boldsymbol{r}_1$, it passes the signal from the second source with the gain of $-\sigma_1\mu/\sigma_2$, while blocking the signals from the other sources uncorrelated with the first source. Consequently, the spatial filter output at $\boldsymbol{r}_1$, $\widehat{s}(\boldsymbol{r}_1,t)$, is given by:

$$\widehat{s}(\boldsymbol{r}_1,t) = s(\boldsymbol{r}_1,t) - (\frac{\sigma_1}{\sigma_2}\mu)s(\boldsymbol{r}_2,t). \tag{9.8}$$

In exactly the same manner, the spatial filter output at $\boldsymbol{r}_2$, $\widehat{s}(\boldsymbol{r}_2,t)$, is given by:

$$\widehat{s}(\boldsymbol{r}_2,t) = s(\boldsymbol{r}_2,t) - (\frac{\sigma_2}{\sigma_1}\mu)s(\boldsymbol{r}_1,t). \tag{9.9}$$

These equations explicitly show that the spatial filter output for the first source contains leakage from the second source and that the output for the second source contains the leakage from the first source. Equations (9.8) and (9.9) are valid for high SNR, but they still hold for considerably lower SNR, as is shown in our numerical experiments.

9.2 Signal cancellation and estimation of source correlation

Leakage from correlated activities not only causes errors in the time-course estimate but also causes intensity reduction of the reconstructed sources as shown below. Using Eqs. (9.8) and (9.9) with the relationship

$$\langle s(\boldsymbol{r}_1,t)s(\boldsymbol{r}_2,t)\rangle = \mu\sigma_1\sigma_2,$$

we obtain

$$\langle \widehat{s}(\boldsymbol{r}_1,t)^2\rangle = \sigma_1^2(1-\mu^2), \tag{9.10}$$
$$\langle \widehat{s}(\boldsymbol{r}_2,t)^2\rangle = \sigma_2^2(1-\mu^2), \tag{9.11}$$

and

$$\langle \widehat{s}(\boldsymbol{r}_1,t)\widehat{s}(\boldsymbol{r}_2,t)\rangle = \sigma_1\sigma_2(\mu^3-\mu). \tag{9.12}$$

Equations (9.10) and (9.11) indicate that the power of the reconstructed sources is reduced by a factor of $1-\mu^2$. This reduction of reconstructed source power due to the source correlation is known as signal cancellation[57][58].

Let us estimate the magnitude of correlation coefficient using the spatial filter outputs. The estimated correlation coefficient between the first and second sources is denoted $\widehat{\mu}$, which is given by:

$$\widehat{\mu} = \frac{\langle \widehat{s}(\boldsymbol{r}_1,t)\widehat{s}(\boldsymbol{r}_2,t)\rangle}{\sqrt{\langle \widehat{s}(\boldsymbol{r}_1,t)^2\rangle\langle \widehat{s}(\boldsymbol{r}_2,t)^2\rangle}}. \tag{9.13}$$

Substituting Eqs. (9.12), (9.10) and (9.11) into Eq. (9.13), we finally obtain

$$|\widehat{\mu}| = \frac{|\sigma_1\sigma_2(\mu^3-\mu)|}{\sqrt{\sigma_1^2(1-\mu^2)\sigma_2^2(1-\mu^2)}} = |\mu|. \tag{9.14}$$

This equation indicates that the correlation coefficient magnitude can be accurately estimated by directly using the spatial filter outputs with Eq. (9.13). This is somewhat surprising because the adaptive spatial filter assumes that the source activities are uncorrelated, so we intuitively think that the output is erroneous if the sources are correlated. However, in the calculation of the correlation coefficient, these errors cancel, and the correct (magnitude) correlation coefficient can be retrieved. It should be noted that the above analysis is valid only when two sources are correlated.

If a third correlated source exists, the accuracy of the estimated correlation coefficient is affected by this source. This influence can be evaluated in the following manner. We assume that the first, second, and third sources are mutually correlated, and that their correlation coefficients are denoted μ_{12}, μ_{13}, and μ_{23}. We further assume that the correlation between the first and second sources is the target of the measurement, and that the third source is interference. According to Eq. (9.3), denoting the location of the third source as $\boldsymbol{r}_3$, the spatial filter outputs at the target source locations $\boldsymbol{r}_1$ and $\boldsymbol{r}_2$ are given by

$$\widehat{s}(\boldsymbol{r}_1,t) = \frac{1}{[\bar{\boldsymbol{R}}_\nu^{-1}]_{1,1}} \left[[\bar{\boldsymbol{R}}_\nu^{-1}]_{1,1} s(\boldsymbol{r}_1,t) + [\bar{\boldsymbol{R}}_\nu^{-1}]_{1,2} s(\boldsymbol{r}_2,t) + [\bar{\boldsymbol{R}}_\nu^{-1}]_{1,3} s(\boldsymbol{r}_3,t) \right], \tag{9.15}$$

and

$$\widehat{s}(\boldsymbol{r}_2,t) = \frac{1}{[\bar{\boldsymbol{R}}_\nu^{-1}]_{2,2}} \left[[\bar{\boldsymbol{R}}_\nu^{-1}]_{2,1} s(\boldsymbol{r}_1,t) + [\bar{\boldsymbol{R}}_\nu^{-1}]_{2,2} s(\boldsymbol{r}_2,t) + [\bar{\boldsymbol{R}}_\nu^{-1}]_{2,3} s(\boldsymbol{r}_3,t) \right], \tag{9.16}$$

where the matrix $\bar{\boldsymbol{R}}_\nu$ is the part of the source covariance matrix that is related to these three sources. This submatrix is given by

$$\bar{\boldsymbol{R}}_\nu = \begin{bmatrix} \sigma_1^2 & \mu_{12}\sigma_1\sigma_2 & \mu_{13}\sigma_1\sigma_3 \\ \mu_{12}\sigma_1\sigma_2 & \sigma_2^2 & \mu_{23}\sigma_2\sigma_3 \\ \mu_{13}\sigma_1\sigma_3 & \mu_{23}\sigma_2\sigma_3 & \sigma_3^2 \end{bmatrix}, \tag{9.17}$$

and, its inverse is expressed as

$$\bar{\boldsymbol{R}}_\nu^{-1} = \frac{1}{|\bar{\boldsymbol{R}}_\nu|} \begin{bmatrix} \sigma_2^2\sigma_3^2(1-\mu_{23}^2) & \sigma_1\sigma_2\sigma_3^2(\mu_{13}\mu_{23}-\mu_{12}) & \sigma_1\sigma_2^2\sigma_3(\mu_{12}\mu_{23}-\mu_{13}) \\ \sigma_1\sigma_2\sigma_3^2(\mu_{13}\mu_{23}-\mu_{12}) & \sigma_1^2\sigma_3^2(1-\mu_{13}^2) & \sigma_1^2\sigma_2\sigma_3(\mu_{12}\mu_{13}-\mu_{23}) \\ \sigma_1\sigma_2^2\sigma_3(\mu_{12}\mu_{23}-\mu_{13}) & \sigma_1^2\sigma_2\sigma_3(\mu_{12}\mu_{13}-\mu_{23}) & \sigma_1^2\sigma_2^2(1-\mu_{12}^2) \end{bmatrix}, \tag{9.18}$$

where $|\bar{\boldsymbol{R}}_\nu|$ is the determinant of $\bar{\boldsymbol{R}}_\nu$ expressed as

$$|\bar{\boldsymbol{R}}_\nu| = \sigma_1^2\sigma_2^2\sigma_3^2(1-\mu_{12}^2-\mu_{13}^2-\mu_{23}^2+2\mu_{12}\mu_{13}\mu_{23}). \tag{9.19}$$

Using Eqs. (9.13), (9.15), (9.16), and (9.18), we finally obtain

$$|\widehat{\mu}_{12}| = \frac{|\mu_{12}-\mu_{13}\mu_{23}|}{\sqrt{(1-\mu_{13}^2)(1-\mu_{23}^2)}}. \tag{9.20}$$

This equation shows how the values of μ_{13} and μ_{23} affect $\widehat{\mu}_{12}$. Clearly, when μ_{13} and μ_{23} are small, $\widehat{\mu}_{12}$ is close to $|\mu_{12}|$.

9.3 Suppression of coherent interferences using the LCMV spatial filter

9.3.1 Weight-vector derivation

As described in Section 4.4, the LCMV spatial filter was originally developed to suppress the influence of highly correlated interferences. The LCMV spatial filter is based on the assumption that the locations of the coherent interferences are known. Let us denote the locations of coherent interferences $\boldsymbol{r}_1, \ldots, \boldsymbol{r}_d$. If these locations are accurately known, Eq. (4.49) gives a weight vector that sets the null sensitivity over these locations. As a result, the spatial filter outputs are free from the leakage that normally occurs when the sources are correlated. In actual applications, however, it is unlikely that we will accurately know the locations of the coherent interferences. Instead, it may be possible to know the approximate locations or some extended region in which coherent interferences could exist. Thus, Equation (4.49) needs to be extended to set the null sensitivity over such an extended region.

This extension can be performed in the following brute-force manner. Let us define Ω_R to be the local region in which the coherent interference sources exist. Then, we divide Ω_R into a total of J voxels whose distances should be comparable to the spatial resolution of the spatial filter. Let us define the locations of these voxels as

$$\boldsymbol{r}_{\Omega_1}, \boldsymbol{r}_{\Omega_2}, \ldots, \boldsymbol{r}_{\Omega_J},$$

and also define an $M \times 3J$ matrix $\boldsymbol{C}_\Omega$ such that

$$\boldsymbol{C}_\Omega = [\boldsymbol{L}(\boldsymbol{r}_{\Omega_1}), \boldsymbol{L}(\boldsymbol{r}_{\Omega_2}), \ldots, \boldsymbol{L}(\boldsymbol{r}_{\Omega_J})]. \tag{9.21}$$

Then, using Eq. (4.49) the weight vector of the LCMV spatial filter is obtained using

$$\boldsymbol{w}(\boldsymbol{r}) = \underset{\boldsymbol{w}(\boldsymbol{r})}{\arg\min}\, \boldsymbol{w}^T(\boldsymbol{r})\boldsymbol{R}\boldsymbol{w}(\boldsymbol{r}), \quad \text{subject to} \quad \boldsymbol{l}^T(\boldsymbol{r})\boldsymbol{w}(\boldsymbol{r}) = 1,$$
$$\text{and} \quad \boldsymbol{C}_\Omega^T\boldsymbol{w}(\boldsymbol{r}) = \begin{bmatrix} 0 \\ \vdots \\ 0 \end{bmatrix}. \tag{9.22}$$

Defining an $M \times (3J+1)$ matrix $\boldsymbol{C}$ such that

$$\boldsymbol{C} = [\boldsymbol{l}(\boldsymbol{r}), \boldsymbol{C}_\Omega],$$

and a $(3J+1) \times 1$ column vector $\boldsymbol{c}$ such that

$$\boldsymbol{c} = \begin{bmatrix} 1 \\ 0 \\ \vdots \\ 0 \end{bmatrix}, \tag{9.23}$$

the resultant weight vector can be expressed as

$$\boldsymbol{w}(\boldsymbol{r}) = \boldsymbol{R}^{-1}\boldsymbol{C}[\boldsymbol{C}^T\boldsymbol{R}^{-1}\boldsymbol{C}]^{-1}\boldsymbol{c}. \tag{9.24}$$

In the weight equation described above, the number of voxels J can be large. However, using a large value for J generally consumes degrees of freedom, defined as $M - 3J - 1$[7]. That is, if a large value of J is used, the degree of freedom, which needs to be sufficiently large in order to formulate the weight vector, approaches zero, and if the degree of freedom is nearly zero, the matrix $[\boldsymbol{C}^T\boldsymbol{R}^{-1}\boldsymbol{C}]$ becomes close to a singular matrix and calculation of its inverse becomes error-prone. Therefore, the weight should be derived from a smaller number of constraints, which still maintain the null sensitivity over Ω_R. Such a weight can be derived in the following manner[59].

We first apply the singular-value decomposition to $\boldsymbol{C}_\Omega$, i.e.,

$$\boldsymbol{C}_\Omega = [\boldsymbol{x}_1, \ldots, \boldsymbol{x}_M] \begin{bmatrix} \varphi_1 & \cdot & \cdots & 0 \\ 0 & \varphi_2 & \cdots & 0 \\ \vdots & \cdot & \ddots & 0 \\ 0 & \cdot & \cdots & \varphi_M \end{bmatrix} [\boldsymbol{y}_1, \ldots, \boldsymbol{y}_M]^T, \tag{9.25}$$

where we assume that the number of sensors M is smaller than $3J$. In the above equation, $\boldsymbol{x}_j$ is an $M \times 1$ singular vector, $\boldsymbol{y}_j$ is a $3J \times 1$ singular vector, and φ_j is the jth singular value. (The singular values are ordered in a decreasing manner.) We then assume that the first D singular values are large compared to the other singular values. Defining $\boldsymbol{\Lambda}_D$ as a diagonal matrix that contains the largest D singular values in its diagonal elements, we get the relationship

$$\boldsymbol{C}_\Omega \approx [\boldsymbol{x}_1, \ldots, \boldsymbol{x}_D]\boldsymbol{\Lambda}_D[\boldsymbol{y}_1, \ldots, \boldsymbol{y}_D]^T. \tag{9.26}$$

Thus, combining Eq. (9.26) with the constraint equation $\boldsymbol{C}_\Omega^T\boldsymbol{w}(\boldsymbol{r}) = 0$ leads to

$$[\boldsymbol{y}_1, \ldots, \boldsymbol{y}_D]\boldsymbol{\Lambda}_D[\boldsymbol{x}_1, \ldots, \boldsymbol{x}_D]^T\boldsymbol{w}(\boldsymbol{r}) \approx \boldsymbol{C}_\Omega^T\boldsymbol{w}(\boldsymbol{r}) = 0. \tag{9.27}$$

Thus, we obtain the relationship

$$[\boldsymbol{x}_1, \ldots, \boldsymbol{x}_D]^T\boldsymbol{w}(\boldsymbol{r}) \approx 0. \tag{9.28}$$

In the above equation, the number of relatively large singular values, D, is generally much smaller than $3J$. The resultant LCMV spatial filter weight vector is expressed as

$$\boldsymbol{w}(\boldsymbol{r}) = \boldsymbol{R}^{-1}\boldsymbol{C}[\boldsymbol{C}^T\boldsymbol{R}^{-1}\boldsymbol{C}]^{-1}\boldsymbol{c}, \tag{9.29}$$

where $\boldsymbol{C}$ is an $M \times (D+1)$ matrix, such that

$$\boldsymbol{C} = [\boldsymbol{l}(\boldsymbol{r}), \boldsymbol{x}_1, \ldots, \boldsymbol{x}_D],$$

and $\boldsymbol{c}$ is a $(D+1) \times 1$ column vector, such that

$$\boldsymbol{c} = \begin{bmatrix} 1 \\ 0 \\ \vdots \\ 0 \end{bmatrix}.$$

The question here is: how can we determine D, the threshold for truncating the singular values? The discussion above regarding the degrees of freedom requires that D should be as small as possible. On the other hand, the truncated SVD matrix in Eq. (9.26) must approximate the matrix $\boldsymbol{C}_\Omega$. Let us denote the error of this approximation ε_D, which is given by:

$$\varepsilon_D = \|\boldsymbol{C}_\Omega - [\boldsymbol{x}_1, \ldots, \boldsymbol{x}_D]\boldsymbol{\Lambda}_D[\boldsymbol{y}_1, \ldots, \boldsymbol{y}_D]^T\|^2 = \sum_{j=D+1}^{M} \varphi_j^2. \tag{9.30}$$

The above equation indicates that small values of D should increase the approximation error. To reduce this error, a larger D should be used. Therefore, we should compromise between the amount of ε_D and the degree of freedom. Unfortunately, there is no rigorous method for this determination and the optimal D should be empirically determined. In addition, to determine the optimal D, a condition for the eigenspace projection described in the next section should also be taken into consideration.

9.3.2 Extension to eigenspace-projected spatial filter

The eigenspace projection described in Section 6.8 cannot be directly applied to the weight of the LCMV spatial filter in Eq. (9.29). This is because the null constraint,

$$\boldsymbol{w}^T(\boldsymbol{r})[\boldsymbol{x}_1, \ldots, \boldsymbol{x}_D] = 0 \tag{9.31}$$

is not preserved through the eigenspace projection. In Section 6.8, we discuss that the eigenspace-projection spatial filter uses the signal-subspace component of the weight vector obtained by $\boldsymbol{E}_S\boldsymbol{E}_S^T\boldsymbol{w}(\boldsymbol{r})$, where $\boldsymbol{E}_S = [\boldsymbol{e}_1, \ldots, \boldsymbol{e}_Q]$ and $\boldsymbol{e}_1, \ldots, \boldsymbol{e}_Q$ are the signal-level eigenvectors of the measurement covariance matrix $\boldsymbol{R}$. However, multiplying the projection operator by the LCMV weight vector in Eq. (9.29) gives

$$(\boldsymbol{E}_S\boldsymbol{E}_S^T\boldsymbol{w}(\boldsymbol{r}))^T[\boldsymbol{x}_1, \ldots, \boldsymbol{x}_D] = \boldsymbol{w}^T(\boldsymbol{r})\boldsymbol{E}_S\boldsymbol{E}_S^T[\boldsymbol{x}_1, \ldots, \boldsymbol{x}_D], \tag{9.32}$$

and since the column vectors, $\boldsymbol{x}_1, \ldots, \boldsymbol{x}_D$, are not necessarily contained in the signal subspace of $\boldsymbol{R}$, multiplying $\boldsymbol{E}_S\boldsymbol{E}_S^T$ to $\boldsymbol{x}_1, \ldots, \boldsymbol{x}_D$ changes these vectors. As a result, we have

$$\boldsymbol{w}^T(\boldsymbol{r})\boldsymbol{E}_S\boldsymbol{E}_S^T[\boldsymbol{x}_1, \ldots, \boldsymbol{x}_D] \neq 0.$$

That is, applying the signal-subspace projector $\boldsymbol{E}_S\boldsymbol{E}_S^T$ does not preserve the null constraints.

One way to solve this problem is to define a modified signal subspace that also contains the subspace spanned by the column vectors $\boldsymbol{x}_1, \ldots, \boldsymbol{x}_D$. We then use the projector onto this modified signal subspace to derive the eigenspace-projection spatial filter[49]. The modified signal subspace is defined as a column span of a matrix,

$$\bar{\boldsymbol{E}}_S = [\boldsymbol{e}_1, \ldots, \boldsymbol{e}_Q, \bar{\boldsymbol{x}}_1, \ldots, \bar{\boldsymbol{x}}_D], \tag{9.33}$$

where the column vectors $\bar{\boldsymbol{x}}_1, \ldots, \bar{\boldsymbol{x}}_D$ are obtained by applying the Gram-Schmidt orthogonalization onto the column vectors $\boldsymbol{e}_1, \ldots, \boldsymbol{e}_Q, \boldsymbol{x}_1, \ldots, \boldsymbol{x}_D$. Therefore, the column vectors of $\bar{\boldsymbol{E}}_S$ form an orthonormal basis. The application of the projection operator $\bar{\boldsymbol{E}}_S \bar{\boldsymbol{E}}_S^T$ onto the LCMV weight vector results in

$$\begin{aligned}
&[\bar{\boldsymbol{E}}_S \bar{\boldsymbol{E}}_S^T \boldsymbol{w}(\boldsymbol{r})]^T [\boldsymbol{x}_1, \ldots, \boldsymbol{x}_D] \\
&\qquad = \boldsymbol{w}^T(\boldsymbol{r}) \bar{\boldsymbol{E}}_S \bar{\boldsymbol{E}}_S^T [\boldsymbol{x}_1, \ldots, \boldsymbol{x}_D] = \boldsymbol{w}^T(\boldsymbol{r}) [\boldsymbol{x}_1, \ldots, \boldsymbol{x}_D] = 0,
\end{aligned} \tag{9.34}$$

because the vectors, $\boldsymbol{x}_1, \ldots, \boldsymbol{x}_D$, lie in the column span of $\bar{\boldsymbol{E}}_S$. Thus, the weight vector of this modified eigenspace LCMV spatial filter preserves the null constraints. However, note that the dimensions of the modified signal subspace $\bar{\boldsymbol{E}}_S$ are greater than the dimensions of the true signal subspace. (The difference between the dimensions of the two subspaces is equal to D.) As discussed in Eq. (6.86), this overestimation of the signal subspace dimension reduces the noise-reduction capability of the eigenspace projection. Therefore, there is also a trade-off between the value of D and the noise-reduction capability of eigenspace projection.

9.4 Imaging magnitude source coherence

There is a growing interest in imaging the coherence of brain activity, because the source coherence is considered a reliable measure of functional connectivity between different brain regions [60][61][62][63][64][65]. In this section, we first show that the magnitude coherence in the frequency domain is estimated by the direct use of the outputs from the frequency-domain minimum-variance spatial filter. We define the Fourier spectrum of the source time course, $s(\boldsymbol{r}, t)$, as $s(\boldsymbol{r}, f)$, and its estimated value as $\widehat{s}(\boldsymbol{r}, f)$. Then, when the first and second sources have a coherence value $\rho(f)$ at frequency f, we obtain the relationships

$$\widehat{s}(\boldsymbol{r}_1, f) = s(\boldsymbol{r}_1, f) - \left[\frac{\sigma_1(f)}{\sigma_2(f)}\rho(f)\right] s(\boldsymbol{r}_2, f), \tag{9.35}$$

$$\text{and} \quad \widehat{s}(\boldsymbol{r}_2, f) = s(\boldsymbol{r}_2, f) - \left[\frac{\sigma_2(f)}{\sigma_1(f)}\rho(f)\right] s(\boldsymbol{r}_1, f), \tag{9.36}$$

where $\sigma_q^2(f)$ is the power spectrum of the qth source at f, i.e.,

$$\sigma_q^2(f) = \langle |s(\boldsymbol{r}_q, f)|^2 \rangle.$$

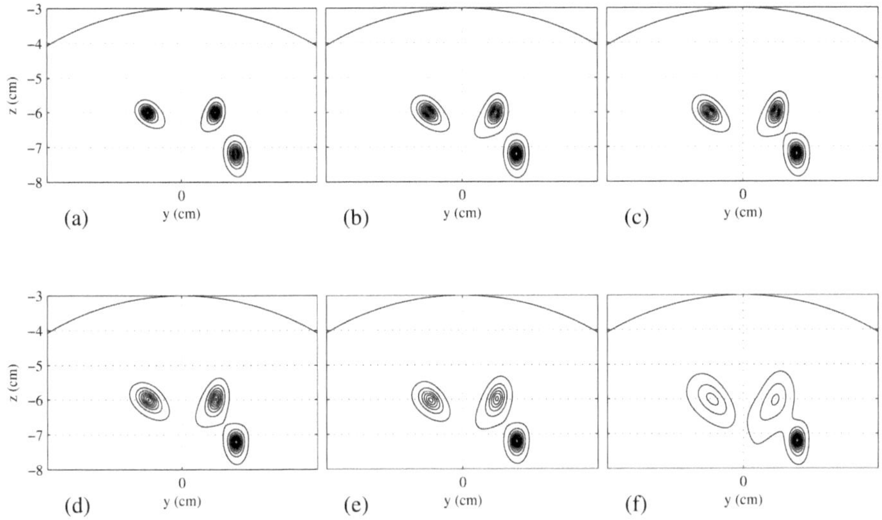

Figure 9.1: Results of the square-root of the power reconstruction, $\sqrt{\langle \widehat{s}(\boldsymbol{r},t)^2 \rangle}$, obtained using the array-gain minimum-variance spatial filter. The total input SNR (α_T) is equal to $16M$; (a) $\mu = 0.08$; (b) $\mu = 0.5$; (c) $\mu = 0.6$; (d) $\mu = 0.7$; (e) $\mu = 0.8$; and (f) $\mu = 0.95$, where μ is the cross-correlation coefficient between the first- and the second-source time courses.

Then, substituting Eqs. (9.35) and (9.36) into

$$\widehat{\rho}(f) = \frac{\langle \widehat{s}(\boldsymbol{r}_1, f)\widehat{s}^H(\boldsymbol{r}_2, f)\rangle}{\sqrt{\langle |\widehat{s}(\boldsymbol{r}_1, f)|^2\rangle \langle |\widehat{s}(\boldsymbol{r}_2, f)|^2\rangle}}, \tag{9.37}$$

leads to

$$|\widehat{\rho}| = \frac{|\sigma_1\sigma_2(\rho^3 - \rho)|}{\sqrt{\sigma_1^2(1-\rho^2)\sigma_2^2(1-\rho^2)}} = |\rho|, \tag{9.38}$$

where $\widehat{\rho}$ is the estimated coherence, and we omit the explicit notation of (f). Here, the superscript H indicates the Hermitian transpose. Equation (9.38) indicates that an accurate estimation of the coherence magnitude can be obtained by substituting the outputs of the minimum-variance spatial filter into Eq. (9.37).

One problem in estimating the coherence is that the interference contained in the measurement is passed through the spatial-filter reconstruction process and gives a pseudo-coherence. This can be understood in the following manner. Let $\boldsymbol{g}(f)$, $\boldsymbol{g}_s(f)$, and $\boldsymbol{g}_I(f)$ be the measurement, signal, and interference in the frequency domain, respectively. Then, $\boldsymbol{g}(f)$ is expressed as

$$\boldsymbol{g}(f) = \boldsymbol{g}_s(f) + \boldsymbol{g}_I(f), \tag{9.39}$$

where we ignore the sensor noise. Considering the relationship

$$\widehat{s}(\boldsymbol{r}_i, f) = \boldsymbol{w}^H(\boldsymbol{r}_i, f)\boldsymbol{g}(f) = \boldsymbol{w}^H(\boldsymbol{r}_i, f)(\boldsymbol{g}_s(f) + \boldsymbol{g}_I(f)),$$

the cross product in the numerator on the right-hand side of Eq. (9.37) is expressed as

$$\begin{aligned}\langle \widehat{s}(\boldsymbol{r}_1, f)\widehat{s}^H(\boldsymbol{r}_2, f)\rangle = \boldsymbol{w}^H(\boldsymbol{r}_i, f)\langle \boldsymbol{g}_s(f)\boldsymbol{g}_s^H(f)\rangle \boldsymbol{w}(\boldsymbol{r}_j, f)\\ + \boldsymbol{w}^H(\boldsymbol{r}_i, f)\langle \boldsymbol{g}_I(f)\boldsymbol{g}_I^H(f)\rangle \boldsymbol{w}(\boldsymbol{r}_j, f),\end{aligned} \tag{9.40}$$

where we assume that the brain signal $\boldsymbol{g}_s(f)$ and the interference $\boldsymbol{g}_I(f)$ are incoherent, i.e., that $\langle \boldsymbol{g}_s(f)\boldsymbol{g}_I^H(f)\rangle = 0$. In Eq. (9.40), the second term on the right-hand side shows the pseudo-correlation caused by the interference. If the interference is region-specific (such as the interference caused by eye blinks), the spatial filter may separate out this pseudo-term from the true coherence (the first term). However, since most of the background interferences are non-region-specific, the pseudo-coherence should also be non-region-specific and it may complicate the interpretation of the final coherence imaging results.

Nolte *et al.* proposed using only the imaginary part of the coherence to avoid this problem[66], because the pseudo-coherence is real valued, and the non-zero imaginary part of the coherence is caused solely from true interactions among brain activities. The rationale can be understood in the following manner. Denoting $\mu(\boldsymbol{r}, f)$ the Fourier spectrum of the background source activity, $\mu(\boldsymbol{r}, t)$, the interference term $\boldsymbol{g}_I(f)$ can be expressed as,

$$\boldsymbol{g}_I(f) = \sum_j \mu(\boldsymbol{r}_j, f)\boldsymbol{l}(\boldsymbol{r}_j), \tag{9.41}$$

where the background sources are represented by discrete independent sources and j is the index for these sources. Then, the cross product in the second term on the right-hand side of Eq. (9.40) is expressed as

$$\begin{aligned}\langle \boldsymbol{g}_I(f)\boldsymbol{g}_I^H(f)\rangle &= \sum_j \sum_{j'} \langle \mu(\boldsymbol{r}_j, f)\mu^*(\boldsymbol{r}_{j'}, f)\rangle \boldsymbol{l}(\boldsymbol{r}_j)\boldsymbol{l}^T(\boldsymbol{r}_{j'})\\ &= \sum_j < |\mu(\boldsymbol{r}_j, f)|^2 > \boldsymbol{l}(\boldsymbol{r}_j)\boldsymbol{l}^T(\boldsymbol{r}_j),\end{aligned} \tag{9.42}$$

where the asterisk $*$ indicates the complex conjugate, and the relationship in Eq. (8.6) is assumed. The above equation shows that $\langle \boldsymbol{g}_I(f)\boldsymbol{g}_I^H(f)\rangle$ is real-valued.

On the other hand, using

$$\boldsymbol{g}_s(f) = \sum_{q=1}^{Q} s(\boldsymbol{r}_q, f)\boldsymbol{l}(\boldsymbol{r}_q),$$

the cross product in the first term on the right-hand side of Eq. (9.40) is expressed as

$$\langle \boldsymbol{g}_s(f)\boldsymbol{g}_s^H(f)\rangle = \sum_{q=1}^{Q} \sum_{q'=1}^{Q} \langle s(\boldsymbol{r}_q, f)s^*(\boldsymbol{r}_{q'}, f)\rangle \boldsymbol{l}(\boldsymbol{r}_q)\boldsymbol{l}^T(\boldsymbol{r}_{q'}) \tag{9.43}$$

Because there are source interactions, the product $\langle s(\boldsymbol{r}_q, f)s^*(\boldsymbol{r}_{q'}, f)\rangle$ can have a non-zero value, and thus the first term on the right-hand side of Eq. (9.40) has both real and imaginary parts. Therefore, the imaginary part of the observed coherence is only caused only by $\langle s(\boldsymbol{r}_q, f)s^*(\boldsymbol{r}_{q'}, f)\rangle$, which is the true brain interaction, but the real part contains not only the coherence from $\langle s(\boldsymbol{r}_q, f)s^*(\boldsymbol{r}_{q'}, f)\rangle$ but also the pseudo-coherence caused by the interference.

In actual implementation, the computation of the magnitude source coherence between the ith voxel and the jth voxel, $\widehat{\rho}_{i,j}$, is carried out based on Eq. (9.37), i.e.,

$$\widehat{\rho}_{i,j} = \frac{|\sum_{k_e} \widehat{s}(\boldsymbol{r}_i, f)\widehat{s}^H(\boldsymbol{r}_j, f)|}{\sqrt{\sum_{k_e} |\widehat{s}(\boldsymbol{r}_i, f)|^2 \sum_{k_e} |\widehat{s}(\boldsymbol{r}_j, f)|^2}}. \tag{9.44}$$

Here, the summation above is carried out over multiple trials and k_e is the trial index. If the data are single-trial data, the whole measurement time course is divided into multiple segments and the summation is performed over these segments. Guggisberg *et al.* proposed[62] to average the imaginary coherence over all voxel connections. That is, they propose to calculate the mean imaginary coherence at the ith voxel, v_i, using

$$v_i = \tanh\left(\frac{1}{N}\sum_{j=1}^{N} \tanh^{-1}(\mathcal{I_M}\{\widehat{\rho}_{i,j}\})\right), \tag{9.45}$$

where $\mathcal{I_M}\{\cdot\}$ indicates the imaginary part of the quantity in the curly braces, and the imaginary coherence is averaged in the Fisher's Z-transformed domain. There is strong evidence that this mean imaginary coherence can detect decreased connectivity among cortical neurons, and can provide useful clinical information on pathological brain regions of patients with brain lesions[62]. One example of this imaginary coherence imaging is presented in Fig. 1.2.

9.5 Numerical examples

Numerical examples are presented to illustrate the results of the arguments in this chapter. We use the same computer-simulation as in Section 4.7. The time courses shown in Fig. 4.3(a) are denoted $s_1(t)$, $s_2(t)$, and $s_3(t)$ from the top to the bottom, respectively. The correlation coefficients are 8×10^{-2} between $s_1(t)$ and $s_2(t)$, 2×10^{-4} between $s_1(t)$ and $s_3(t)$, and 4×10^{-3} between $s_2(t)$ and $s_3(t)$. We assign $s_2(t)$ and $s_3(t)$ to the time courses of the second and the third sources. The new time course calculated using

$$s_1'(t) = (1-\omega)s_1(t) + \omega s_2(t), \tag{9.46}$$

is assigned to the first source where parameter ω controls the degree of correlation between $s_1'(t)$ and $s_2(t)$. The simulated MEG recordings are calculated at 400 time points, and the sampling interval is assumed to be 1 ms. The powers of the three sources were set in order for their input SNR to be equal, and the total input SNR α_T $(= \alpha_1 + \alpha_2 + \alpha_3)$ was set to $16M$. We first investigate the effects of the signal cancellation caused by source correlation. We generate six data sets with the correlation coefficients between the first and second sources, μ, of 0.08, 0.5, 0.6, 0.7, 0.8, and 0.95. The results of square root of the power reconstruction for the six values of μ are shown in Fig. 9.1. Here, the array-gain-constraint minimum-variance spatial filter was used. These results show that, compared to the intensity of the third source, the intensities of the first and second sources are reduced according to their degree of correlation, thus demonstrating the signal cancellation phenomenon.

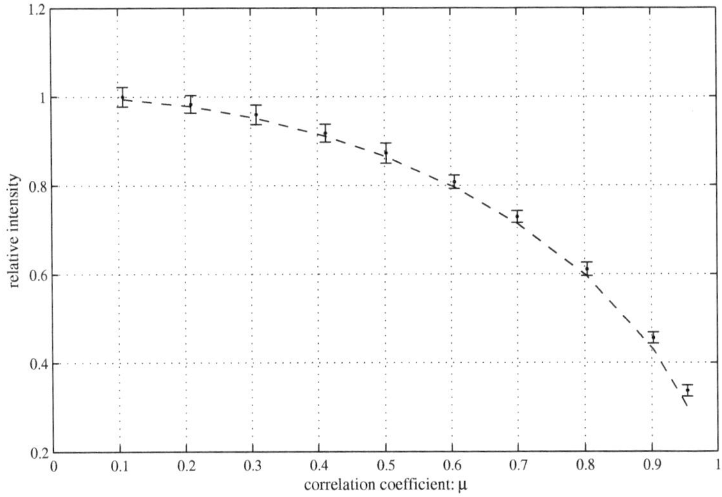

Figure 9.2: The relative intensity of the first source with respect to the correlation between the first and second sources. The broken line indicates the theoretical relationship $\sqrt{1-\mu^2}$. In these Monte Carlo simulations, the mean intensity of the first source is calculated from 100 Monte Carlo results. The error bars indicate a range of ± 2 standard deviations. The total input SNR, α_T, is set to $4M$.

This intensity reduction is theoretically predicted in Eqs. (9.10) and (9.11), which indicate that the intensities of the reconstructed sources are reduced by a factor of $\sqrt{(1-\mu^2)}$. Monte Carlo-type experiments are performed to check whether the reconstructed signal intensity changes according to $\sqrt{(1-\mu^2)}$ even when SNR is quite low. One hundred sets of simulated magnetic recordings are generated with the same SNR but with different noise realizations for ten different values of μ. The mean reconstructed intensity of the first source is plotted in Fig. 9.2. The total input SNR α_T is set to $4M$. The theoretical trend $\sqrt{(1-\mu^2)}$ is plotted with the broken line, and the error bars represent the range of ± 2 standard deviations of the Monte Carlo results. In Fig. 9.2, although a small discrepancy is observed when μ approaches 1, the theoretical trend (the broken line) generally overlaps with the plots from the Monte Carlo experiments, indicating that the theoretical relationship holds well, even in such low SNR situation as α_T/M equal to 4.

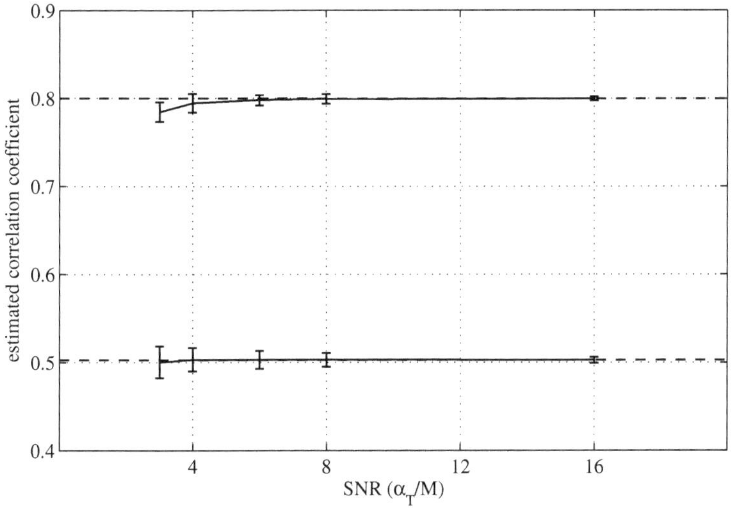

Figure 9.3: Estimated correlation coefficient between the first and second sources. Monte Carlo simulations, which generated 100 sets of simulated MEG recordings, are performed for the five cases of α_T, i.e., $\alpha_T = 3, 4, 6, 8$ and 16. The average values of $\widehat{\mu}$ across 100 Monte Carlo results are plotted. The experiments are performed for the two cases of μ: $\mu = 0.5$ and $\mu = 0.8$. The abscissa is expressed as α_T/M. The error bar shows the range of ± 2 standard deviations.

According to the trend $\sqrt{(1-\mu^2)}$, the intensity reduction is less than twenty percent, unless the correlation coefficient exceeds 0.6. Likewise, sixty percent of the original source intensity is still maintained when the correlation reaches 0.8. Therefore, as far as the signal cancellation is concerned, no serious influences arise from sources with weak or medium degrees of correlation ($\mu \le 0.6$). That is, there is no large difference between the reconstructed sources when $\mu = 0$ (Fig. 9.1(a)) and $\mu = 0.6$ (Fig. 9.1(c)).

The correlation coefficient between the first and second sources is estimated from Eq. (9.13) using spatial filter outputs $\widehat{s}(\boldsymbol{r}_1, t)$ and $\widehat{s}(\boldsymbol{r}_2, t)$. To investigate the influence of noise on the accuracy of the estimated correlation coefficient, we estimate the correlation coefficient between the reconstructed time courses of the first and second sources by averaging the correlation coefficients from one hundred Monte Carlo trials, where each trial uses simulated magnetic recordings having a specified SNR and a unique noise realization. Such experiments are repeated for five values of the total input SNR (α_T), and the results are plotted in Fig. 9.3 for

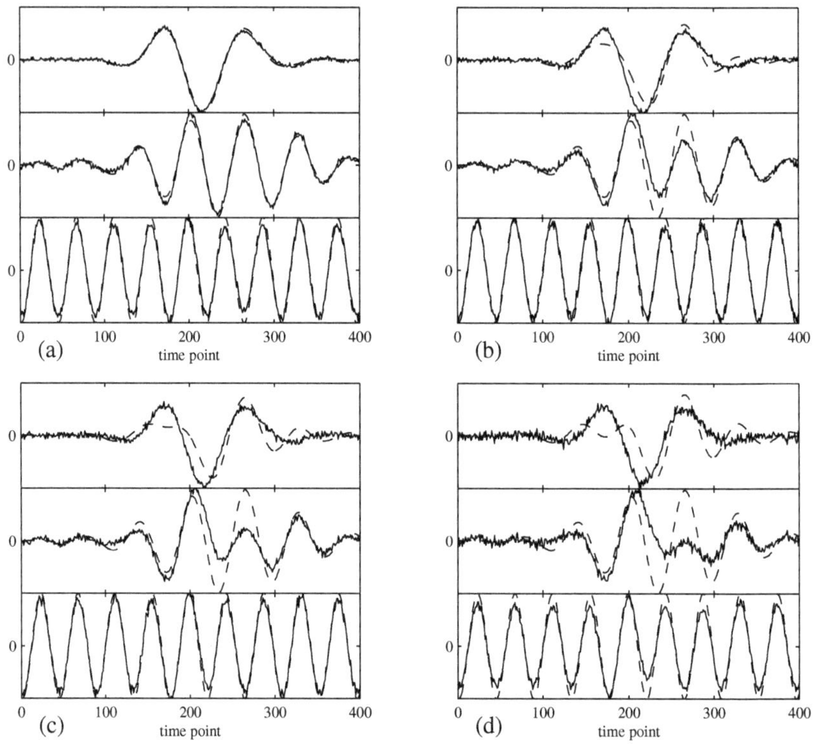

Figure 9.4: Time course outputs for the three sources when $\alpha_T = 4M$ and (a) $\mu = 0.08$, (b) $\mu = 0.4$, (c) $\mu = 0.6$, and (d) $\mu = 0.8$, where μ is the correlation coefficient between the first and second source time courses. The solid lines indicate the spatial filter outputs and the broken lines indicate the true time courses. The eigenspace-projection array-gain minimum-variance spatial filter was used with the signal subspace dimension Q set to 3.

the correlation coefficients μ equal to 0.5 and 0.8. In this figure, the error bars show the range of ± 2 standard deviations. This plot shows that the influence of the noise on the estimated correlation coefficient is generally very small. The estimated correlation coefficient is biased so that the average of the estimates is slightly smaller than the true value when α_T equals three. However, this bias is small and is less than 5% for the whole SNR range used in the experiments.

Reconstructed time courses of the three sources are shown in Fig. 9.4. The total input SNR (α_T) was set to $4M$ in this experiment. The results in (a), (b), (c), and (d) respectively correspond to the cases of $\mu = 0.08$, $\mu = 0.4$, $\mu = 0.6$, and $\mu = 0.8$. The solid lines indicate the spatial filter outputs, and the broken lines indicate the original time courses assumed in the computer simulation. When μ is small, the original time courses can be retrieved as spatial filter outputs.

However, as μ increases, the difference between the reconstructed and true time courses becomes large because of the interference from the other correlated source. This time-course distortion due to leakage is evident when $\mu = 0.6$ and $\mu = 0.8$; but it is also discernible when $\mu = 0.4$.

We next present an example of the scenario where a third correlated source exists, and check how the accuracy of the estimated correlation coefficient is affected by this source. This influence can be evaluated using Eq. (9.20). When $\mu_{13} \approx \mu_{23}$, this equation can be further simplified to

$$\widehat{\mu}_{12} = \frac{|\mu_{12} - \mu_I^2|}{(1 - \mu_I^2)}, \tag{9.47}$$

where $\widehat{\mu}_{12}$ is the estimate of the correlation coefficient between the target sources, and μ_I $(= \mu_{13} = \mu_{23})$ is the correlation between one of the target sources (the first or second sources) and the interference source (the third source). This equation is plotted for the three values of the target correlation, μ_{12}, in Fig. 9.5. The figure shows that even when an additional correlated source exists, the error due to this interference is less than ten percent, unless μ_I exceeds eighty percent of μ_{12}. Therefore, it is generally true that if the correlation between the interference source and either target source is not as strong as the correlation between the target sources, a reasonably accurate estimate of the target-source correlation coefficient can be obtained.

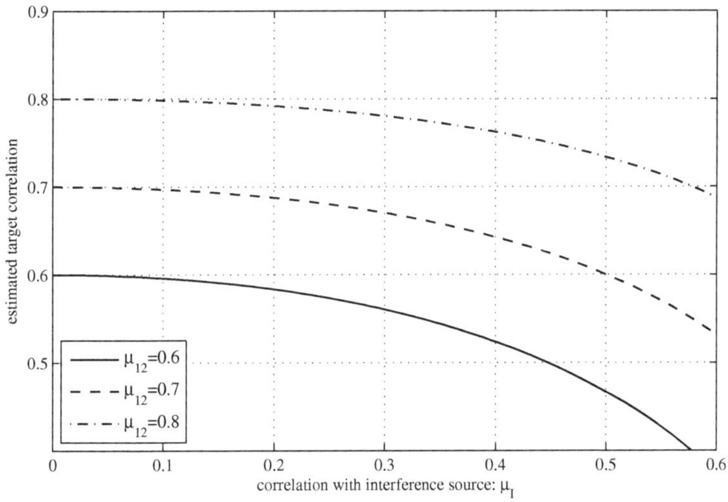

Figure 9.5: Effects of a third correlated source on the estimated target correlation coefficient $\widehat{\mu}_{12}$. Dotted, broken, and solid lines correspond to the estimate, $\widehat{\mu}_{12}$, when $\mu_{12} = 0.8$, $\mu_{12} = 0.7$, and $\mu_{12} = 0.6$, respectively where μ_{12} is the true correlation coefficient.

We next perform numerical experiments for coherent interference suppression as described in Section 9.3. We assumed three sources exist at $(0, -1.7, -6)$, $(0, 1.7, -6)$, and $(0, 1.6, -7.2)$ on the plane $x = 0$, and assigned the three time courses shown in Fig. 4.3(a) to the three sources. The power reconstruction obtained using the array-gain constraint minimum-variance spatial filter is shown in Fig. 9.6(a). Here, we set the SNR α_T to $16M$. We then assumed that the first source located in the left-hand region is an interference and the second and the third sources located in the right-hand region are the sources of interest, and only the right-hand region is reconstructed. The reconstruction results, with and without imposing null-sensitivity, are shown in Fig. 9.6(b)–(d).

The results of the source reconstruction when the first and the second sources are uncorrelated are shown in Fig. 9.6 (b). These results are exactly the same as the right-hand side of the results in 9.6(a). The results of the source reconstruction obtained when the correlation coefficient is set to 0.99 are shown in Fig. 9.6(c). Since the signal cancellation effects are so large, the second source disappears. Then, we applied the LCMV spatial filter reconstruction to the same data. We applied null-sensitivity to a 2-cm cubic region containing the first source, the region shown by the square in Fig. 9.6(a). The results of this reconstruction are shown in Fig. 9.6(d). Despite the high correlation between the first and the second sources, the intensity of the second source is retrieved, demonstrating the effectiveness of the coherent interference suppression method.

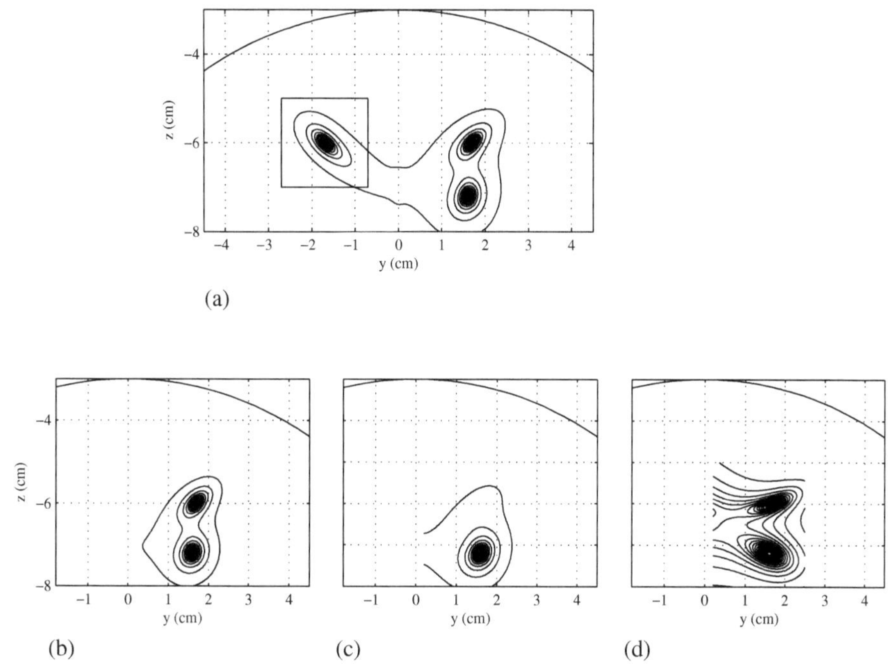

Figure 9.6: (a) Results of source-power reconstruction for the entire field of view. The array-gain constraint minimum-variance spatial filter was used. The solid square indicates the region where null-sensitivity was imposed to obtain the reconstructed results in (d). (b)–(d)Reconstruction for only the right-hand region of the field of view. (b) Reconstructed results obtained when the first- and second-source time courses are uncorrelated ($\mu = 0.07$). (c) The reconstructed results obtained when the first- and second-source time courses are highly correlated ($\mu = 0.99$). (d) The results obtained from the LCMV spatial filter when we imposed a null-sensitivity over the 2-cm cubic region shown by the square in (a).

Chapter 10

Effects of using the sample covariance matrix

Throughout most of our arguments in the previous chapters, we have assumed that the theoretical covariance matrix is known. However, in practical applications of adaptive spatial filters, the covariance matrix should be estimated from the data, and usually we estimate the covariance matrix using the data samples. That is, denoting the spatio-temporal measurement as $\boldsymbol{B}$:

$$\boldsymbol{B} = [\boldsymbol{b}(t_1), \boldsymbol{b}(t_2), \ldots, \boldsymbol{b}(t_K)],$$

the sample covariance matrix $\widehat{\boldsymbol{R}}$ is estimated using

$$\widehat{\boldsymbol{R}} = \frac{1}{K}\boldsymbol{B}\boldsymbol{B}^T = \frac{1}{K}\sum_{k=1}^{K}\boldsymbol{b}(t_k)\boldsymbol{b}^T(t_k), \tag{10.1}$$

where K is the number of time points used for the covariance estimation. Naturally, when this number is small, the estimated sample covariance is likely erroneous and such an error may affect the output of adaptive spatial-filters. This chapter discusses the influence of using the sample covariance matrix on the source reconstruction results from minimum-variance spatial filters.

10.1 Sample covariance matrix: the maximum-likelihood estimate of the true covariance matrix

We first discuss the rationale of using the sample covariance matrix, and show that the sample covariance matrix is the maximum-likelihood estimate of the true covariance matrix. The fundamental assumption is that the measured data are a realization of a multivariate Gaussian process, where the probability distribution of the measurement vector $\boldsymbol{b}(t)$ is expressed as

$$p(\boldsymbol{b}(t)) = \frac{1}{(2\pi)^{(M/2)}|\boldsymbol{R}|^{1/2}} \exp[-\frac{1}{2}\boldsymbol{b}^T(t)\boldsymbol{R}^{-1}\boldsymbol{b}(t)]. \tag{10.2}$$

Assuming that the Gaussian process for each time point is independent and identically distributed, the probability for the spatio-temporal data set $\boldsymbol{B}$ is given by

$$p(\boldsymbol{B}) = \prod_{k=1}^{K} \frac{1}{(2\pi)^{(M/2)}|\boldsymbol{R}|^{1/2}} \exp[-\frac{1}{2}\boldsymbol{b}^T(t_k)\boldsymbol{R}^{-1}\boldsymbol{b}(t_k)]. \tag{10.3}$$

Neglecting an additive constant, the log-likelihood function is then expressed as

$$\log \mathcal{L}(\boldsymbol{B}) = \log p(\boldsymbol{B}) = -\frac{K}{2}\log|\boldsymbol{R}| - \frac{1}{2}\sum_{k=1}^{K} \boldsymbol{b}^T(t_k)\boldsymbol{R}^{-1}\boldsymbol{b}(t_k). \tag{10.4}$$

To derive the optimum estimate of $\boldsymbol{R}$, $\widehat{\boldsymbol{R}}$, the likelihood function is maximized with respect to $\boldsymbol{R}$. Calculating

$$\frac{\partial \log \mathcal{L}(\boldsymbol{B})}{\partial \boldsymbol{R}} = \frac{K}{2}\boldsymbol{R}^{-1} - \frac{1}{2}\boldsymbol{R}^{-1}\sum_{k=1}^{K} \boldsymbol{b}(t_k)\boldsymbol{b}^T(t_k)\boldsymbol{R}^{-1}, \tag{10.5}$$

and setting $\partial \log \mathcal{L}(\boldsymbol{B})/\partial \boldsymbol{R}$ to zero give the optimum estimate of the covariance matrix, $\widehat{\boldsymbol{R}}$:

$$\widehat{\boldsymbol{R}} = \frac{1}{K}\sum_{k=1}^{K} \boldsymbol{b}(t_k)\boldsymbol{b}^T(t_k) = \frac{1}{K}\boldsymbol{B}\boldsymbol{B}^T, \tag{10.6}$$

where we use the formulae[43]

$$\frac{\partial \log|\boldsymbol{R}|}{\partial \boldsymbol{R}} = \boldsymbol{R}^{-1}, \tag{10.7}$$

and

$$\frac{\partial}{\partial \boldsymbol{R}}\boldsymbol{b}^T\boldsymbol{R}^{-1}\boldsymbol{b} = \boldsymbol{R}^{-1}\boldsymbol{b}\boldsymbol{b}^T\boldsymbol{R}^{-1}. \tag{10.8}$$

Discussion above shows that the sample covariance matrix, $\widehat{\boldsymbol{R}}$, is the maximum likelihood estimate of the covariance matrix.

10.2 Effects of using sample covariance matrices on the minimum-variance filters

Because the measured data are one realization of a random process, the sample covariance matrix $\widehat{\boldsymbol{R}}$ is a random variable. Assuming again that the data $\boldsymbol{b}(t)$ are drawn from the M-dimensional Gaussian process with mean equal to zero, the probability density distribution of the matrix

$$\widehat{\boldsymbol{C}} = K\widehat{\boldsymbol{R}} = \sum_{k=1}^{K} \boldsymbol{b}(t_k)\boldsymbol{b}^T(t_k), \tag{10.9}$$

is known to be a Whishart distribution $W_M(\boldsymbol{R}, K)$, which is expressed as[67][68]

$$p(\widehat{\boldsymbol{C}}) = \frac{|\widehat{\boldsymbol{C}}|^{(K-M-1)/2} \exp[-\frac{1}{2}\mathrm{tr}(\boldsymbol{R}^{-1}\widehat{\boldsymbol{C}})]}{2^{\frac{1}{2}KM}\pi^{\frac{1}{4}M(M-1)}|\boldsymbol{R}|^{K/2}\prod_{j=1}^{M}\Gamma[(K+1-j)/2]}, \tag{10.10}$$

where $\Gamma[\cdot]$ is the Gamma function. Note that $\boldsymbol{R}$ is the true covariance matrix. The Whishart distribution is an important tool for analyzing the stochastic property of the sample covariance matrix. Although a discussion regarding the various properties of the Whishart distribution is beyond the scope of this book, we introduce one useful formula. That is, if $\widehat{\boldsymbol{C}}$ obeys the Whishart distribution, defining $\boldsymbol{x}$ as any $M \times 1$ fixed (non-stochastic) vector, we have the relationship[68],

$$\frac{\boldsymbol{x}^T\boldsymbol{R}^{-1}\boldsymbol{x}}{\boldsymbol{x}^T\widehat{\boldsymbol{C}}^{-1}\boldsymbol{x}} \sim \chi^2_{K-M+1}, \tag{10.11}$$

where χ^2_{K-M+1} is the chi-squared distribution with $K-M+1$ degrees of freedom. In the equation above, the notation "$\sim$" indicates that the stochastic variable on the left-hand side is distributed according to the probability distribution shown on the right-hand side. Considering $\widehat{\boldsymbol{C}} = K\widehat{\boldsymbol{R}}$, we can obtain

$$(\boldsymbol{x}^T\widehat{\boldsymbol{R}}^{-1}\boldsymbol{x})^{-1} \sim \frac{1}{K}(\boldsymbol{x}^T\boldsymbol{R}^{-1}\boldsymbol{x})^{-1}\chi^2_{K-M+1}. \tag{10.12}$$

This equation can be used for discussing how the sample covariance matrix affects the output power of the minimum-variance filter. That is, let us use $\widehat{P}_{\widehat{\boldsymbol{R}}}(\boldsymbol{r})$ to denote the output power of the unit-gain minimum-variance filter obtained using the sample covariance and $\widehat{P}_{\boldsymbol{R}}(\boldsymbol{r})$ to denote the output power obtained using the true (theoretical) covariance matrix. These two kinds of output powers are respectively given by

$$\widehat{P}_{\widehat{\boldsymbol{R}}}(\boldsymbol{r}) = \frac{1}{[\boldsymbol{l}^T(\boldsymbol{r})\widehat{\boldsymbol{R}}^{-1}\boldsymbol{l}(\boldsymbol{r})]}, \tag{10.13}$$

$$\text{and} \quad \widehat{P}_{\boldsymbol{R}}(\boldsymbol{r}) = \frac{1}{[\boldsymbol{l}^T(\boldsymbol{r})\boldsymbol{R}^{-1}\boldsymbol{l}(\boldsymbol{r})]}. \tag{10.14}$$

Then, using Eq. (10.12), it is straightforward to obtain

$$\widehat{P}_{\widehat{\boldsymbol{R}}}(\boldsymbol{r}) = \widehat{P}_{\boldsymbol{R}}(\boldsymbol{r})\frac{1}{K}\chi^2_{K-M+1}. \tag{10.15}$$

It is apparent that the above relationship also holds for the output power of the array-gain constraint minimum-variance filter. The above equation clearly shows that $\widehat{P}_{\widehat{\boldsymbol{R}}}(\boldsymbol{r})$ is a random variable and its probability distribution is the chi-squared distribution with degrees of freedom of $K - M + 1$. Therefore, we can derive the relationship, first derived by Capon and Goodman[69],

$$\langle \widehat{P}_{\widehat{\boldsymbol{R}}}(\boldsymbol{r}) \rangle = \widehat{P}_{\boldsymbol{R}}(\boldsymbol{r}) \frac{1}{K} \langle \chi^2_{K-M+1} \rangle = \frac{K - M + 1}{K} \widehat{P}_{\boldsymbol{R}}(\boldsymbol{r}), \tag{10.16}$$

where $\langle \cdot \rangle$ indicates the expectation operator and we use $\langle \chi^2_{K-M+1} \rangle = K - M + 1$. The equation above indicates that the average output power obtained using the sample covariance matrix is smaller than the output power obtained using the true covariance matrix. The intensity reduction is characterized by the factor $(K - M - 1)/K$. Therefore, if the number of time points K is close to the number of sensors, M, a severe intensity reduction should occur.

10.3 Recovering from the sample covariance effects: Beamspace processing

The intensity reduction of the output power due to the use of sample covariance matrices can partly be overcome by diagonal loading, which is discussed in Section 6.6, where we argue that diagonal loading can reduce the influence of the array mismatch. In the same manner, we can show that diagonal loading is also effective in recovering the signal intensity that is degraded by the use of a sample covariance matrix. The effectiveness of diagonal loading is, however, achieved at the sacrifice of the spatial resolution, as was discussed in Section 6.6. In the following, we describe a different method, called beamspace processing, to reduce the sample-covariance influence.

The basic idea of beamspace processing is to reduce the dimensionality of the data. That is, defining an $M \times J_R$ (where $M > J_R$) matrix $\boldsymbol{T}$, the dimensionality of the data $\boldsymbol{b}(t)$ is reduced by applying

$$\boldsymbol{\beta}(t) = \boldsymbol{T}^T \boldsymbol{b}(t), \tag{10.17}$$

where $\boldsymbol{\beta}(t)$ is a set of "transformed" data whose dimension is equal to J_R. We apply the same transformation to the lead field vector, such that

$$\boldsymbol{l}_\beta(\boldsymbol{r}) = \boldsymbol{T}^T \boldsymbol{l}(\boldsymbol{r}),$$

and define the covariance matrix of the transformed data as

$$\boldsymbol{R}_\beta = \langle \boldsymbol{\beta}(t) \boldsymbol{\beta}^T(t) \rangle = \boldsymbol{T}^T \langle \boldsymbol{b}(t) \boldsymbol{b}^T(t) \rangle \boldsymbol{T} = \boldsymbol{T}^T \boldsymbol{R} \boldsymbol{T}.$$

The weight of the beamspace (unit-gain) minimum variance filter is thus obtained using

$$\boldsymbol{w}(\boldsymbol{r}) = \arg\min_{\boldsymbol{w}(\boldsymbol{r})} \boldsymbol{w}^T(\boldsymbol{r}) \boldsymbol{R}_\beta \boldsymbol{w}(\boldsymbol{r}), \quad \text{subject to} \quad \boldsymbol{w}^T(\boldsymbol{r}) \boldsymbol{l}_\beta(\boldsymbol{r}) = 1. \tag{10.18}$$

The resultant weight is expressed as

$$\boldsymbol{w}(\boldsymbol{r}) = \frac{\boldsymbol{R}_\beta^{-1}\boldsymbol{l}_\beta(\boldsymbol{r})}{[\boldsymbol{l}_\beta^T(\boldsymbol{r})\boldsymbol{R}_\beta^{-1}\boldsymbol{l}_\beta(\boldsymbol{r})]}, \tag{10.19}$$

and the filter output is given by

$$\widehat{s}(\boldsymbol{r},t) = \frac{\boldsymbol{l}_\beta^T(\boldsymbol{r})\boldsymbol{R}_\beta^{-1}\boldsymbol{\beta}(t)}{[\boldsymbol{l}_\beta^T(\boldsymbol{r})\boldsymbol{R}_\beta^{-1}\boldsymbol{l}_\beta(\boldsymbol{r})]} = \frac{\boldsymbol{l}^T(\boldsymbol{r})\boldsymbol{T}(\boldsymbol{T}^T\boldsymbol{R}\boldsymbol{T})^{-1}\boldsymbol{T}^T\boldsymbol{b}(t)}{[\boldsymbol{l}^T(\boldsymbol{r})\boldsymbol{T}(\boldsymbol{T}^T\boldsymbol{R}\boldsymbol{T})^{-1}\boldsymbol{T}^T\boldsymbol{l}(\boldsymbol{r})]}. \tag{10.20}$$

Comparing the above equation with Eq. (4.6), we can see that the beamspace minimum variance filter is implemented by simply replacing $\boldsymbol{R}^{-1}$ in the regular minimum-variance filter formulation with

$$\boldsymbol{T}(\boldsymbol{T}^T\boldsymbol{R}\boldsymbol{T})^{-1}\boldsymbol{T}^T.$$

The dimension of the beamspace covariance matrix $\boldsymbol{R}_\beta$ is J_R, so the intensity reduction caused by the use of the sample covariance is characterized by $(K-(J_R-1))/K$ where K is the number of time samples used for the covariance calculation. Therefore, when $K \gg J_R$, the intensity bias can be significantly reduced.

We next argue how to find a reasonable $\boldsymbol{T}$. Limpiti and Van Veen[70] proposed to derive such a $\boldsymbol{T}$ by restricting the source space to a smaller region Ω_R. That is, we first calculate $\boldsymbol{G}_R$, which is the gram matrix over Ω_R, such that

$$\boldsymbol{G}_R = \int_{\Omega_R} \boldsymbol{L}(\boldsymbol{r})\boldsymbol{L}^T(\boldsymbol{r})\, d^3r. \tag{10.21}$$

This $\boldsymbol{G}_R$ is called the local gram matrix in this book. Defining the eigenvectors corresponding to the J_R largest eigenvalues of $\boldsymbol{G}_R$ as $\boldsymbol{x}_1, \ldots, \boldsymbol{x}_{J_R}$, we obtain $\boldsymbol{T}$ such that

$$\boldsymbol{T} = [\boldsymbol{x}_1, \boldsymbol{x}_2, \ldots, \boldsymbol{x}_{J_R}]. \tag{10.22}$$

The method here is very similar to the method described in Section 9.3. We first define the error produced when the column span of $\boldsymbol{T}$ represents the lead-field $\boldsymbol{L}(\boldsymbol{r})$ as

$$e_T^2(\boldsymbol{r}) = \|(\boldsymbol{I}-\boldsymbol{T}\boldsymbol{T}^T)\boldsymbol{L}(\boldsymbol{r})\|^2 = \mathrm{tr}\{(\boldsymbol{I}-\boldsymbol{T}\boldsymbol{T}^T)\boldsymbol{L}(\boldsymbol{r})\boldsymbol{L}^T(\boldsymbol{r})(\boldsymbol{I}-\boldsymbol{T}\boldsymbol{T}^T)\}. \tag{10.23}$$

Then, the total error over the region Ω_R is given by

$$\int_{\Omega_R} e_T^2(\boldsymbol{r})\, d^3r = \mathrm{tr}\{(\boldsymbol{I}-\boldsymbol{T}\boldsymbol{T}^T)\boldsymbol{G}_R(\boldsymbol{I}-\boldsymbol{T}\boldsymbol{T}^T)\} = \sum_{j=J_R+1}^{M} \varpi_j, \tag{10.24}$$

where ϖ_j is the jth eigenvalue of $\boldsymbol{G}_R$. Thus, the relative value of this total error is defined as

$$E_R = \sum_{j=J_R+1}^{M} \varpi_j \Big/ \sum_{j=1}^{M} \varpi_j. \tag{10.25}$$

The equation above indicates that a smaller J_R gives a greater E_R, although a smaller J_R is preferable. The rule of thumb is to check the eigenvalues ϖ_j ($j = 1, \ldots, M$) and to choose a reasonably small J_R that gives a reasonably small E_R. This beamspace processing is a method that reduces the data dimensionality by incorporating prior knowledge regarding the possible source locations and configurations. If we know in advance the region in which the sources exist, this knowledge can be used to reduce the data dimensionality through beamspace processing.

10.4 Numerical examples

10.4.1 Effects of using sample covariance matrices

In this section, we perform numerical experiments to illustrate the results of our discussion so far. We used the same computer simulation scheme depicted in Fig. 5.2, except that the coordinates of the single source are $(0, -1, -6)$. The time course shown in the top panel in Fig. 4.3(a) was assigned to this source. The signal MEG recordings, $\boldsymbol{b}_s(t)$, were computed between 0 and 400 ms, and Gaussian noise was added to generate the array outputs, $\boldsymbol{b}(t)$, where the input SNR was set equal to $2M$. In this computer simulation, we use six values of the sampling frequency equal to 0.5, 1, 2, 4, 8, and 16 kHz, which correspond to the number of time points K equal to 200, 400, 800, 1600, 3200, and 6400, respectively. The sample covariance matrix $\widehat{\boldsymbol{R}}$ was calculated using Eq. (10.1) with the whole data window between 0 and 400 ms. We performed Monte-Carlo-type experiments in which fifty sets of $\widehat{P}_{\widehat{\boldsymbol{R}}}(\boldsymbol{r})$ were calculated from fifty generated sets of $\boldsymbol{b}(t)$ having the same sampling frequency, the same noise statistics and different noise realizations. We then calculated the average across fifty Monte Carlo results of $\widehat{P}_{\widehat{\boldsymbol{R}}}(\boldsymbol{r})$ for each sampling frequency.

The reconstruction results for the six sampling frequencies are shown in Fig. 10.1 (a). Here, the cross-sectional profiles of the source at $(0, -1, -6)$ along the line defined as $x = 0$ and $z = -6$ are shown for $-3 \le y \le 1$. In this figure, we can clearly see that the intensity decrease depends on the sampling frequency. We plot the peak amplitude of these reconstructed profiles with respect to the number of time points K. The results are shown in Fig. 10.1 (b). In this figure, the theoretical results obtained using Eq. (10.16) are plotted with the broken line. The theoretical plot almost entirely overlaps with the plot from the computer simulation, indicating that the theoretical and computer-simulation results agree very well.

10.4.2 Recovering from the sample covariance effects

We next applied diagonal loading to the same data. The diagonal-loading factor was determined as $\epsilon = \varepsilon_R \lambda_{max}\{\widehat{\boldsymbol{R}}\}$, where $\lambda_{max}\{\widehat{\boldsymbol{R}}\}$ indicates the maximum eigenvalue of $\widehat{\boldsymbol{R}}$ and ε_R is a user-defined constant. The peak intensity of the reconstructed profiles is plotted with respect to the number of samples for four values of ε_R. The results are shown in Fig. 10.2(a). It can be seen that when K is less than 1000, the intensity reduction is significant for small values of ε_R, such as 10^{-4} and 10^{-5}. However, such an intensity reduction becomes negligible when ε_R is set equal to 10^{-2}. Setting ε_R equal to 10^{-2} makes it possible to recover 90% of the asymptotic source intensity even for K equal to 200. However, such a large diagonal-loading factor should cause a large spatial blur. This is shown by the reconstructed profiles in Figs. 10.2(b) where the case of K equal to 1600 is shown. In this figure, a large blur is caused when ε_R is set to 10^{-2}.

We then applied the beamspace processing. The regional source space Ω_R was set to a 4-cm cubic region with its center equal to the source location, $(0, -1, -6)$. We calculated the local gram matrix $\boldsymbol{G}_R$ using Eq. 10.21, and performed the eigendecomposition of $\boldsymbol{G}_R$. The eigenvalue spectrum in this case is shown in Fig. 10.3. According to this plot, a reasonable choice for the beamspace dimension, J_R, is between $J_R = 3$ and $J_R = 6$. The peak intensity is plotted with respect to the number of samples for the four cases of J_R. The results are shown in Fig. 10.4(a). These plots show that the true intensity is almost perfectly recovered and more than 95% of the asymptotic value is recovered even when K equals 200. These plots also show that the effectiveness of the beamspace technique is not significantly affected by the choice of J_R. The reconstructed profiles of the point source obtained from the data with K equal to 200 for the four choices of J_R are shown in Fig. 10.4(b), which shows that the beamspace technique does not cause a large image blur and the profiles are approximately equal for all four cases.

We next changed the size of Ω_R to check how the size affects the performance of the beamspace technique. We define Ω_R as the h_w-cm^3 cubic region centered at the true source location and we measure the performance of the beamspace technique for four cases of h_w: $h_w = 2$, 4, 6, and 8. The beamspace dimension J_R was chosen as the smallest value that satisfies the criterion that the error E_R be less than 10^{-4}. Using this criterion, J_R was determined as 5, 5, 7, and 10 when h_w is equal to 2, 4, 6, and 8 cm, respectively. The peak intensity plotted with respect to the number of samples for these four cases of h_w is shown in Fig. 10.5 (a). This plot shows that, in this particular example, the influence of the size of Ω_R on the effectiveness of the beamspace technique is small, and any choice of h_w gives a nearly perfect recovery of the source intensity. The reconstruction profiles of the point source obtained when K equals 200 are shown for the four h_w values in Fig. 10.5(b). This figure shows that the image blur is nearly equal for all four cases and the spatial resolution is not affected by the size of Ω_R.

10.4.3 Effects of using sample covariance matrices on unit-noise-gain minimum-variance filter

We next investigate the effects of using the sample covariance matrix on the output of the unit-noise-gain minimum-variance filter. Unfortunately, since there are no theoretical arguments for the effects of using sample-covariance matrices on this type of adaptive spatial filter, we can not derive an equation corresponding to Eq. (10.16). Thus, we must explore the sample-covariance influence on the unit-noise-gain filter solely by numerical experiments. We performed exactly the same Monte-Carlo-type computer simulation. The reconstructed cross-sectional profiles of the point source for six sampling frequencies (corresponding to six different values of K) are shown in Fig. 10.6 (a). We can see that, compared to the results in Fig. 10.1(a), the unit-noise-gain minimum-variance filter is much more sensitive to small sample sizes than the unit-gain and array-gain minimum-variance filters. The plot of the peak intensity with respect to the number of time points, K, is shown in Fig. 10.1(b), which shows that the peak intensity decreases almost linearly with respect to the number of time points K.

We tested the effectiveness of the diagonal loading to recover the source intensity. The results are shown in Fig. 10.7(a). It can be seen that diagonal loading has almost no effects when ε_R is set below 10^{-3}, and some improvements can be observed when ε_R is set to 10^{-2}. The true intensity is nearly perfectly recovered when ε_R is set to 10^{-1}. However, using such a large diagonal-loading factor should cause a large spatial blur. This fact can be seen in Fig. 10.7(b), which shows the results for K equal to 1600. A very large blur is caused when ε_R is set to 10^{-1}.

We then tested the beamspace technique, and the results are shown in Fig. 10.8(a). To obtain these results, Ω_R was set to a 4-cm^3 cubic region with its center at the source location. We used four different values of J_R: 3, 4, 5, and 6. The plotted results show that the effectiveness of the beamspace technique is rather modest, and the intensity recovery is small, compared to the case in Fig. 10.4(a). To check the spatial resolution, the reconstruction profiles of the point source when $K = 200$ are plotted for the four J_R choices in Fig. 10.8(b). This figure shows that the beamspace technique does not cause an image blur and the profiles are approximately equal for all four cases.

In summary, our numerical experiments show that the inaccuracy in the sample covariance matrices much more severely affects the unit-noise-gain spatial filter than the unit-gain and the array-gain minimum-variance filters. Therefore, unless the data contain a sufficient number of time samples, it may be safer not to use the unit-noise-gain minimum-variance spatial filter.

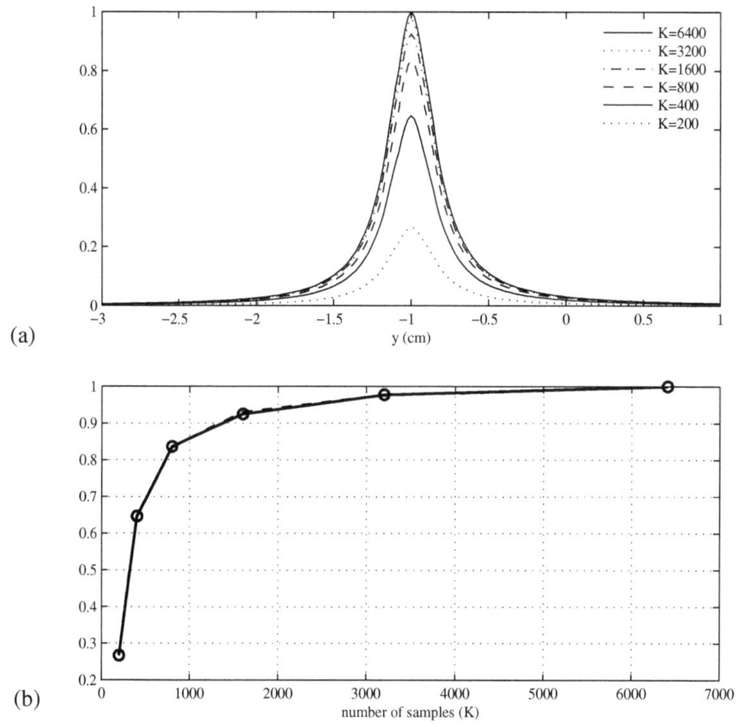

Figure 10.1: (a) The cross-sectional profiles of the reconstructed source at $(0, -1, -6)$ for six values of the number of time samples, K, used for computing the sample covariance matrix. Each profile was obtained by averaging fifty Monte Carlo results, with the input SNR equal to $2M$. (b) The plot of the peak intensities of the reconstructed profiles with respect to the number of time samples, K. The theoretical change in the peak-intensity with respect to the number of samples is plotted with the broken line. (The theoretical plot almost completely overlaps with the plot from the computer simulation.) These theoretical results were obtained using Eq. (10.16).

(a) 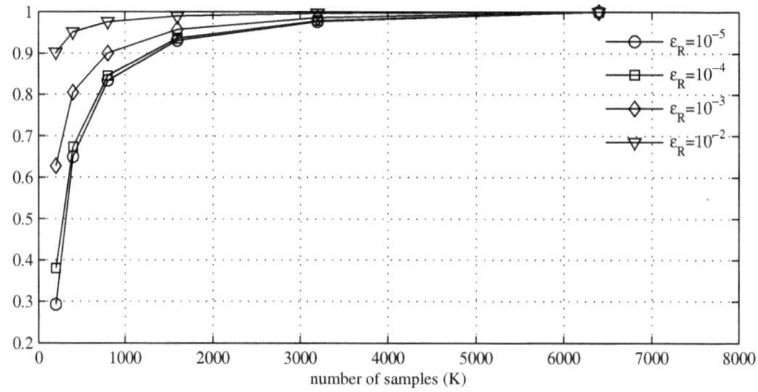

(b) 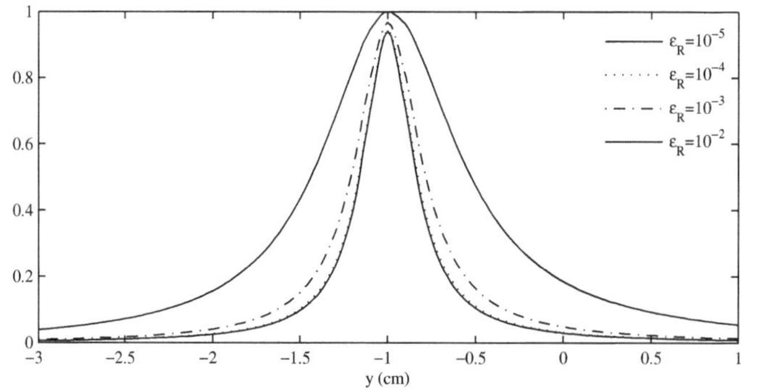

Figure 10.2: (a) The peak intensity of the reconstructed profiles obtained using the array-gain minimum-variance filter with diagonal loading. The peak intensity is plotted with respect to the number of samples, K, with four values of ε_R: $\varepsilon_R = 10^{-2}$, $\varepsilon_R = 10^{-3}$, $\varepsilon_R = 10^{-4}$, and $\varepsilon_R = 10^{-5}$. (b) The cross-sectional profiles of the reconstructed point source at $(0, -1, -6)$ for the four values of ϵ_R. The data with K equal to 1600 were used. These profiles were obtained by averaging fifty Monte-Carlo results of $\widehat{P}_{\widehat{\boldsymbol{R}}}(\boldsymbol{r})$.

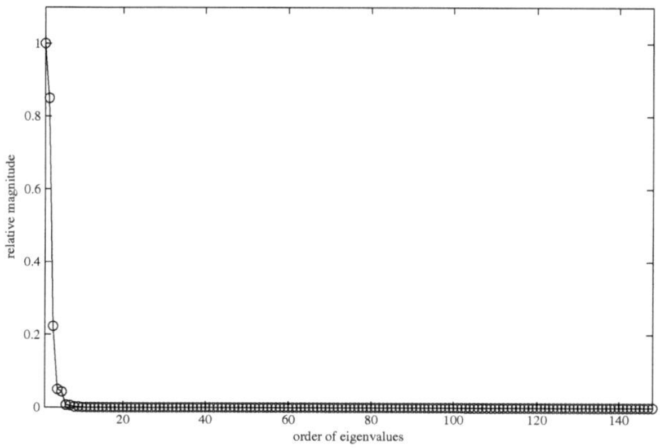

Figure 10.3: Eigenvalue spectrum of $\boldsymbol{G}_R$ calculated using Eq. (10.21) with Ω_R set equal to a 4-cm cubic region with its center at the source location.

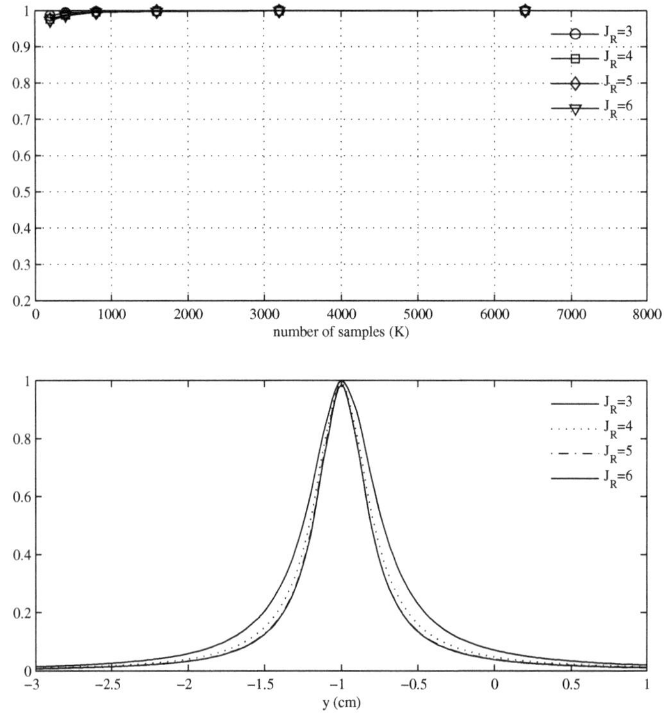

Figure 10.4: (a) The peak intensities of the reconstructed profiles with respect to the number of samples. The beamspace technique was used with four choices of J_R ($J_R = 3$, 4, 5, and 6). (b) The cross-sectional profiles of the reconstructed point source at $(0, -1, -6)$. The peak profiles were obtained with the beamspace technique for these four choices of J_R. The data with K equal to 200 were used. These profiles were obtained by averaging fifty Monte-Carlo results of $\widehat{P}_{\widehat{\boldsymbol{R}}}(\boldsymbol{r})$.

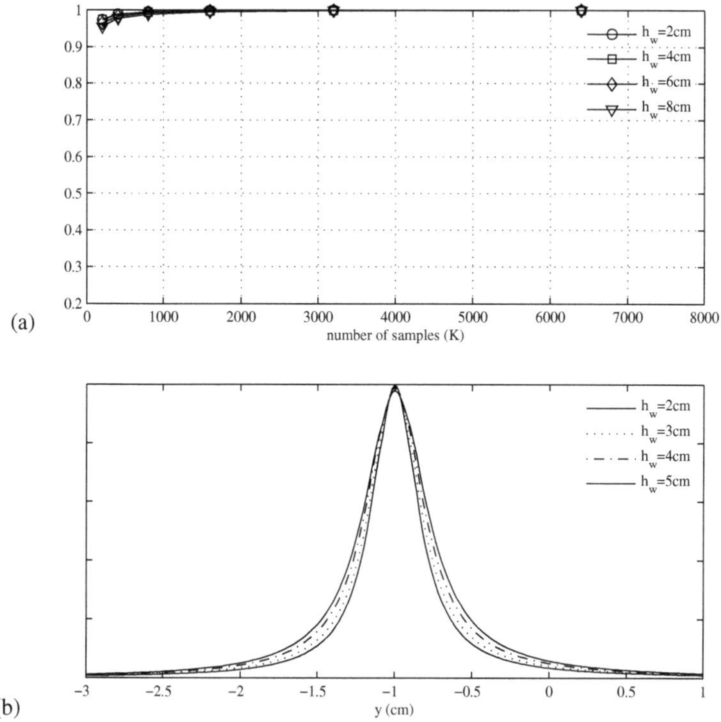

Figure 10.5: (a) The peak intensity of the reconstructed profiles with respect to the number of samples. The beamspace technique was used with four values of h_w: 2, 3, 4, and 5, where h_w is the length in cm of each side of the cubic region Ω_R. (b) The cross-sectional profiles of the reconstructed point source at $(0, -1, -6)$. The peak profiles were obtained with the beamspace technique with these four values of h_w. The data with K equal to 200 were used. These profiles were obtained by averaging fifty Monte-Carlo results.

(a)

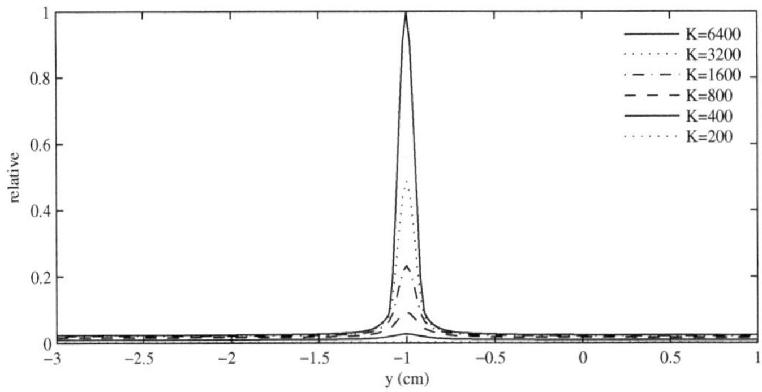

(b)

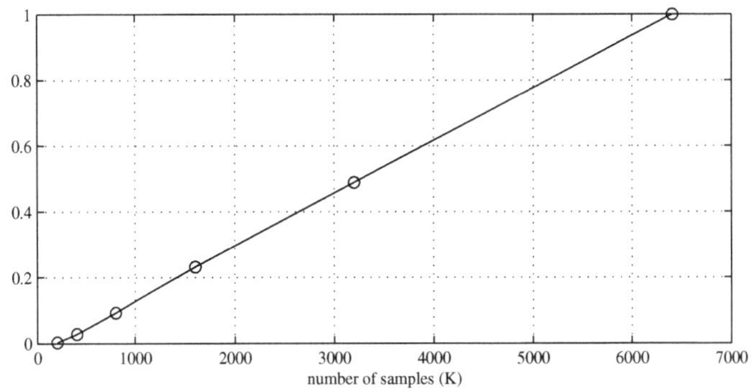

Figure 10.6: Results for the unit-noise-gain minimum-variance filter. (a) The cross-sectional profiles of the reconstructed point source at $(0, -1, -6)$ for the six sampling frequencies. The peak profiles were obtained by averaging fifty Monte-Carlo results, and the input SNR was set to $2M$. (b) The plot of the peak intensities of the reconstructed profiles with respect to the number of time points K.

(a)
(b)

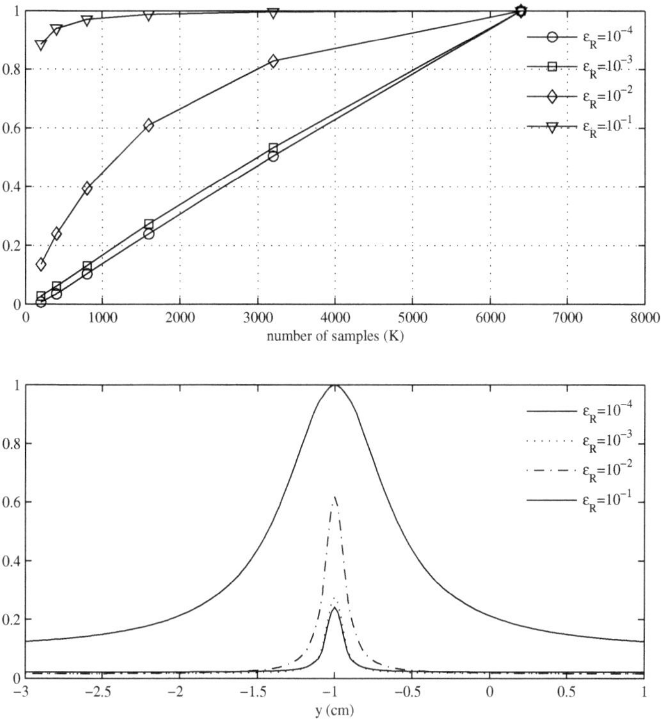

Figure 10.7: Results for the unit-noise-gain minimum-variance filter with diagonal loading. (a) The peak intensities of the reconstructed profiles plotted with respect to the number of samples, K, with four values of ε_R: $\varepsilon_R = 10^{-2}$, $\varepsilon_R = 10^{-3}$, $\varepsilon_R = 10^{-4}$, and $\varepsilon_R = 10^{-5}$. (b) The cross-sectional profiles of the reconstructed point source at $(0, -1, -6)$ with these four values of ε_R. These profiles were obtained by averaging fifty Monte-Carlo results. The data with K equal to 1600 were used.

(a)

(b)

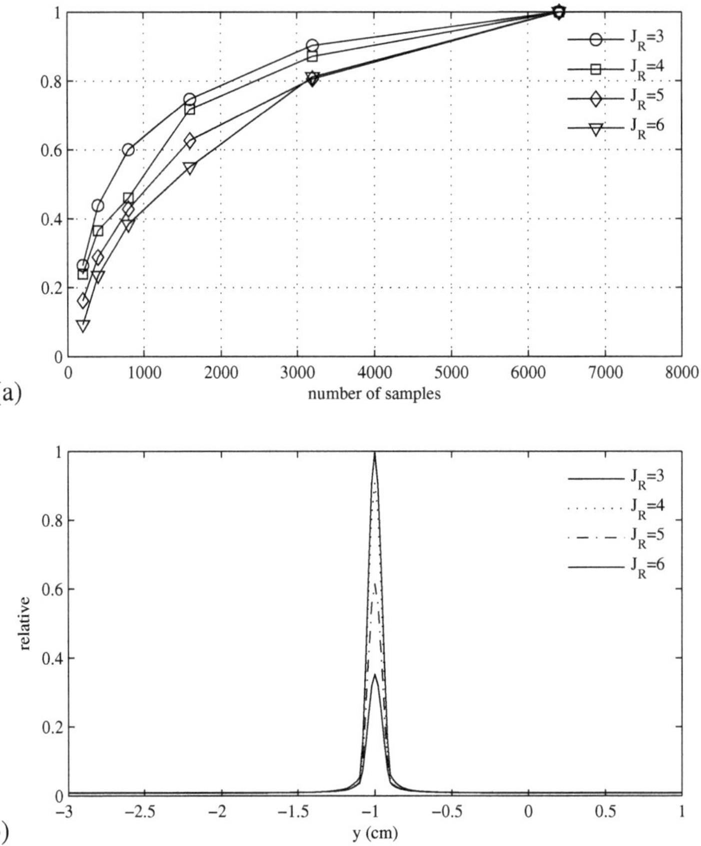

Figure 10.8: Results for the unit-noise-gain minimum-variance filter with beamspace processing. (a) The peak intensities of the reconstructed profiles with respect to the number of samples. The beamspace technique was used with four choices of J_R ($J_R = 3$, 4, 5, and 6) (b) The cross-sectional profiles of the reconstructed point source at $(0, -1, -6)$ with these four choices of J_R. These profiles were obtained by averaging fifty Monte-Carlo results. The data with K equal to 200 were used.

Chapter 11

Statistical evaluation of the spatial filter output

This chapter describes statistical methods for assessing the statistical significance of spatial filter output. Following a brief description of a method based on the Gaussianity assumption, we introduce methods based on non-parametric statistics. These methods first derive, at each voxel location, an empirical probability distribution of the spatial filter output. A statistical threshold is then derived from the empirical distributions obtained from all voxel locations. Here, we describe two methods that can address the multiple comparison problem.

11.1 Problem with Gaussian-distribution-based methods

The statistical significance of the spatial filter output can be evaluated using a method that is based on the Gaussian distribution[71][25], by assuming that the measurement consists of deterministic signal and Gaussian noise, i.e.,

$$\boldsymbol{b}(t) = \boldsymbol{b}_s(t) + \boldsymbol{n}(t), \tag{11.1}$$

where $\boldsymbol{b}_s(t)$ is the signal of interest, i.e., the signal generated by the brain sources that are the target of the investigation. Here, $\boldsymbol{n}(t)$ is the noise vector and each element of $\boldsymbol{n}(t)$ is assumed to be distributed as $\mathcal{N}(0, \sigma_0^2)$, which indicates a Gaussian distribution with zero mean and a variance of σ_0^2. The spatial filter output, $\widehat{s}(\boldsymbol{r}, t)$, is expressed as

$$\widehat{s}(\boldsymbol{r}, t) = \boldsymbol{w}^T(\boldsymbol{r})\boldsymbol{b}(t) = \boldsymbol{w}^T(\boldsymbol{r})\boldsymbol{b}_s(t) + \boldsymbol{w}^T(\boldsymbol{r})\boldsymbol{n}(t). \tag{11.2}$$

Therefore, assuming these Gaussian processes are uncorrelated between different sensor recordings, the spatial-filter output $\widehat{s}(\boldsymbol{r}, t)$ is distributed as $\mathcal{N}(\boldsymbol{w}^T(\boldsymbol{r})\boldsymbol{b}_s(t), \sigma_0^2\|\boldsymbol{w}(\boldsymbol{r})\|^2)$, which is a Gaussian distribution with a mean of

$\boldsymbol{w}^T(\boldsymbol{r})\boldsymbol{b}_s(t)$ and a variance of $\sigma_0^2\|\boldsymbol{w}(\boldsymbol{r})\|^2$. Actually, since σ_0^2 must be estimated from the measured data, the distribution of $\widehat{s}(\boldsymbol{r},t)$ is not exactly represented by a Gaussian distribution but by a t distribution. The statistical evaluation can be performed by testing the null hypothesis that there is no signal source activity at each voxel location. That is, the t score under the null hypothesis, $\widehat{s}(\boldsymbol{r},t)/(\sigma_0\|\boldsymbol{w}(\boldsymbol{r})\|)$, is calculated and compared to $t_{\alpha/2}$, which is the two-tailed t score corresponding to a significance level of α, which is equal to the probability of a Type I error. This procedure is performed at each voxel location, and if the calculated t score is higher than $t_{\alpha/2}$, the estimated source activity, $\widehat{s}(\boldsymbol{r},t)$, is considered to be statistically significant. We can proceed with a multiple comparisons step by using the false discovery rate[72][73], which is described in Section 11.2.4.

The problem with the above approach is that the signal and noise model expressed in Eq. (11.1) is, in general, insufficient to express the actual measurements. As discussed in Chapter 8, the major cause of errors in electromagnetic brain imaging is not the Gaussian sensor noise but background interferences, which are the brain signals generated by sources other than the sources of interest. By denoting such interferences $\boldsymbol{b}_I(t)$ (as denoted in Chapter 8), the measured data is expressed as

$$\boldsymbol{b}(t) = \boldsymbol{b}_s(t) + \boldsymbol{b}_I(t) + \boldsymbol{n}(t) = \boldsymbol{b}_s(t) + \boldsymbol{b}_{i+n}(t), \tag{11.3}$$

where $\boldsymbol{b}_{i+n}(t) = \boldsymbol{b}_I(t) + \boldsymbol{n}(t)$. The spatial filter output obtained from $\boldsymbol{b}(t)$ is expressed as

$$\widehat{s}(\boldsymbol{r},t) = \boldsymbol{w}^T(\boldsymbol{r})\boldsymbol{b}_s(t) + \boldsymbol{w}^T(\boldsymbol{r})\boldsymbol{b}_{i+n}(t) = \widehat{s}_0(\boldsymbol{r},t) + \widehat{s}_{i+n}(\boldsymbol{r},t), \tag{11.4}$$

where

$$\widehat{s}_0(\boldsymbol{r},t) = \boldsymbol{w}^T(\boldsymbol{r})\boldsymbol{b}_s(t), \tag{11.5}$$

$$\text{and} \quad \widehat{s}_{i+n}(\boldsymbol{r},t) = \boldsymbol{w}^T(\boldsymbol{r})\boldsymbol{b}_{i+n}(t). \tag{11.6}$$

Here, $\widehat{s}_0(\boldsymbol{r},t)$ is the estimated signal-source activity of interest, and $\widehat{s}_{i+n}(\boldsymbol{r},t)$ is the spatial filter output due to the background interference and the sensor noise. The null hypothesis is that there is no signal source activity, and the statistical significance of the filter output, $\widehat{s}(\boldsymbol{r},t)$, is assessed using the probability distribution of $\widehat{s}_{i+n}(\boldsymbol{r},t)$ under this null hypothesis. The problem here is that the probability distribution of $\widehat{s}_{i+n}(\boldsymbol{r},t)$ is generally unknown. This has lead us to develop methods based on nonparametric statistics, which are described in the following sections.

11.2 Evaluation of statistical significance using nonparametric statistics

11.2.1 Voxel-by-voxel statistical significance test

Here, we describe methods based on nonparametric statistics in which the probability distribution at each voxel is determined from the data, and a priori assumption of Gaussianity is not needed. The prerequisite of these methods is that multiple measurements of $\boldsymbol{b}_{i+n}(t)$ be available. These multiple measurements of $\boldsymbol{b}_{i+n}(t)$ are denoted with an index β, as $\boldsymbol{b}_{i+n}^{\beta}(t)$ where $\beta = 1, \ldots, K_\beta$, and K_β is the total number of measurements. We can then obtain multiple $\widehat{s}_{i+n}(\boldsymbol{r}, t)$ such that

$$\widehat{s}_{i+n}^{\beta}(\boldsymbol{r}, t) = \boldsymbol{w}^T(\boldsymbol{r})\boldsymbol{b}_{i+n}^{\beta}(t) \quad \beta = 1, \ldots, K_\beta. \tag{11.7}$$

We can now derive the empirical cumulative null distribution of $\widehat{s}_{i+n}(\boldsymbol{r}, t)$ using these $\widehat{s}_{i+n}^{\beta}(\boldsymbol{r}, t)$. That is, we calculate $\widehat{F}(x)$, which is the empirical cumulative distribution of $|\widehat{s}_{i+n}^{\beta}(\boldsymbol{r}, t)|$, such that $\widehat{F}(x) = \sharp\{|\widehat{s}_{i+n}^{\beta}(\boldsymbol{r}, t)| \le x\}/K_\beta$ where $\sharp\{|\widehat{s}_{i+n}^{\beta}(\boldsymbol{r}, t)| \le x\}$ indicates the number of $|\widehat{s}_{i+n}^{\beta}(\boldsymbol{r}, t)|$ that is less than or equal to x. This procedure is repeated and the empirical distribution is calculated at all voxel locations. Since $\widehat{F}(x)$ is obtained at each voxel location $\boldsymbol{r}$ and each time point t, $\widehat{F}(x)$ is rewritten as $\widehat{F}(x|\boldsymbol{r}, t)$ in the following.

Using $\widehat{F}(x|\boldsymbol{r}, t)$, we could obtain the statistical threshold, $\Sigma(\boldsymbol{r}, t)$, such that $\Sigma(\boldsymbol{r}, t) = \widehat{F}^{-1}(1 - \alpha|\boldsymbol{r}, t)$ where α is the level of significance. In practice, the inverse of the empirical cumulative distribution can be calculated by first sorting $|\widehat{s}_{i+n}^{\beta}(\boldsymbol{r}, t)|$ in increasing order:

$$|\widehat{s}_{i+n}^{(1)}(\boldsymbol{r}, t)| \le |\widehat{s}_{i+n}^{(2)}(\boldsymbol{r}, t)| \le \cdots \le |\widehat{s}_{i+n}^{(K_\beta)}(\boldsymbol{r}, t)|, \tag{11.8}$$

where $|\widehat{s}_{i+n}^{(\beta)}(\boldsymbol{r}, t)|$ is the βth smallest value among $|\widehat{s}_{i+n}^{1}(\boldsymbol{r}, t)|, \ldots, |\widehat{s}_{i+n}^{K_\beta}(\boldsymbol{r}, t)|$. Then, we could use $|\widehat{s}_{i+n}^{(q)}(\boldsymbol{r}, t)|$ as the threshold value where $q = \lfloor(1 - \alpha)K_\beta\rfloor$, and $\lfloor\cdot\rfloor$ indicates the maximum integer that does not exceed the value in the parentheses. However, the statistical threshold obtained in this manner does not take multiple comparisons into consideration, and it generally leads to a situation in which many false-positive voxels arise, i.e., many voxels that do not contain brain activation are found to be active. That is, denoting the total number of voxels N, αN false-positive voxels are, on average, found to be statistically significant. To avoid this problem, the statistical significance is determined using a procedure that takes multiple comparisons into account. We describe two procedures: one is a procedure based on the maximum statistics described in the following section, and the other is a procedure based on the false-discovery rate, which is described in Section 11.2.4.

11.2.2 Multiple comparisons using maximum statistics

Here, we use maximum statistics [74][75][76] to address the multiple comparison problem. To utilize maximum statistics, we first standardize the empirical cumulative distribution of $|\widehat{s}_{i+n}^{\beta}(\boldsymbol{r}, t)|$ by calculating $T^\beta(\boldsymbol{r}, t)$ such that

$$T^\beta(\boldsymbol{r}, t) = \frac{|\widehat{s}_{i+n}^{\beta}(\boldsymbol{r}, t)| - \langle|\widehat{s}_{i+n}^{\beta}(\boldsymbol{r}, t)|\rangle_\beta}{\widehat{\sigma}_\beta(\boldsymbol{r}, t)}. \tag{11.9}$$

Here,

$$\widehat{\sigma}_\beta^2(\boldsymbol{r}, t) = \langle\widehat{s}_{i+n}^{\beta}(\boldsymbol{r}, t)^2\rangle_\beta - \langle|\widehat{s}_{i+n}^{\beta}(\boldsymbol{r}, t)|\rangle_\beta^2,$$

and $\langle\cdot\rangle_\beta$ indicates the average over the index β, i.e.,

$$\langle \widehat{s}^{\beta}_{i+n}(\boldsymbol{r},t)^2\rangle_\beta = \frac{1}{K_\beta}\sum_{\beta=1}^{K_\beta} \widehat{s}^{\beta}_{i+n}(\boldsymbol{r},t)^2,$$

$$\text{and} \quad \langle |\widehat{s}^{\beta}_{i+n}(\boldsymbol{r},t)|\rangle_\beta = \frac{1}{K_\beta}\sum_{\beta=1}^{K_\beta} |\widehat{s}^{\beta}_{i+n}(\boldsymbol{r},t)|.$$

We then calculate the maximum $T^\beta(\boldsymbol{r},t)$ among $T^1(\boldsymbol{r},t),\ldots,T^{K_\beta}(\boldsymbol{r},t)$ at each voxel location and at each time point. This maximum value at the ith voxel location and at the kth time point is denoted $T^{i,k}_{max}$, where $i = 1,\ldots,N$, and $k = 1,\ldots,K$. Here, N again indicates the total number of voxels and K indicates the total number of time points. We next obtain the empirical distribution of $T^{i,k}_{max}$, $\widehat{H}(x)$, such that $\widehat{H}(x) = \sharp\{T^{i,k}_{max} \le x\}/(NK)$, where $\sharp\{T^{i,k}_{max} \le x\}$ is the number of $T^{i,k}_{max}$ values which are less than or equal to x. We can then obtain the threshold of the $T^{i,k}_{max}$ value for the α-significance level, T^{th}_{max}, such that $T^{th}_{max} = \widehat{H}^{-1}(1-\alpha)$. The inverse of this empirical cumulative distribution can be calculated by first sorting $T^{i,k}_{max}$ in increasing order:

$$T^{(1)}_{max} \le T^{(2)}_{max} \le \cdots \le T^{(NK)}_{max}, \tag{11.10}$$

and choosing $T^{(p)}_{max}$ as T^{th}_{max} where $p = \lfloor(1-\alpha)NK\rfloor$. We finally obtain the statistical threshold for the spatial-filter reconstruction, $\Sigma(\boldsymbol{r},t)$, by converting T^{th}_{max} into a source activity value, that is,

$$\Sigma(\boldsymbol{r},t) = T^{th}_{max}\widehat{\sigma}_\beta(\boldsymbol{r},t) + \langle |\widehat{s}^{\beta}_{i+n}(\boldsymbol{r},t)|\rangle_\beta. \tag{11.11}$$

We evaluate the statistical significance of the spatial filter output by comparing the output $|\widehat{s}(\boldsymbol{r},t)|$ with $\Sigma(\boldsymbol{r},t)$. When $|\widehat{s}(\boldsymbol{r},t)| \ge \Sigma(\boldsymbol{r},t)$, the output $\widehat{s}(\boldsymbol{r},t)$ is considered to be statistically significant. Note that $\widehat{s}(\boldsymbol{r},t)$ is obtained using $\widehat{s}(\boldsymbol{r},t) = \boldsymbol{w}^T(\boldsymbol{r})\boldsymbol{b}(t)$ in Eq. (11.4).

11.2.3 Modification for power image

A slight modification of the above-mentioned method gives the statistical threshold for the power image $\widehat{P}(\boldsymbol{r}) = \langle \widehat{s}(\boldsymbol{r},t)^2\rangle$. To derive the threshold for $\widehat{P}(\boldsymbol{r})$, we calculate a power reconstruction using $\boldsymbol{b}^{\beta}_{i+n}(t)$, such that

$$\widehat{P}^{\beta}_{i+n}(\boldsymbol{r}) = \langle(\boldsymbol{w}^T(\boldsymbol{r})\boldsymbol{b}^{\beta}_{i+n}(t))^2\rangle. \tag{11.12}$$

We then standardize $\widehat{P}^{\beta}_{i+n}(\boldsymbol{r})$ using

$$T^\beta(\boldsymbol{r}) = \frac{\widehat{P}^{\beta}_{i+n}(\boldsymbol{r}) - \langle \widehat{P}^{\beta}_{i+n}(\boldsymbol{r})\rangle_\beta}{\widehat{\sigma}_\beta(\boldsymbol{r})}, \tag{11.13}$$

where

$$\widehat{\sigma}^2_\beta(\boldsymbol{r}) = \langle \widehat{P}^\beta_{i+n}(\boldsymbol{r})^2 \rangle_\beta - \langle \widehat{P}^\beta_{i+n}(\boldsymbol{r}) \rangle^2_\beta,$$

$$\langle \widehat{P}^\beta_{i+n}(\boldsymbol{r})^2 \rangle_\beta = \frac{1}{K_\beta} \sum_{\beta=1}^{K_\beta}, \widehat{P}^\beta_{i+n}(\boldsymbol{r})^2, \tag{11.14}$$

$$\text{and} \quad \langle \widehat{P}^\beta_{i+n}(\boldsymbol{r}) \rangle_\beta = \frac{1}{K_\beta} \sum_{\beta=1}^{K_\beta} \widehat{P}^\beta_{i+n}(\boldsymbol{r}).$$

We then calculate the maximum $T^\beta(\boldsymbol{r})$ among $\beta = 1, \ldots, K_\beta$ at each voxel location. This maximum value at the ith voxel location is denoted T^i_{max} where $i = 1, \ldots, N$ and N again indicates the total number of voxels. We next obtain the empirical cumulative distribution of T^i_{max}, $\widehat{H}(x)$, such that $\widehat{H}(x) = \sharp\{T^i_{max} \le x\}/N$, where $\sharp\{T^i_{max} \le x\}$ is the number of T^i_{max} values which are less than or equal to x. We can then obtain the threshold of the T^i_{max} value for the α-significance level, T^{th}_{max}, such that $T^{th}_{max} = \widehat{H}^{-1}(1-\alpha)$. The inverse of this empirical distribution can be calculated by first sorting T^i_{max} in increasing order:

$$T^{(1)}_{max} \le T^{(2)}_{max} \le \cdots \le T^{(N)}_{max}, \tag{11.15}$$

and choosing $T^{(p)}_{max}$ as T^{th}_{max} where $p = \lfloor (1-\alpha)N \rfloor$. We finally obtain the statistical threshold for the spatial-filter reconstruction, $\Sigma(\boldsymbol{r})$, by converting T^{th}_{max} into a source activity value, that is,

$$\Sigma(\boldsymbol{r}) = T^{th}_{max} \widehat{\sigma}_\beta(\boldsymbol{r}) + \langle \widehat{P}^\beta_{i+n}(\boldsymbol{r}) \rangle_\beta. \tag{11.16}$$

11.2.4 Multiple comparisons using the false discovery rate

The multiple-comparison problem can also be addressed using the false discovery rate[72], and this method can replace the use of the maximum statistics for multiple comparisons. To explain the false discovery rate, we make some definitions as follows:

K_{aa} : the numbers of voxels that are truly active and discovered to be active,
K_{ai} : the numbers of voxels that are truly active but discovered to be inactive,
K_{ia} : the numbers of voxels that are truly inactive but discovered to be active,
K_{ii} : the numbers of voxels that are truly inactive and discovered to be inactive.

The false discovery rate, F_D, is then defined such that

$$F_D = \frac{K_{ia}}{K_{aa} + K_{ia}}. \tag{11.17}$$

That is, the false discovery rate is the ratio of the number of voxels that are truly inactive but discovered to be active to the total number of voxels discovered to be active. The procedure described below, which was introduced by Benjamini and

Hochberg[73], guarantees the expectation of this false discovery rate, F_D, to be less than a user-specified value $\mathcal{A}$. That is, the procedure guarantees that

$$\langle F_D \rangle \leq \frac{K_{ia} + K_{ii}}{N} \mathcal{A} \leq \mathcal{A}, \tag{11.18}$$

where $\langle F_D \rangle$ denotes the expected value of F_D and N is again the total number of voxels. Note that the ratio $(K_{ia} + K_{ii})/N$ is less than but close to 1, because it is the ratio of the number of inactive voxels to the total number of voxels. What is guaranteed in Eq. (11.18) is that if one replicates the same experiment many times, the average of F_D over those replications is less than the user-specified value $\mathcal{A}$. For any particular case, the obtained F_D might be larger than $\mathcal{A}$.

The procedure using the false discovery rate is summarized as follows.

(I) Specify a desired value for $\mathcal{A}$ between 0 and 1.

(II) Calculate, at each voxel location, the probability for the event that the voxel value exceeds $|\widehat{s}(\boldsymbol{r}, t)|$. Assuming that we have the empirical distribution of $|\widehat{s}^{\beta}_{i+n}(\boldsymbol{r}, t)|$ ($\beta = 1, \ldots, K_\beta$), this probability can be obtained first by sorting $|\widehat{s}(\boldsymbol{r}, t)|$ and $|\widehat{s}^{\beta}_{i+n}(\boldsymbol{r}, t)|$ in increasing order, such that

$$|\widehat{s}^{(1)}_{i+n}(\boldsymbol{r}, t)| \leq \cdots \leq |\widehat{s}^{(m)}_{i+n}(\boldsymbol{r}, t)| \leq |\widehat{s}(\boldsymbol{r}, t)| \leq |\widehat{s}^{(m+1)}_{i+n}(\boldsymbol{r}, t)| \leq \cdots \leq |\widehat{s}^{(K_\beta)}_{i+n}(\boldsymbol{r}, t)|, \tag{11.19}$$

where again $|\widehat{s}^{(\beta)}_{i+n}(\boldsymbol{r}, t)|$ is the βth smallest $|\widehat{s}_{i+n}(\boldsymbol{r}, t)|$. Here, we assume that $|\widehat{s}(\boldsymbol{r}, t)|$ is greater than $|\widehat{s}^{(m)}_{i+n}(\boldsymbol{r}, t)|$ but smaller than $|\widehat{s}^{(m+1)}_{i+n}(\boldsymbol{r}, t)|$. If $|\widehat{s}(\boldsymbol{r}, t)|$ is smaller than $|\widehat{s}^{(1)}_{i+n}(\boldsymbol{r}, t)|$, we set m to 1. Then, the probability for the event that the voxel value exceeds $|\widehat{s}(\boldsymbol{r}, t)|$, p, is obtained as

$$p = \frac{(K_\beta - m) + 1}{K_\beta + 1}. \tag{11.20}$$

(III) Sort the p values from all voxel locations in increasing order, such that

$$p_{(1)} \leq p_{(2)} \leq \cdots \leq p_{(N)}, \tag{11.21}$$

where $p_{(j)}$ indicates the jth smallest p value.

(IV) Find the largest j that satisfies the relationship

$$p_{(j)} \leq \frac{j}{N} \mathcal{A}.$$

Denoting such j as j^{max}, $p_{(j^{max})}$ is the threshold value, i.e., the voxels whose p value is smaller than $p_{(j^{max})}$ are determined to be active, and the voxels whose p value is greater than $p_{(j^{max})}$ are determined to be inactive.

11.3 Deriving a voxel-wise empirical null distribution

11.3.1 Method when the signal is time-locked and the interference is non-time-locked to the stimulus

We mentioned that the prerequisite of the nonparametric method described in the preceding sections is that multiple baseline measurements, $\boldsymbol{b}^{\beta}_{i+n}(t)$, (where $\beta = 1, 2, \ldots, K_\beta$) be available. Here, we describe a method of creating multiple $\boldsymbol{b}^{\beta}_{i+n}(t)$ from a single-subject data set. The method is applicable to event-related measurements where multiple raw epochs are measured and averaged. The key assumption here is that the target brain activity and thus the signal measurements, $\boldsymbol{b}_s(t)$, are time-locked to the stimulus, but the non-target brain activity, including $\boldsymbol{b}_I(t)$, are not time-locked to the stimulus. Therefore, in event-related measurements, the signal measurements generated by time-locked activity are obtained by averaging raw-epoch measurements. On the other hand, the background non-time-locked interference, $\boldsymbol{b}_{i+n}(t)$, can be estimated by calculating the plus/minus averages of the raw epochs. That is, half of the epochs are multiplied by -1 before averaging over the epochs. By doing so, the time-locked signals are averaged out, whereas the signals not time-locked to the stimulus are not.

Denoting the raw epoch measurements as $\{\boldsymbol{h}_1(t), \ldots, \boldsymbol{h}_{K_E}(t)\}$ where K_E is the number of raw epochs, we have

$$\boldsymbol{b}(t) = \frac{1}{K_E} \sum_{k=1}^{K_E} \boldsymbol{h}_k(t). \tag{11.22}$$

When K_E is very large or when the interference is very small, this $\boldsymbol{b}(t)$ is approximately equal to $\boldsymbol{b}_s(t)$. However, in all other cases, a considerable amount of the interference still exists in the averaged data, $\boldsymbol{b}(t)$. The non-time-locked interference component existing in the averaged data, $\boldsymbol{b}_{i+n}(t)$, can be estimated from

$$\boldsymbol{b}_{i+n}(t) \approx \frac{1}{K_E} \sum_{k=1}^{K_E} \epsilon_k \boldsymbol{h}_k(t), \tag{11.23}$$

where the coefficients $\epsilon_1, \ldots, \epsilon_{K_E}$ have a value of either -1 or 1. We assign -1 or 1 to $\epsilon_1, \ldots, \epsilon_{K_E}$ by drawing -1 or 1 randomly and without replacement from a pool of a total $K_E/2$ of -1 and a total $K_E/2$ of 1. As a result, one half of $\epsilon_1, \ldots, \epsilon_{K_E}$ have a value of -1 and the other half of $\epsilon_1, \ldots, \epsilon_{K_E}$ have a value of 1[77]. Here, since there are many ways to assign -1 or 1 to $\epsilon_1, \ldots, \epsilon_{K_E}$ (the number of ways is equal to $K_E!/(\frac{K_E}{2}!)^2$), we can obtain many different $\boldsymbol{b}_{i+n}(t)$[78]. We denote each realization $\boldsymbol{b}^{\beta}_{i+n}(t)$, which can be expressed as

$$\boldsymbol{b}^{\beta}_{i+n}(t) = \frac{1}{K_E} \sum_{k=1}^{K_E} \epsilon^*_k \boldsymbol{h}_k(t), \tag{11.24}$$

where $\epsilon_1^*, \ldots, \epsilon_{K_E}^*$ is one realization of the random assignments of -1 or 1. In Eq. (11.24), β is an index for the different ways of assigning -1 and 1, and $\beta = 1, \ldots, K_\beta$, where K_β indicates the total number of different assignments. Once $\boldsymbol{b}_{i+n}^{\beta}(t)$ are obtained for $\beta = 1, \ldots, K_\beta$, we can then obtain multiple $\widehat{s}_{i+n}^{\beta}(\boldsymbol{r}, t)$ using Eq. (11.7), and we can now derive the empirical cumulative distribution of $\widehat{s}_{i+n}(\boldsymbol{r}, t)$ using these $\widehat{s}_{i+n}^{\beta}(\boldsymbol{r}, t)$.

11.3.2 Method when both the signal and the interference are non-time-locked to the stimulus

With a slight modification based on the permutation test[74][79], the above-mentioned method can be applied to the case of imaging induced source activity. Because an induced activity is not time locked to the stimulus, we cannot use waveform-based averaging. Instead, we should use covariance-based averaging. In Section 8.4 we describe two methods, the F-image method and the prewhitening-based method, to reconstruct such induced activities using averaged covariance matrices. To explain the method here, we use the F-image method in order to simplify the explanation. The method is applicable to the prewhitening method in exactly the same manner. The Fourier transforms of the raw-epoch measurements are denoted $\{\boldsymbol{g}_1(f), \ldots, \boldsymbol{g}_{K_E}(f)\}$ for task data sets and $\{\boldsymbol{g}_1^c(f), \ldots, \boldsymbol{g}_{K_E}^c(f)\}$ for control data sets. As mentioned in Section 8.4, the frequency-specific task and control covariance matrices, $\widehat{\boldsymbol{\Gamma}}(F_w)$ and $\widehat{\boldsymbol{\Gamma}}_c(F_w)$, are calculated using

$$\widehat{\boldsymbol{\Gamma}}(F_w) = \frac{1}{K_E} \sum_{k=1}^{K_E} \widehat{\gamma}_k(F_w), \tag{11.25}$$

where $\widehat{\gamma}_k(F_w) = \sum_{f \in F_w} \boldsymbol{g}_k(f)(\boldsymbol{g}_k(f))^H$, and

$$\widehat{\boldsymbol{\Gamma}}_c(F_w) = \frac{1}{K_E} \sum_{k=1}^{K_E} \widehat{\gamma}_k^c(F_w), \tag{11.26}$$

where $\widehat{\gamma}_k^c(F_w) = \sum_{f \in F_w} \boldsymbol{g}_k^c(f)(\boldsymbol{g}_k^c(f))^H$. Here, $\widehat{\gamma}_k(F_w)$ and $\widehat{\gamma}_k^c(F_w)$ are the covariance matrices obtained from the kth epoch data. The task and control power images are obtained using Eqs. (8.46) and (8.47), which are repeated here:

$$\langle \widehat{s}(\boldsymbol{r}, F_w)^2 \rangle = \boldsymbol{w}^T(\boldsymbol{r}, F_w) \widehat{\boldsymbol{\Gamma}}(F_w) \boldsymbol{w}(\boldsymbol{r}, F_w), \tag{11.27}$$

$$\text{and} \quad \langle \widehat{s}_C(\boldsymbol{r}, F_w)^2 \rangle = \boldsymbol{w}^T(\boldsymbol{r}, F_w) \widehat{\boldsymbol{\Gamma}}_c(F_w) \boldsymbol{w}(\boldsymbol{r}, F_w), \tag{11.28}$$

where the weight $\boldsymbol{w}(\boldsymbol{r}, , F_w)$ is obtained using Eq. (8.44). These task and control power images are then used for calculating the F-image such that

$$F(\boldsymbol{r}, F_w) = \frac{\langle \widehat{s}(\boldsymbol{r}, F_w)^2 \rangle - \langle \widehat{s}_C(\boldsymbol{r}, F_w)^2 \rangle}{\langle \widehat{s}_C(\boldsymbol{r}, F_w)^2 \rangle}.$$

Multiple F-images can be created to derive the voxel-wise cumulative distribution using the permutation test in the following manner. That is, under the null hypothesis that there is no activation in the task period, there is no difference between the task and control data sets. Therefore, to calculate the task covariance matrix, we could use

$$\widehat{\boldsymbol{\Gamma}}(F_w) = \frac{1}{K_E}\sum_{k=1}^{K_E} \widehat{\gamma}_k^*(F_w), \tag{11.29}$$

where $\{\widehat{\gamma}_1^*(F_w), \ldots, \widehat{\gamma}_{K_E}^*(F_w)\}$ are drawn randomly and without replacement from the $2K_E$ sets, $\{\widehat{\gamma}_1(t), \ldots, \widehat{\gamma}_{K_E}(t), \widehat{\gamma}_1^c(t), \ldots, \widehat{\gamma}_{K_E}^c(t)\}$. The remaining K_E sets of the epoch-based covariance matrices are denoted $\widehat{\gamma}_1^{c*}(F_w), \ldots, \widehat{\gamma}_{K_E}^{c*}(F_w)$, and they are used to calculate $\widehat{\boldsymbol{\Gamma}}_c(F_w)$, such that

$$\widehat{\boldsymbol{\Gamma}}_c(F_w) = \frac{1}{K_E}\sum_{k=1}^{K_E} \widehat{\gamma}_k^{c*}(F_w). \tag{11.30}$$

Since there are many ways to choose $\{\widehat{\gamma}_1^*(t), \ldots, \widehat{\gamma}_{K_E}^*(t)\}$ and $\{\widehat{\gamma}_1^{c*}(t), \ldots, \widehat{\gamma}_{K_E}^{c*}(t)\}$ ($K_E!/(\frac{K_E}{2}!)^2$ ways, in fact), we have multiple sets of $\widehat{\boldsymbol{\Gamma}}(F_w)$ and $\widehat{\boldsymbol{\Gamma}}_c(F_w)$, which are denoted $\widehat{\boldsymbol{\Gamma}}^{\beta}(F_w)$ and $\widehat{\boldsymbol{\Gamma}}_c^{\beta}(F_w)$, where $\beta = 1, \ldots, K_\beta$ and K_β is the total number of created data sets. Using $\widehat{\boldsymbol{\Gamma}}^{\beta}(F_w)$ and $\widehat{\boldsymbol{\Gamma}}_c^{\beta}(F_w)$, we can calculate multiple F images, $F^\beta(\boldsymbol{r}, F_w)$, where $\beta = 1, \ldots, K_\beta$. These multiple $F^\beta(\boldsymbol{r}, F_w)$ form an empirical null distribution at each voxel location.

11.4 Non-parametric method using reconstructed voxel time courses

The nonparametric method described in Section 11.2 can provide the spatio-temporal threshold, $\Sigma(\boldsymbol{r}, t)$. However, the method requires that multi-epoch measurements be performed and stored in memory. In a practical application, such requirements are not always fulfilled. In addition, the method needs to process a five-dimensional data set and the resultant computational load and memory requirements may be prohibitively large. Here, we present a simpler method in which a time-independent threshold, $\Sigma(\boldsymbol{r})$, is obtained[80]. The prerequisite for the method described here is that a single control measurement be available. In general, this assumption is approximately fulfilled on various occasions.

The method calculates an empirical cumulative distribution at each voxel location using the time-course reconstruction during the control period, $\widehat{s}_c(\boldsymbol{r}, t_k)$ ($k = 1, \ldots, K_c$), where $t_1, \ldots, t_{K_c}$ are the discrete time points during the control period. The assumption here is that the time course reconstruction, $\widehat{s}_c(\boldsymbol{r}, t_k)$, is identically and independently distributed over the control time period, and the probability distribution of $\widehat{s}_c(\boldsymbol{r}, t_k)$ is approximately equal to the distribution of $\widehat{s}_{i+n}(\boldsymbol{r}, t)$ in the task period. Thus, we can use the empirical distribution of $\widehat{s}_c(\boldsymbol{r}, t_k)$ as the probability distribution of $\widehat{s}_{i+n}(\boldsymbol{r}, t)$ for assessing the statistical significance of the spatial filter output, $\widehat{s}(\boldsymbol{r}, t)$, in the task period.

To utilize the maximum statistics, we first standardize the empirical distribution of $|\widehat{s}_c(\boldsymbol{r}, t_k)|$ by calculating $T(\boldsymbol{r}, t_k)$ such that

$$T(\boldsymbol{r}, t_k) = \frac{|\widehat{s}_c(\boldsymbol{r}, t_k)| - \langle|\widehat{s}_c(\boldsymbol{r}, t_k)|\rangle_c}{\widehat{\sigma}_c(\boldsymbol{r})}. \tag{11.31}$$

Here,

$$\widehat{\sigma}_c^2(\boldsymbol{r}) = \langle\widehat{s}_c(\boldsymbol{r}, t_k)^2\rangle_C - \langle|\widehat{s}_c(\boldsymbol{r}, t_k)|\rangle_C^2,$$

and $\langle\cdot\rangle_C$ indicates the time average over the control period, i.e.,

$$\langle\widehat{s}_c(\boldsymbol{r}, t_k)^2\rangle_C = \frac{1}{K_c}\sum_{k=1}^{K_c}\widehat{s}_c(\boldsymbol{r}, t_k)^2,$$

$$\text{and} \quad \langle|\widehat{s}_c(\boldsymbol{r}, t_k)|\rangle_C = \frac{1}{K_c}\sum_{k=1}^{K_c}|\widehat{s}_c(\boldsymbol{r}, t_k)|.$$

We next calculate the maximum T value at each voxel location. The maximum T value at the ith voxel location is denoted T^i_{max}, where $i = 1, \ldots, N$ and N again indicates the total number of voxels. We then obtain the empirical cumulative distribution of T^i_{max}, $\widehat{H}(x)$, such that $\widehat{H}(x) = \sharp\{T^i_{max} \leq x\}/N$, where $\sharp\{T^i_{max} \leq x\}$ is the number of T^i_{max} values which are less than or equal to x. We can then obtain the threshold of the T^i_{max} value for the α-significance level, T^{th}_{max}, such that $T^{th}_{max} = \widehat{H}^{-1}(1-\alpha)$. The inverse of this empirical distribution can be calculated by first sorting T^i_{max} in increasing order:

$$T^{(1)}_{max} \leq T^{(2)}_{max} \leq \cdots \leq T^{(N)}_{max}, \tag{11.32}$$

and choosing $T^{(p)}_{max}$ as T^{th}_{max} where $p = \lfloor(1-\alpha)N\rfloor$. We finally obtain the statistical threshold for the spatial-filter output, $\Sigma(\boldsymbol{r})$, by converting T^{th}_{max} into a source activity value, that is,

$$\Sigma(\boldsymbol{r}) = T^{th}_{max}\widehat{\sigma}_c(\boldsymbol{r}) + \langle|\widehat{s}_c(\boldsymbol{r}, t_k)|\rangle_C. \tag{11.33}$$

We evaluate the statistical significance of the spatial filter output by comparing the output $|\widehat{s}(\boldsymbol{r}, t)|$ with $\Sigma(\boldsymbol{r})$. Then, the output $\widehat{s}(\boldsymbol{r}, t)$ for which $|\widehat{s}(\boldsymbol{r}, t)| \geq \Sigma(\boldsymbol{r})$ are considered to be statistically significant.

It should be mentioned that once we calculate T^i_{max} at each voxel location, the distributional properties of the process don't matter any more and the assumption that the time-course reconstruction is identically and independently distributed is not needed. Instead, this multiple comparison procedure imposes a new assumption that the shape of the distribution of $|\widehat{s}_c(\boldsymbol{r}, t_k)|$ is the same at all voxels, although its scale may be different.

11.5 Numerical examples

We conducted numerical experiments to illustrate some results for the statistical thresholding described in this chapter. We used the same experimental scheme

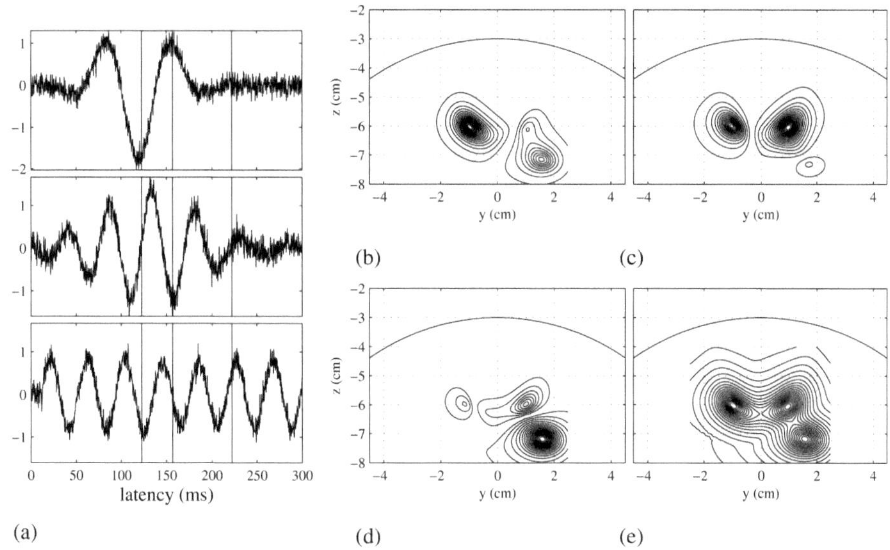

Figure 11.1: Source reconstruction results using the eigenspace spatial filter. (a) The reconstructed time courses at the first source location (top), the second source location (middle), and the third source location (bottom). The three vertical lines indicate the time points for 123 ms, 157 ms, and 222 ms. Reconstructed source activity at (b) 123 ms, (c) 157 ms, and (d) 222 ms. (e) the square root of the power reconstruction, $\sqrt{\langle \widehat{s}(\boldsymbol{r},t)^2 \rangle}$.

used in Section 4.7, and generated, at 600 time points, one hundred raw epochs having the total input SNR, α_T, equal to 0.3. The sampling interval is assumed to be 0.5 ms. We then averaged these one-hundred epochs to obtain the average recordings with α_T equal to 3. The array-gain eigenspace-projected spatial filter was applied to the averaged recordings with the signal-subspace dimension set to 3. The reconstructed time courses at the three source locations are shown in Fig. 11.1(a), and source reconstruction results are shown in Fig. 11.1(b)–(e). We then calculated the statistical threshold using the randomized plus/minus averages of the raw epochs described in Section 11.3.1. The reconstructed time courses with the spatio-temporal threshold are shown in Fig. 11.2(a). The thresholded reconstruction results are shown in Fig. 11.2(b)–(e). In Fig. 11.2(a), the two dotted horizontal lines show the statistical threshold at the three sources locations, $\pm\Sigma(\boldsymbol{r}_1,t)$, $\pm\Sigma(\boldsymbol{r}_2,t)$, and $\pm\Sigma(\boldsymbol{r}_3,t)$. In these figures, source activities greater than the threshold values are considered statistically significant. Reconstructed source activities lower than these threshold values (for example, the first and second sources at 222 ms) were set equal to zero.

Next, we used the method described in Section 11.4 to calculate the statistical threshold. The reconstructed time courses with obtained threshold values are

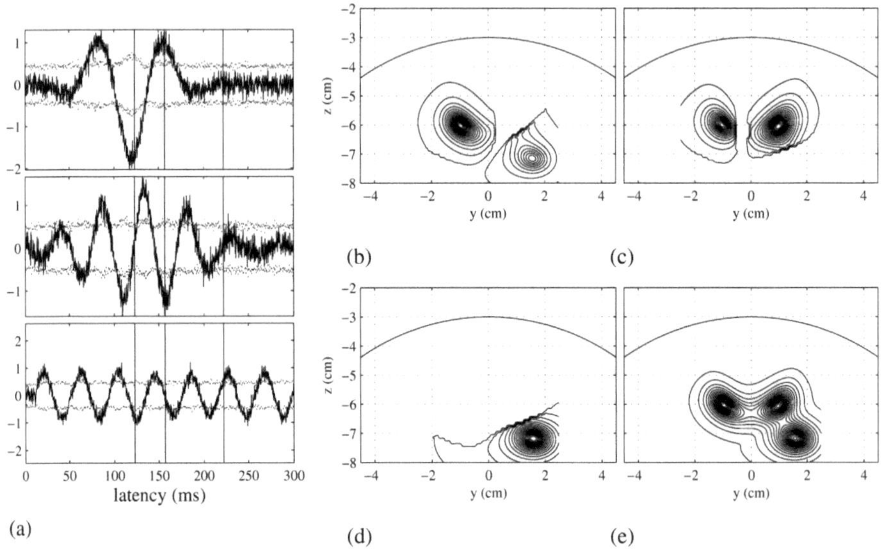

Figure 11.2: (a) Reconstructed time courses and statistical threshold at the first source location (top), the second source location (middle), and the third source location (bottom). The three vertical lines indicate the time points for 123 ms, 157 ms, and 222 ms. The two (fluctuating) dotted horizontal lines show the spatio-temporal statistical threshold $\pm\Sigma(\boldsymbol{r}, t)$. Thresholded source reconstruction results at (b) 123 ms, (c) 157 ms, and (d) 222 ms. (e) the square root of the thresholded power reconstruction, $\sqrt{\langle \widehat{s}(\boldsymbol{r}, t)^2 \rangle}$. Statistical threshold was obtained using the nonparametric method in Section 11.2 with the randomized plus/minus averages described in Section 11.3.1.

shown in Fig. 11.3(a). In this case, the threshold values are time-independent so they are shown as straight horizontal lines in this figure. The thresholded reconstruction results are shown in Fig. 11.3(b)–(d).

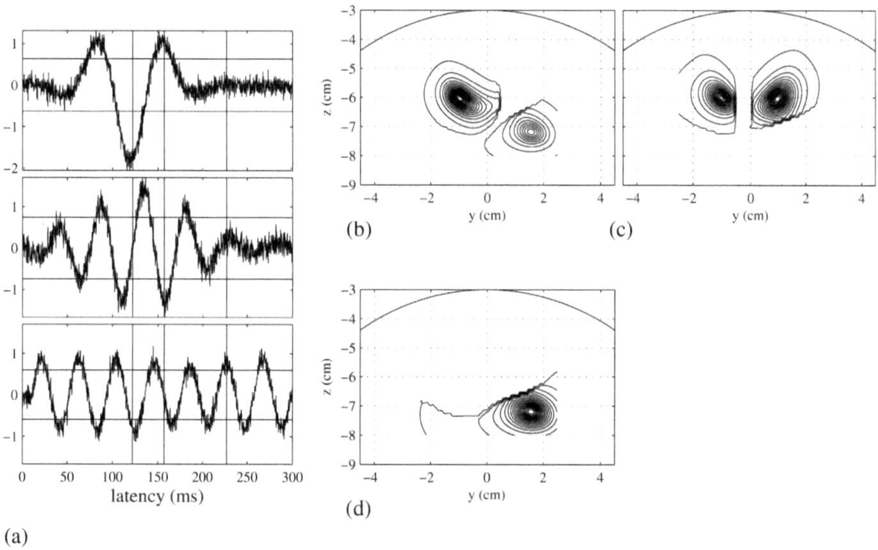

Figure 11.3: (a) Reconstructed time courses and statistical threshold at the first source location (top), the second source location (middle), and the third source location (bottom). The two horizontal lines show the statistical threshold, $\pm\Sigma(\boldsymbol{r})$. Thresholded source activity at (b) 123 ms, (c) 157 ms, and (d) 222 ms. The time-independent statistical threshold were obtained using the method described in Section 11.4.

Chapter 12

Methods related to adaptive spatial filters

This chapter describes methods closely related to adaptive spatial filters. The first one is the Wiener filter, from which the formulation for the minimum-variance spatial filter can be directly derived. The second one is the multiple signal classification (MUSIC) algorithm. This algorithm explicitly exploits the orthogonality relationship between the noise subspace and the sensor lead field at the source locations, and it can localize the multiple sources through a three-dimensional search. We will discuss the fact that the algorithm has a close relationship with adaptive spatial filters. The third one is the generalized likelihood ratio test (GLRT) scanning method. We will show the similarity between the final form of the GLRT scanning function and adaptive spatial-filter power reconstruction.

12.1 Wiener filter

12.1.1 Minimum-mean-squared-error criterion

The Wiener filter formulation can be obtained by minimizing the cost function in Eq. (3.13) with $\boldsymbol{\varUpsilon}_A$ equal to the noise covariance matrix, $\boldsymbol{R}_n$, and $\boldsymbol{\varUpsilon}_B$ equal to the source covariance matrix, $\boldsymbol{R}_\nu$. We simplify the arguments in this section by assuming that the source orientation is predetermined, and define the composite lead-field matrix at all voxel locations as $\boldsymbol{L}_\mathcal{V}$, such that

$$\boldsymbol{L}_\mathcal{V} = [\boldsymbol{l}(\boldsymbol{r}_1), \boldsymbol{l}(\boldsymbol{r}_2), \ldots, \boldsymbol{l}(\boldsymbol{r}_N)]. \tag{12.1}$$

Note that the columns of $\boldsymbol{L}_\mathcal{V}$ are the lead-field vectors in the predetermined orientations[1]. Then, by setting $\epsilon\boldsymbol{\varUpsilon}_A = \boldsymbol{R}_n$, $\boldsymbol{\varUpsilon}_B = \boldsymbol{R}_{vox}$, and $\boldsymbol{L}_V = \boldsymbol{L}_\mathcal{V}$ in Eq. (3.14),

[1]The composite lead-field matrix, $\boldsymbol{L}_\mathcal{V}$, is different from the composite lead-field matrix, $\boldsymbol{L}_V$, in Eq. (3.2), because in $\boldsymbol{L}_V$ the columns are the lead-field vectors in the x, y, and z directions, whereas in $\boldsymbol{L}_\mathcal{V}$ the columns are the lead-field vectors in predetermined directions.

the formula of the Wiener estimate can be obtained as

$$\widehat{\boldsymbol{\nu}}_{vox}(t) = \boldsymbol{R}_{vox}\boldsymbol{L}_{\mathcal{V}}^T(\boldsymbol{L}_{\mathcal{V}}\boldsymbol{R}_{vox}\boldsymbol{L}_{\mathcal{V}}^T + \boldsymbol{R}_n)^{-1}\boldsymbol{b}(t), \tag{12.2}$$

where $\widehat{\boldsymbol{\nu}}_{vox}(t)$ is the estimated source distribution at voxel locations defined in Eq. (3.5), which is

$$\widehat{\boldsymbol{\nu}}_{vox}(t) = \begin{bmatrix} \widehat{s}(\boldsymbol{r}_1, t) \\ \widehat{s}(\boldsymbol{r}_2, t) \\ \vdots \\ \widehat{s}(\boldsymbol{r}_N, t) \end{bmatrix}.$$

The voxel-based source covariance matrix $\boldsymbol{R}_{vox}$ is expressed as

$$\boldsymbol{R}_{vox} = \begin{bmatrix} \langle s(\boldsymbol{r}_1,t)^2\rangle & \cdots & \langle s(\boldsymbol{r}_1,t)s(\boldsymbol{r}_N,t)\rangle \\ \vdots & \ddots & \vdots \\ \langle s(\boldsymbol{r}_N,t)s(\boldsymbol{r}_1,t)\rangle & \cdots & \langle s(\boldsymbol{r}_N,t)^2\rangle \end{bmatrix}, \tag{12.3}$$

where the bracket $\langle \cdot \rangle$ indicates the expectation operator.

The formula in Eq. (12.2) can also be obtained by minimizing the squared error between the true and the estimated source spatial distributions[81][82]. We define the weight matrix for an inverse filter that is used to estimate $\widehat{\boldsymbol{\nu}}_{vox}(t)$ as $\boldsymbol{W}_{\mathrm{inv}}$, i.e.,

$$\widehat{\boldsymbol{\nu}}_{vox}(t) = \boldsymbol{W}_{\mathrm{inv}}\boldsymbol{b}(t), \tag{12.4}$$

The square error between $\boldsymbol{\nu}_{vox}(t)$ and $\widehat{\boldsymbol{\nu}}_{vox}(t)$ is given by

$$\begin{aligned}\mathcal{F} &= \langle[\boldsymbol{\nu}_{vox}(t) - \widehat{\boldsymbol{\nu}}_{vox}(t)]^T[\boldsymbol{\nu}_{vox}(t) - \widehat{\boldsymbol{\nu}}_{vox}(t)]\rangle \\ &= \langle[\boldsymbol{\nu}_{vox}(t) - \boldsymbol{W}_{\mathrm{inv}}\boldsymbol{b}(t)]^T[\boldsymbol{\nu}_{vox}(t) - \boldsymbol{W}_{\mathrm{inv}}\boldsymbol{b}(t)]\rangle. \end{aligned} \tag{12.5}$$

To derive the $\boldsymbol{W}_{\mathrm{inv}}$ that minimizes this cost function, we first calculate the derivative of $\mathcal{F}$ with respect to $\boldsymbol{W}_{\mathrm{inv}}$,

$$\frac{\partial \mathcal{F}}{\partial \boldsymbol{W}_{\mathrm{inv}}} = \langle -2[\boldsymbol{\nu}_{vox}(t) - \boldsymbol{W}_{\mathrm{inv}}\boldsymbol{b}(t)]\boldsymbol{b}^T(t)\rangle = -2\langle\boldsymbol{\nu}_{vox}(t)\boldsymbol{b}^T(t)\rangle + 2\boldsymbol{W}_{\mathrm{inv}}\langle\boldsymbol{b}(t)\boldsymbol{b}^T(t)\rangle. \tag{12.6}$$

Here, using $\boldsymbol{b}(t) = \boldsymbol{L}_{\mathcal{V}}\boldsymbol{\nu}_{vox}(t)$, we can derive

$$\langle\boldsymbol{\nu}_{vox}(t)\boldsymbol{b}^T(t)\rangle = \langle\boldsymbol{\nu}_{vox}(t)[\boldsymbol{L}_{\mathcal{V}}\boldsymbol{\nu}_{vox}(t)]^T\rangle = \langle\boldsymbol{\nu}_{vox}(t)\boldsymbol{\nu}_{vox}^T(t)\rangle\boldsymbol{L}_{\mathcal{V}}^T = \boldsymbol{R}_{vox}\boldsymbol{L}_{\mathcal{V}}^T. \tag{12.7}$$

Substituting this relationship and $\boldsymbol{R} = \langle\boldsymbol{b}(t)\boldsymbol{b}^T(t)\rangle$ into Eq. (12.6), and setting the right-hand side of this equation to zero, we can obtain the expression for the inverse filter $\boldsymbol{W}_{\mathrm{inv}}$, such that

$$\boldsymbol{W}_{\mathrm{inv}} = \boldsymbol{R}_{vox}\boldsymbol{L}_{\mathcal{V}}^T\boldsymbol{R}^{-1} = \boldsymbol{R}_{vox}\boldsymbol{L}_{\mathcal{V}}^T(\boldsymbol{L}_{\mathcal{V}}\boldsymbol{R}_{vox}\boldsymbol{L}_{\mathcal{V}}^T + \boldsymbol{R}_n)^{-1}, \tag{12.8}$$

where we use

$$\boldsymbol{R} \approx \boldsymbol{L}_{\mathcal{V}}\boldsymbol{R}_{vox}\boldsymbol{L}_{\mathcal{V}}^T + \boldsymbol{R}_n.$$

Substituting Eq. (12.8) into (12.4), we can derive the Wiener estimate (Eq. (12.2)), which is also called the minimum-mean-squared-error estimate.

12.1.2 Derivation of the minimum-variance spatial filter

The minimum-variance spatial filter can be derived from the Wiener filter formulation in Eq. (12.2) with the assumption that sources are uncorrelated[83][84]. With this assumption, the voxel-based source covariance matrix $\boldsymbol{R}_{vox}$ is expressed as

$$\boldsymbol{R}_{vox} = \begin{bmatrix} \langle s(\boldsymbol{r}_1,t)^2\rangle & 0 & \cdots & 0 \\ 0 & \langle s(\boldsymbol{r}_2,t)^2\rangle & \cdots & 0 \\ \vdots & . & \ddots & \vdots \\ 0 & 0 & \cdots & \langle s(\boldsymbol{r}_N,t)^2\rangle \end{bmatrix}. \tag{12.9}$$

Substituting Eqs. (12.1) and (12.9) into (12.2), we obtain

$$\begin{bmatrix} \widehat{s}(\boldsymbol{r}_1,t) \\ \vdots \\ \widehat{s}(\boldsymbol{r}_n,t) \\ \vdots \\ \widehat{s}(\boldsymbol{r}_N,t) \end{bmatrix} = \begin{bmatrix} \langle s(\boldsymbol{r}_1,t)^2\rangle & 0 & \cdots & 0 \\ 0 & \langle s(\boldsymbol{r}_2,t)^2\rangle & \cdots & 0 \\ \vdots & . & \ddots & \vdots \\ 0 & 0 & \cdots & \langle s(\boldsymbol{r}_N,t)^2\rangle \end{bmatrix} \boldsymbol{L}_\mathcal{V}^T \boldsymbol{R}^{-1}\boldsymbol{b}(t)$$

$$= \begin{bmatrix} \langle s(\boldsymbol{r}_1,t)^2\rangle \boldsymbol{l}^T(\boldsymbol{r}_1) \\ \vdots \\ \langle s(\boldsymbol{r}_n,t)^2\rangle \boldsymbol{l}^T(\boldsymbol{r}_n) \\ \vdots \\ \langle s(\boldsymbol{r}_N,t)^2\rangle \boldsymbol{l}^T(\boldsymbol{r}_N) \end{bmatrix} \boldsymbol{R}^{-1}\boldsymbol{b}(t). \tag{12.10}$$

Thus, the expression for the nth voxel at $\boldsymbol{r}_n$ is given by

$$\widehat{s}(\boldsymbol{r}_n,t) = \langle s(\boldsymbol{r}_n,t)^2\rangle \boldsymbol{l}^T(\boldsymbol{r}_n)\boldsymbol{R}^{-1}\boldsymbol{b}(t). \tag{12.11}$$

From this equation, we get

$$\begin{aligned} \langle \widehat{s}(\boldsymbol{r}_n,t)^2\rangle &= (\langle s(\boldsymbol{r}_n,t)^2\rangle)^2 \boldsymbol{l}^T(\boldsymbol{r}_n)\boldsymbol{R}^{-1}\langle \boldsymbol{b}(t)\boldsymbol{b}^T(t)\rangle \boldsymbol{R}^{-1}\boldsymbol{l}(\boldsymbol{r}_n) \\ &= (\langle s(\boldsymbol{r}_n,t)^2\rangle)^2 \boldsymbol{l}^T(\boldsymbol{r}_n)\boldsymbol{R}^{-1}\boldsymbol{l}(\boldsymbol{r}_n). \end{aligned} \tag{12.12}$$

Setting $\langle \widehat{s}(\boldsymbol{r}_n,t)^2\rangle$ equal to $\langle s(\boldsymbol{r}_n,t)^2\rangle$, we obtain

$$\langle s(\boldsymbol{r}_n,t)^2\rangle = \frac{1}{\boldsymbol{l}^T(\boldsymbol{r}_n)\boldsymbol{R}^{-1}\boldsymbol{l}(\boldsymbol{r}_n)}, \tag{12.13}$$

and substituting the equation above back into Eq. (12.11), we finally get

$$\widehat{s}(\boldsymbol{r}_n,t) = \boldsymbol{w}^T(\boldsymbol{r}_n)\boldsymbol{b}(t), \tag{12.14}$$

and

$$\boldsymbol{w}(\boldsymbol{r}_n) = \frac{\boldsymbol{R}^{-1}\boldsymbol{l}(\boldsymbol{r}_n)}{\boldsymbol{l}^T(\boldsymbol{r}_n)\boldsymbol{R}^{-1}\boldsymbol{l}(\boldsymbol{r}_n)}. \tag{12.15}$$

This is exactly equal to the weight expression for the unit-gain minimum-variance spatial filter in which the filter is pointing at $\boldsymbol{r}_n$[84].

12.2 MUSIC algorithm

12.2.1 Single- and multi-dipole search

Before introducing the well-known MUSIC algorithm, we briefly review the conventional single- and multi-dipole search algorithms. These algorithms are based on the least-squares principle, where the optimum estimate of the signal vector $\boldsymbol{b}_s(t)$, $\widehat{\boldsymbol{b}}_s(t)$, is obtained by minimizing the error function[1],

$$\mathcal{F} = \|\boldsymbol{b}(t) - \widehat{\boldsymbol{b}}_s(t)\|^2.$$

Let us first consider the case where all locations of the Q sources, $\boldsymbol{r}_1, \ldots, \boldsymbol{r}_Q$, are known. In such cases, the source magnitudes $\boldsymbol{\nu}(t)$ (defined in Eq. (2.28)) are estimated using

$$\widehat{\boldsymbol{\nu}}(t) = \underset{\boldsymbol{\nu}(t)}{\arg\min}\mathcal{F} = \underset{\boldsymbol{\nu}(t)}{\arg\min}\|\boldsymbol{b}(t) - \boldsymbol{L}_D\boldsymbol{\nu}(t)\|^2, \tag{12.16}$$

where the composite lead-field matrix is defined in Eq. (2.32). To derive $\widehat{\boldsymbol{\nu}}(t)$, we differentiate $\mathcal{F}$ with respect to $\boldsymbol{\nu}(t)$ and this gives

$$\begin{aligned}\frac{\partial \mathcal{F}}{\partial \boldsymbol{\nu}(t)} &= \frac{\partial}{\partial \boldsymbol{\nu}(t)}(\boldsymbol{b}(t)\boldsymbol{b}^T(t) - \boldsymbol{b}^T(t)\boldsymbol{L}_D\boldsymbol{\nu}(t) - \boldsymbol{\nu}^T(t)\boldsymbol{L}_D^T\boldsymbol{b}(t) + \boldsymbol{\nu}^T\boldsymbol{L}_D^T\boldsymbol{L}_D\boldsymbol{\nu}(t)) \\ &= -2\boldsymbol{L}_D^T\boldsymbol{b}(t) + 2\boldsymbol{L}_D^T\boldsymbol{L}_D\boldsymbol{\nu}(t). \end{aligned} \tag{12.17}$$

Setting the right-hand side of the above equation to zero, we obtain

$$\widehat{\boldsymbol{\nu}}(t) = (\boldsymbol{L}_D^T\boldsymbol{L}_D)^{-1}\boldsymbol{L}_D^T\boldsymbol{b}(t). \tag{12.18}$$

Substituting the above expression back into $\mathcal{F} = \|\boldsymbol{b}(t) - \boldsymbol{L}_D\boldsymbol{\nu}(t)\|^2$ gives

$$\mathcal{F} = \|\boldsymbol{b}(t) - \boldsymbol{L}_D(\boldsymbol{L}_D^T\boldsymbol{L}_D)^{-1}\boldsymbol{L}_D^T\boldsymbol{b}(t)\|^2. \tag{12.19}$$

Therefore, the least-squares estimate of the signal vector is obtained as

$$\widehat{\boldsymbol{b}}_s(t) = \boldsymbol{L}_D(\boldsymbol{L}_D^T\boldsymbol{L}_D)^{-1}\boldsymbol{L}_D^T\boldsymbol{b}(t). \tag{12.20}$$

The matrix $\boldsymbol{L}_D(\boldsymbol{L}_D^T\boldsymbol{L}_D)^{-1}\boldsymbol{L}_D^T$ is a projector that projects an M-dimensional vector onto the column span of $\boldsymbol{L}_D$, which is the signal subspace. The estimated signal vector $\widehat{\boldsymbol{b}}_s(t)$ is, therefore, a projection of the data vector, $\boldsymbol{b}(t)$, onto the signal subspace $\mathcal{E}_S$. This is the geometric interpretation of the least-squares estimate $\widehat{\boldsymbol{b}}_s(t)$.

So far, we have assumed that the locations of all sources, $\boldsymbol{r}_1, \ldots, \boldsymbol{r}_Q$, are known, i.e., that the basis vectors of the signal subspace, $\boldsymbol{l}(\boldsymbol{r}_1), \ldots, \boldsymbol{l}(\boldsymbol{r}_Q)$, are known. However, the source locations, $\boldsymbol{r}_1, \ldots, \boldsymbol{r}_Q$, are generally unknown, and they should be estimated from the data vector $\boldsymbol{b}(t)$. The geometric interpretation given above still holds even when the source locations are unknown. The source locations $\boldsymbol{r}_1, \ldots, \boldsymbol{r}_Q$ are estimated by searching for those that make

$\boldsymbol{L}_D(\boldsymbol{L}_D^T\boldsymbol{L}_D)^{-1}\boldsymbol{L}_D^T$ equal to a signal-subspace projector where $\widehat{\boldsymbol{b}}_s(t)$ is the projection of $\boldsymbol{b}(t)$ onto the signal subspace. When $\boldsymbol{L}_D(\boldsymbol{L}_D^T\boldsymbol{L}_D)^{-1}\boldsymbol{L}_D^T$ becomes the signal-subspace projector, the cost function $\mathcal{F}$ in Eq. (12.19) is minimized[85]. This is the basic principle of the least-squares-based single- or multi-dipole search methods. The major problem with the multi-dipole search method is that it requires a $3Q$-dimensional search, which is generally difficult to implement when $Q >= 2$ due to the computational complexity, and there is no guarantee that a global solution can be obtained, unless initial estimates can be set reasonably close to the true solution. We next describe the multiple signal classification (MUSIC) algorithm that can avoid a high-dimensional search by making use of the properties of the noise subspace[14][15][16].

12.2.2 Making use of the noise subspace–the MUSIC algorithm

As stated previously, the lead-field vectors at the true source locations are orthogonal to the noise subspace, and the source locations, $\boldsymbol{r}_1, \ldots, \boldsymbol{r}_Q$, can be obtained by making use of this orthogonality relationship, if we can obtain the basis vectors of the noise subspace. The question then is how we can obtain the basis vectors of the noise subspace. In Section 2.4, we already mentioned that the eigenvectors that correspond to the noise-level eigenvalues of $\boldsymbol{R}$ form basis vectors of the noise subspace, and Section 13.1 in the Appendix provides the proof that the noise-level eigenvectors of the sample data covariance are the maximum likelihood estimates of the basis vectors of the noise subspace. According to Eq. (2.57), the eigenvectors $\boldsymbol{e}_{Q+1}, \ldots, \boldsymbol{e}_M$ are orthogonal to $\boldsymbol{L}_D$. Therefore, the source locations, $\boldsymbol{r}_1, \ldots, \boldsymbol{r}_Q$, can be estimated by exploiting this orthogonality.

The orthogonality between the lead-field vector and the noise subspace can be evaluated by calculating $\mathcal{J}(\boldsymbol{r})$ as

$$\mathcal{J}(\boldsymbol{r}) = \frac{1}{\widetilde{\boldsymbol{l}}(\boldsymbol{r})^T \boldsymbol{E}_N \boldsymbol{E}_N^T \widetilde{\boldsymbol{l}}(\boldsymbol{r})} = \frac{\boldsymbol{l}(\boldsymbol{r})^T \boldsymbol{l}(\boldsymbol{r})}{\boldsymbol{l}(\boldsymbol{r})^T \boldsymbol{E}_N \boldsymbol{E}_N^T \boldsymbol{l}(\boldsymbol{r})}, \tag{12.21}$$

where $\widetilde{\boldsymbol{l}}(\boldsymbol{r})$ is the normalized lead-field vector. The MUSIC algorithm calculates this $\mathcal{J}(\boldsymbol{r})$, called the MUSIC localizer, in the source space and determines the source locations as $\boldsymbol{r}$ at which $\mathcal{J}(\boldsymbol{r})$ reaches large values. The algorithm uses only a three-dimensional search to detect multiple sources, regardless of the number of sources.

It should be pointed out that a significant similarity exists between the minimum-variance spatial filter and the MUSIC algorithm, if we compare the MUSIC localizer in Eq. (12.21) and the output power of the array-gain minimum-variance spatial filter in Eq. (4.10). Using Eq. (2.63), Eq. (4.10) can be rewritten as

$$\langle \widehat{s}(\boldsymbol{r},t)^2 \rangle = \frac{\boldsymbol{l}(\boldsymbol{r})^T \boldsymbol{l}(\boldsymbol{r})}{\boldsymbol{l}(\boldsymbol{r})^T \boldsymbol{E}_S \boldsymbol{\Lambda}_S^{-1} \boldsymbol{E}_S^T \boldsymbol{l}(\boldsymbol{r}) + \boldsymbol{l}(\boldsymbol{r})^T \boldsymbol{E}_N \boldsymbol{\Lambda}_N^{-1} \boldsymbol{E}_N^T \boldsymbol{l}(\boldsymbol{r})}. \tag{12.22}$$

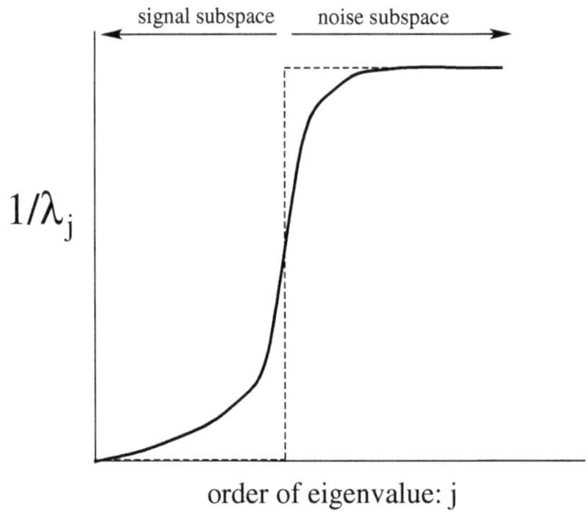

Figure 12.1: Typical shape of the plot of the inverse eigenvalues, $1/\lambda_j$ (solid line), and the distorted shape used to derive the music localizer (broken line). The eigenvalues λ_j are numbered in decreasing order.

Comparing the equation above with Eq. (12.21), one finds that the localizer in Eq. (12.21) is obtained by setting the inverse of the signal-level eigenvalues to zero and setting the noise-level eigenvalues to one in Eq. (12.22). Namely, the MUSIC localizer can be obtained by distorting the eigenvalue spectrum to emphasize the noise and signal subspace separation, as depicted in Fig. 12.1. In the MUSIC localizer expression in Eq. (12.21), when $\boldsymbol{r}$ approaches one of the source locations, the term $\boldsymbol{E}_N^T\boldsymbol{l}(\boldsymbol{r})$ becomes nearly zero and $\mathcal{J}(\boldsymbol{r})$ forms a sharp peak as a function $\boldsymbol{r}$. On the other hand, the minimum-variance output power in Eq. (12.22) includes the first term, $\boldsymbol{l}(\boldsymbol{r})^T\boldsymbol{E}_S\boldsymbol{\Lambda}_S^{-1}\boldsymbol{E}_S^T\boldsymbol{l}(\boldsymbol{r})$ in its denominator. Thus, when $\boldsymbol{r}$ approaches one of the source locations, $\langle\widehat{s}(\boldsymbol{r},t)^2\rangle$ does not become unreasonably large, but forms a peak whose height corresponds to the source magnitude. Namely, this first term gives the quantitative reliability of the voxel values in the adaptive spatial-filter reconstruction.

12.3 Scanning with the generalized-likelihood-ratio test function

Here, we introduce a method fairly similar to adaptive spatial filters. This method also uses a test function that scans over voxels in order to obtain a three-dimensional source reconstruction. This scanning function represents the likelihood ratio between the null and alternative hypotheses, where the null hypothesis posits that there is no source at the scanning location. The method is called the generalized-likelihood-ratio test (GLRT) scanning[86][87].

12.3.1 Data model

When we have Q total signal sources, the data, $\boldsymbol{b}(t)$, is expressed as

$$\boldsymbol{b}(t) = \sum_{j=1}^{Q} \boldsymbol{L}(\boldsymbol{r}_j)\boldsymbol{s}(\boldsymbol{r}_j, t) + \boldsymbol{b}_I(t) + \boldsymbol{n}(t), \tag{12.23}$$

where $\boldsymbol{b}_I(t)$ is the interference and $\boldsymbol{n}(t)$ is the sensor noise. When the scanning location $\boldsymbol{r}$ is equal to one of the source locations, for example, the location of the first source $\boldsymbol{r}_1$, the measured data are expressed as

$$\begin{aligned}\boldsymbol{b}(t) = \boldsymbol{L}(\boldsymbol{r}_1)\boldsymbol{s}(\boldsymbol{r}_1, t) + \sum_{j=2}^{Q} \boldsymbol{L}(\boldsymbol{r}_j)\boldsymbol{s}(\boldsymbol{r}_j, t) + \boldsymbol{b}_I(t) + \boldsymbol{n}(t) \\ = \boldsymbol{L}(\boldsymbol{r}_1)\boldsymbol{s}(\boldsymbol{r}_1, t) + \boldsymbol{b}_{I+n}^{(\mathrm{ex})}(t),\end{aligned} \tag{12.24}$$

where

$$\boldsymbol{b}_{I+n}^{(\mathrm{ex})}(t) = \sum_{j=2}^{Q} \boldsymbol{L}(\boldsymbol{r}_j)\boldsymbol{s}(\boldsymbol{r}_j, t) + \boldsymbol{b}_I(t) + \boldsymbol{n}(t). \tag{12.25}$$

This $\boldsymbol{b}_{I+n}^{(\mathrm{ex})}(t)$ is the extended interference plus sensor noise. The data model above is based on the same idea as that for deriving Eqs. (6.2) and (6.3), i.e., on the idea that, when the method attempts to detect the source activity at $\boldsymbol{r}_1$, all other sources are considered interference. When the scanning location $\boldsymbol{r}$ is not equal to any of the locations of the signal sources, we have

$$\boldsymbol{b}_{I+n}^{(\mathrm{ex})}(t) = \sum_{j=1}^{Q} \boldsymbol{L}(\boldsymbol{r}_j)\boldsymbol{s}(\boldsymbol{r}_j, t) + \boldsymbol{b}_I(t) + \boldsymbol{n}(t) = \boldsymbol{b}(t). \tag{12.26}$$

That is, when no signal source exists at the scanning location $\boldsymbol{r}$, the extended interference plus sensor noise, $\boldsymbol{b}_{I+n}^{(\mathrm{ex})}(t)$, is equal to the measured signal.

We further assume that the extended interference plus noise, $\boldsymbol{b}_{I+n}^{(\mathrm{ex})}(t)$, is represented by a multi-variate Gaussian process,

$$\boldsymbol{b}_{I+n}^{(\mathrm{ex})}(t) \sim \mathcal{N}(0, \boldsymbol{\Sigma}), \tag{12.27}$$

where $\boldsymbol{\Sigma}$ is the covariance matrix of $\boldsymbol{b}_{I+n}^{(\mathrm{ex})}(t)$. Note that since $\boldsymbol{b}_{I+n}^{(\mathrm{ex})}(t)$ depends on $\boldsymbol{r}$, $\boldsymbol{\Sigma}$ also depends on the scanning location $\boldsymbol{r}$. To derive the scanning function, we have to define the null and alternative hypotheses. The null hypothesis, H_0, posits that there is no source activity at the scanning location $\boldsymbol{r}$. The alternative hypothesis, H_1, posits that there is a source activity at $\boldsymbol{r}$. The method described here performs this hypothesis testing at each voxel location by scanning the concentrated data likelihood.

12.3.2 Deriving the scanning function

Likelihood of the null hypothesis

We first derive the data likelihood when the null hypothesis holds, i.e., when there is no source at the scanning location, In this case, according to the Gaussianity assumption in Eq. (12.27) and the relationship in Eq. (12.26), the probability distribution of the measurement $\boldsymbol{b}(t)$ given $\boldsymbol{\Sigma}$ is expressed as

$$p(\boldsymbol{b}(t)) = \frac{1}{(2\pi)^{(M/2)}|\boldsymbol{\Sigma}|^{1/2}} \exp[-\frac{1}{2}\boldsymbol{b}^T(t)\boldsymbol{\Sigma}^{-1}\boldsymbol{b}(t)]. \tag{12.28}$$

Defining the spatio-temporal measurement as $\boldsymbol{B}$: $\boldsymbol{B} = [\boldsymbol{b}(t_1), \boldsymbol{b}(t_2), \ldots, \boldsymbol{b}(t_K)]$, and assuming that the Gaussian processes are independent and identically distributed for these time points, the probability for this spatio-temporal data set is expressed as

$$p(\boldsymbol{B}) = \prod_{k=1}^{K} \frac{1}{(2\pi)^{(M/2)}|\boldsymbol{\Sigma}|^{1/2}} \exp[-\frac{1}{2}\boldsymbol{b}^T(t_k)\boldsymbol{\Sigma}^{-1}\boldsymbol{b}(t_k)]. \tag{12.29}$$

The log-likelihood function is then expressed as

$$\log \mathcal{L}(\boldsymbol{B}) = -\frac{K}{2}\log|\boldsymbol{\Sigma}| - \frac{1}{2}\sum_{k=1}^{K}\boldsymbol{b}^T(t_k)\boldsymbol{\Sigma}^{-1}\boldsymbol{b}(t_k) = -\frac{K}{2}\log|\boldsymbol{\Sigma}| - \frac{1}{2}\mathrm{tr}\{\boldsymbol{\Sigma}^{-1}\boldsymbol{B}\boldsymbol{B}^T\}, \tag{12.30}$$

where we ignore additive constants that are not related to the arguments here. Using exactly the same derivation as in Section 10.1, we can derive the maximum likelihood estimate of $\boldsymbol{\Sigma}$, such that

$$\boldsymbol{\Sigma} = \frac{1}{K}\sum_{k=1}^{K}\boldsymbol{b}(t_k)\boldsymbol{b}^T(t_k) = \frac{1}{K}\boldsymbol{B}\boldsymbol{B}^T = \widehat{\boldsymbol{R}}. \tag{12.31}$$

Equation (12.31) shows that $\boldsymbol{\Sigma}$ is equal to the measurement sample covariance matrix, $\widehat{\boldsymbol{R}}$, when H_0 holds. The maximized log-likelihood function is obtained by substituting the equation above into Eq. (12.30), resulting in

$$\max_{\boldsymbol{\Sigma}} \log \mathcal{L}(\boldsymbol{B}) = -\frac{K}{2}\log|\widehat{\boldsymbol{R}}| - \frac{K}{2}\mathrm{tr}\{(\boldsymbol{B}\boldsymbol{B}^T)^{-1}\boldsymbol{B}\boldsymbol{B}^T\} = -\frac{K}{2}\log|\widehat{\boldsymbol{R}}| - \frac{KM}{2}. \tag{12.32}$$

We again ignore the additive constant $-KM/2$ that is not related to the following arguments. Finally, we attain

$$\max_{\boldsymbol{\Sigma}} \log \mathcal{L}(\boldsymbol{B}) = -\frac{K}{2}\log|\widehat{\boldsymbol{R}}|. \tag{12.33}$$

Likelihood for the alternative hypothesis

To derive the likelihood for the alternative hypothesis, we assume that the source time course is given by a weighted sum of temporal basis functions, $\phi_j(t)$, where

$j = 1, \ldots, \mathcal{D}$ and $\mathcal{D}$ is the total number of basis functions. Therefore, using a $3 \times \mathcal{D}$ matrix $\boldsymbol{X}$, the time course of a source at $\boldsymbol{r}$ is expressed as

$$\boldsymbol{s}(\boldsymbol{r}, t) = \begin{bmatrix} s_x(\boldsymbol{r}, t) \\ s_y(\boldsymbol{r}, t) \\ s_z(\boldsymbol{r}, t) \end{bmatrix} = \boldsymbol{X} \begin{bmatrix} \phi_1(t) \\ \vdots \\ \phi_{\mathcal{D}}(t) \end{bmatrix} \tag{12.34}$$

Defining a vector $\varphi(t)$ such that

$$\varphi(t) = \begin{bmatrix} \phi_1(t) \\ \vdots \\ \phi_{\mathcal{D}}(t) \end{bmatrix}, \tag{12.35}$$

Eq. (12.24) is rewritten as

$$\boldsymbol{b}(t) = \boldsymbol{L}(\boldsymbol{r})\boldsymbol{X}(\boldsymbol{r})\varphi(t) + \boldsymbol{b}_{I+n}^{(\mathrm{ex})}(t), \tag{12.36}$$

where the scanning location is denoted $\boldsymbol{r}$. When the alternative hypothesis holds, the probability distribution of the measurement, $\boldsymbol{B}$, given $\boldsymbol{X}$ and $\boldsymbol{\Sigma}$, is expressed such that

$$p(\boldsymbol{B}) = \prod_{k=1}^{K} \frac{1}{(2\pi)^{(M/2)}|\boldsymbol{\Sigma}|^{1/2}} \exp\left[-\frac{1}{2}[\boldsymbol{b}(t_k) - \boldsymbol{L}\boldsymbol{X}\varphi(t_k)]^T \boldsymbol{\Sigma}^{-1}[\boldsymbol{b}(t_k) - \boldsymbol{L}\boldsymbol{X}\varphi(t_k)]\right], \tag{12.37}$$

where we omit the explicit notation of $(\boldsymbol{r})$ from $\boldsymbol{L}(\boldsymbol{r})$ and $\boldsymbol{X}(\boldsymbol{r})$ for simplicity. The log-likelihood function is then expressed as

$$\begin{aligned} \log \mathcal{L}(\boldsymbol{B}) &= -\frac{K}{2}\log|\boldsymbol{\Sigma}| - \frac{1}{2}\sum_{k=1}^{K}[\boldsymbol{b}(t_k) - \boldsymbol{L}\boldsymbol{X}\varphi(t_k)]^T \boldsymbol{\Sigma}^{-1}[\boldsymbol{b}(t_k) - \boldsymbol{L}\boldsymbol{X}\varphi(t_k)] \\ &= -\frac{K}{2}\log|\boldsymbol{\Sigma}| - \frac{1}{2}\mathrm{tr}\{\boldsymbol{\Sigma}^{-1}(\boldsymbol{B} - \boldsymbol{L}\boldsymbol{X}\boldsymbol{\Phi})(\boldsymbol{B} - \boldsymbol{L}\boldsymbol{X}\boldsymbol{\Phi})^T\}, \end{aligned} \tag{12.38}$$

where we again neglect additive constants that are not related to the arguments here. In the equation above, the matrix $\boldsymbol{\Phi}$ is defined as

$$\boldsymbol{\Phi} = [\varphi(t_1), \varphi(t_2), \ldots, \varphi(t_K)]. \tag{12.39}$$

We then maximize the log-likelihood function in Eq. (12.38) with respect to $\boldsymbol{\Sigma}$, and we get the following relationship by calculating $\partial \log \mathcal{L}/\partial \boldsymbol{\Sigma}$ and setting it to zero:

$$\begin{aligned} \boldsymbol{\Sigma} &= \frac{1}{K}\sum_{k=1}^{K}[\boldsymbol{b}(t_k) - \boldsymbol{L}\boldsymbol{X}\varphi(t_k)][\boldsymbol{b}(t_k) - \boldsymbol{L}\boldsymbol{X}\varphi(t_k)]^T \\ &= \frac{1}{K}(\boldsymbol{B} - \boldsymbol{L}\boldsymbol{X}\boldsymbol{\Phi})(\boldsymbol{B} - \boldsymbol{L}\boldsymbol{X}\boldsymbol{\Phi})^T. \end{aligned} \tag{12.40}$$

The maximized log-likelihood function is obtained by substituting the equation above into Eq. (12.38) and by ignoring the additive constant $-KM/2$, resulting in

$$\max_{\boldsymbol{\Sigma}} \log \mathcal{L}(\boldsymbol{B}) = -\frac{K}{2} \log |\boldsymbol{\Sigma}|. \tag{12.41}$$

Note that the above $\log \mathcal{L}(\boldsymbol{B})$ is a function of $\boldsymbol{X}$ and we also need to maximize the above $\log \mathcal{L}(\boldsymbol{B})$ with respect to $\boldsymbol{X}$.

GLRT scanning function

The scanning function that determines whether there is a source activity at $\boldsymbol{r}$ is derived by calculating the ratio of the concentrated likelihood for H_1 to that for H_0. According to the discussion above, the (non-log) likelihood function for H_0 is expressed as

$$\mathcal{L}(\boldsymbol{B}) = |\widehat{\boldsymbol{R}}|^{-K/2}, \tag{12.42}$$

and that for H_1 as

$$\mathcal{L}(\boldsymbol{B}) = \max_{\boldsymbol{X}} |\boldsymbol{\Sigma}|^{-K/2} = \max_{\boldsymbol{X}} |\frac{1}{K}(\boldsymbol{B} - \boldsymbol{L}\boldsymbol{X}\boldsymbol{\Phi})(\boldsymbol{B} - \boldsymbol{L}\boldsymbol{X}\boldsymbol{\Phi})^T|^{-K/2}. \tag{12.43}$$

Therefore, the scanning function is the ratio of these likelihood functions, such that

$$\frac{\max_{\boldsymbol{X}} |\boldsymbol{\Sigma}|^{-K/2}}{|\widehat{\boldsymbol{R}}|^{-K/2}}.$$

However, the scanning function above is generally hard to compute, because it contains the exponent $K/2$, which is generally a very large value. Therefore, removing this power calculation, we define the scanning function equivalent to the one above as

$$\mathcal{S}(\boldsymbol{r}) = \frac{|\widehat{\boldsymbol{R}}|}{\min_{\boldsymbol{X}} |\boldsymbol{\Sigma}|} = \frac{|\boldsymbol{B}\boldsymbol{B}^T|}{\min_{\boldsymbol{X}} |(\boldsymbol{B} - \boldsymbol{L}\boldsymbol{X}\boldsymbol{\Phi})(\boldsymbol{B} - \boldsymbol{L}\boldsymbol{X}\boldsymbol{\Phi})^T|}. \tag{12.44}$$

Through a rather tedious derivation, which is described in Section 13.6 of the Appendix, we show that the scanning function in Eq (12.44) can finally be rewritten as

$$\mathcal{S}(\boldsymbol{r}) = \frac{|\boldsymbol{L}^T(\boldsymbol{r})\boldsymbol{A}^{-1}\boldsymbol{L}(\boldsymbol{r})|}{|\boldsymbol{L}^T(\boldsymbol{r})\widehat{\boldsymbol{R}}^{-1}\boldsymbol{L}(\boldsymbol{r})|}, \tag{12.45}$$

where $\boldsymbol{A}$ is given by

$$\boldsymbol{A} = \widehat{\boldsymbol{R}} - \boldsymbol{R}_{\Phi B}^T \boldsymbol{R}_{\Phi\Phi}^{-1} \boldsymbol{R}_{\Phi B}, \tag{12.46}$$

$$\boldsymbol{R}_{\Phi B} = \frac{1}{K}\boldsymbol{\Phi}\boldsymbol{B}^T, \tag{12.47}$$

$$\text{and} \quad \boldsymbol{R}_{\Phi\Phi} = \frac{1}{K}\boldsymbol{\Phi}\boldsymbol{\Phi}^T. \tag{12.48}$$

We can observe the considerable similarity between the GLRT scanning function in Eq. (12.45) and the source power reconstruction such as in Eq. (4.10). The obvious difference is that the GLRT scanning function contains $\boldsymbol{A}$, which depends on the choice of the set of temporal basis functions, $\varphi(t)$. Consequently, the performance of the GLRT method depends on the choice of $\varphi(t)$. In the following, we show a numerical example of the GLRT reconstruction in which the GLRT method outperforms the adaptive spatial filter.

12.3.3 Numerical examples

Here, we present numerical examples of the GLRT scanning results. In this computer simulation, we used the discrete prolate spheroidal sequences (DPSS)[88][89] as the temporal basis functions, $\phi_j(t)$, according to papers by Baryshnikov *et al.* [90] and Limpiti *et al.* [70]. The computer simulation scheme as in the previous chapters was used, except that the source locations were set to $(0, -2, -6)$, $(0, 2, -6)$, and $(0, 1.5, -7.5)$. The uncorrelated source time courses shown in Fig. 4.3(a) are assigned to the three sources. The GLRT scanning results were presented in Fig. 12.2 (a). The results in Fig. 12.2 (b) show the source power reconstruction from the array-gain minimum-variance spatial filter. The results of the GLRT scanning are almost the same as those of the minimum-variance spatial filter.

We next generated the simulated data by assigning correlated time courses to the first and second sources with Eq. (9.46). The GLRT scanning results when the source correlation coefficient equals 0.8 and 0.98 are shown in (c) and (e), respectively. The corresponding results from the array-gain minimum-variance filter are shown in (d) and (f). In these minimum-variance results, we can observe the effect of signal cancellation so that the reconstructed intensities of the first and second sources are significantly reduced. This is particularly true in (f), which are the results when the correlation coefficient equals 0.98. In these results, the first and the second sources almost disappear. On the other hand, the GLRT scanning results are significantly robust to the source correlation. Even when the correlation coefficient is equal to 0.98 in (e), the method can reconstruct the three sources, although we can observe some spatial blur, which is likely the effect of the source correlation.

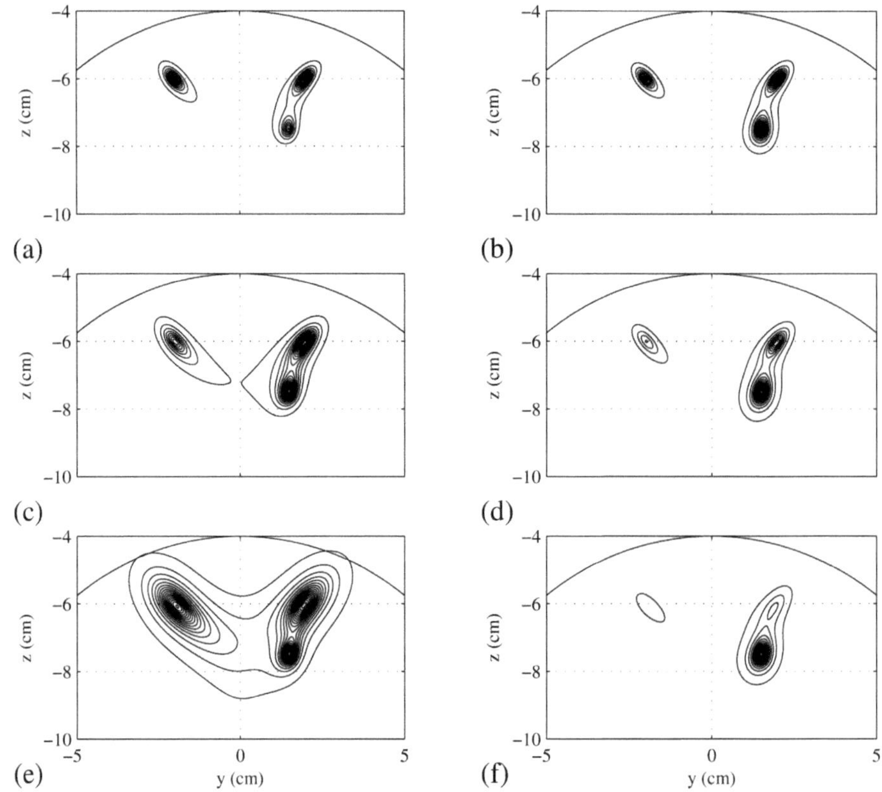

Figure 12.2: Results of GLRT scanning and the source power reconstruction from the minimum-variance spatial filter. (a) The GLRT scanning results when the source correlation is almost zero. (b) Results from the array-gain minimum-variance spatial filter with the same data as for (a). (c) The results of GLRT scanning when the correlation coefficient between the first and second source time courses was set to 0.8. (d) Results from the minimum-variance spatial filter with the same data as for (c). (e) The results of GLRT scanning when the correlation coefficient was set to 0.98. (f) Results from the minimum-variance spatial filter with the same data as for (e). The GLRT scanning results were obtained using the discrete prolate spheroidal sequences (DPSSs) used as the temporal basis functions.

Chapter 13

Appendices

13.1 Maximum-likelihood estimation of noise and signal subspaces

In this appendix, we show that when the sample measurement covariance matrix $\widehat{\boldsymbol{R}}$ is used, the column spans of $\boldsymbol{E}_S$ and $\boldsymbol{E}_N$ provide the maximum likelihood estimates of the noise and the signal subspaces of $\boldsymbol{R}$. The proof here is according to Scharf[91]. We start from the basic equation of the measurements, Eq. (2.27). Denoting $\boldsymbol{b}_s(t)$ as

$$\boldsymbol{b}_s(t) = \sum_{q=1}^{Q} s(\boldsymbol{r}_q, t)\boldsymbol{l}(\boldsymbol{r}_q), \tag{13.1}$$

Eq. (2.27) becomes

$$\boldsymbol{b}(t) = \boldsymbol{b}_s(t) + \boldsymbol{n}(t). \tag{13.2}$$

According to the noise-and-signal-subspace arguments in Section 2.4.1, there are a total of $(M-Q)$ linearly independent vectors $\boldsymbol{x}_j$ $(j = 1, \ldots, M-Q)$, which are orthogonal to $\boldsymbol{b}_s(t)$, i.e.,

$$\boldsymbol{x}_j^T \boldsymbol{b}_s(t) = 0, \quad j = 1, \ldots, M-Q. \tag{13.3}$$

We assume that the noise is multi-variate Gaussian with the covariance matrix of $\sigma_0^2\boldsymbol{I}$, and that the noise probability distributions are independent and identically distributed for all time points. Then, denoting the spatio-temporal measurement, $\boldsymbol{B}$, as

$$\boldsymbol{B} = [\boldsymbol{b}(t_1), \ldots, \boldsymbol{b}(t_K)],$$

the log-likelihood function for $\boldsymbol{B}$, $\log \mathcal{L}(\boldsymbol{B})$, is expressed as

$$\log \mathcal{L}(\boldsymbol{B}) = -\frac{1}{2\sigma_0^2} \sum_{k=1}^{K} [\boldsymbol{b}(t_k) - \boldsymbol{b}_s(t_k)]^T [\boldsymbol{b}(t_k) - \boldsymbol{b}_s(t_k)], \tag{13.4}$$

where additive constants are ignored.

If $\boldsymbol{x}_j$ is known, the maximum-likelihood estimate of the signal vector, $\boldsymbol{b}_s(t)$, is obtained by maximizing the log-likelihood function in Eq. (13.4) under the constraint of Eq. (13.3). We change this constrained maximization problem to an unconstrained minimization problem by introducing the Lagrange multiplier. We define the Lagrangian such that

$$\mathbb{L}(\boldsymbol{b}_s(t_k),\boldsymbol{\kappa}(t_k)) = \sum_{k=1}^{K}[\boldsymbol{b}(t_k)-\boldsymbol{b}_s(t_k)]^T[\boldsymbol{b}(t_k)-\boldsymbol{b}_s(t_k)] + 2\sum_{j=1}^{M-Q}\sum_{k=1}^{K}\kappa_j(t_k)\boldsymbol{x}_j^T\boldsymbol{b}_s(t_k), \tag{13.5}$$

where each $\kappa_j(t_k)$ is a Lagrange multiplier, and we define

$$\boldsymbol{\kappa}(t_k) = [\kappa_1(t_k),\ldots,\kappa_{M-Q}(t_k)]^T.$$

Defining

$$\boldsymbol{A} = [\boldsymbol{x}_1,\ldots,\boldsymbol{x}_{M-Q}],$$

and calculating the derivative of the right-hand side of Eq. (13.5) with respect to $\boldsymbol{b}_s(t_k)$, we obtain

$$\frac{\partial \mathbb{L}(\boldsymbol{b}_s(t_k),\boldsymbol{\kappa}(t_k))}{\partial \boldsymbol{b}_s(t_k)} = -2[\boldsymbol{b}(t_k)-\boldsymbol{b}_s(t_k)] + 2\boldsymbol{A}\boldsymbol{\kappa}(t_k). \tag{13.6}$$

Setting the left-hand side to zero, the maximum-likelihood solution of $\boldsymbol{b}_s(t_k)$ is

$$\widehat{\boldsymbol{b}}_s(t_k) = \boldsymbol{b}(t_k) - \boldsymbol{A}\boldsymbol{\kappa}(t_k). \tag{13.7}$$

Here, substituting the equation above into Eq. (13.3), we get

$$\boldsymbol{A}^T[\boldsymbol{b}(t_k) - \boldsymbol{A}\boldsymbol{\kappa}(t_k)] = 0, \tag{13.8}$$

and

$$\boldsymbol{\kappa}(t_k) = (\boldsymbol{A}^T\boldsymbol{A})^{-1}\boldsymbol{A}^T\boldsymbol{b}(t_k). \tag{13.9}$$

Substituting this equation back into Eq. (13.7), the final maximum-likelihood solution of $\boldsymbol{b}_s(t_k)$ is obtained as

$$\widehat{\boldsymbol{b}}_s(t_k) = \boldsymbol{b}(t_k) - \boldsymbol{A}(\boldsymbol{A}^T\boldsymbol{A})^{-1}\boldsymbol{A}^T\boldsymbol{b}(t_k) = (\boldsymbol{I} - \boldsymbol{\Pi}_A)\boldsymbol{b}(t_k), \tag{13.10}$$

where

$$\boldsymbol{\Pi}_A = \boldsymbol{A}(\boldsymbol{A}^T\boldsymbol{A})^{-1}\boldsymbol{A}^T.$$

This $\boldsymbol{\Pi}_A$ is the projection matrix onto the column span of $\boldsymbol{A}$, namely, onto the noise subspace of $\boldsymbol{R}$.

The above solution, $\widehat{\boldsymbol{b}}_s(t_k)$, is the maximum likelihood solution when $\boldsymbol{A}$ is known. When $\boldsymbol{A}$ is unknown, the maximum likelihood estimate of $\boldsymbol{A}$ is obtained as the $\boldsymbol{A}$ that maximizes the log-likelihood in Eq. (13.4). When $\widehat{\boldsymbol{b}}_s(t_k)$ is expressed as in Eq. (13.10), the residual of the likelihood function is given by

$$\log \mathcal{L}(\boldsymbol{B}) = -\frac{1}{2\sigma_0^2}\sum_{k=1}^{K} \boldsymbol{b}^T(t_k)\boldsymbol{\Pi}_A\boldsymbol{b}(t_k) = -\frac{K}{2\sigma_0^2}\mathrm{tr}\{\boldsymbol{\Pi}_A\widehat{\boldsymbol{R}}\}, \tag{13.11}$$

where $\mathrm{tr}\{\cdot\}$ indicates the trace operation, and $\widehat{\boldsymbol{R}}$ is the sample measurement covariance matrix defined in Eq. (10.1). When deriving Eq. (13.11), the properties of the projection matrix, $\boldsymbol{\Pi}_A^T = \boldsymbol{\Pi}_A$, and $\boldsymbol{\Pi}_A^2 = \boldsymbol{\Pi}_A$, are used. The log-likelihood function in Eq. (13.11) is known to have an upper bound, i.e., the relationship

$$\log \mathcal{L}(\boldsymbol{B}) = -\frac{K}{2\sigma_0^2}\mathrm{tr}\{\boldsymbol{\Pi}_A\widehat{\boldsymbol{R}}\} \le -\frac{K}{2\sigma_0^2}\sum_{j=Q+1}^{M}\widehat{\lambda}_j \tag{13.12}$$

holds, where $\widehat{\lambda}_j$ is the jth eigenvalue of $\widehat{\boldsymbol{R}}$ and the eigenvalues are sorted in decreasing order. The upper bound is known to be achieved when the relationship

$$\boldsymbol{\Pi}_A = [\widehat{\boldsymbol{e}}_{Q+1},\ldots,\widehat{\boldsymbol{e}}_M][\widehat{\boldsymbol{e}}_{Q+1},\ldots,\widehat{\boldsymbol{e}}_M]^T = \widehat{\boldsymbol{E}}_N\widehat{\boldsymbol{E}}_N^T \tag{13.13}$$

holds, where $\widehat{\boldsymbol{e}}_j$ is the eigenvector corresponding to $\widehat{\lambda}_j$, and $\widehat{\boldsymbol{E}}_N$ is a matrix whose columns consist of the eigenvectors corresponding to the $Q+1$ smallest eigenvalues of $\widehat{\boldsymbol{R}}$. That is, the maximum-likelihood estimate of the noise subspace, $\widehat{\mathcal{E}}_N$, is the column span of $\widehat{\boldsymbol{E}}_N$, i.e.,

$$\widehat{\mathcal{E}}_N = span\{\widehat{\boldsymbol{e}}_{Q+1},\ldots,\widehat{\boldsymbol{e}}_M\}. \tag{13.14}$$

Since the noise and signal subspaces are orthogonal complements, it is obvious that the maximum-likelihood estimates of the signal subspace $\widehat{\mathcal{E}}_S$ is the column span of $\widehat{\boldsymbol{E}}_S$, i.e.,

$$\widehat{\mathcal{E}}_S = span\{\widehat{\boldsymbol{e}}_1,\ldots,\widehat{\boldsymbol{e}}_Q\}. \tag{13.15}$$

13.2 Additional topics related to non-adaptive spatial filters

13.2.1 Determination of the optimum orientation for scalar non-adaptive spatial filters

The scalar spatial filter formulation requires determination of the optimum orientation, $\boldsymbol{\eta}_{opt}(\boldsymbol{r})$, using Eq. (2.70), unless the voxel orientation is predetermined by other means. The formulae to compute this $\boldsymbol{\eta}_{opt}(\boldsymbol{r})$ for adaptive spatial filters are presented in Section 4.3. This appendix presents formulae to calculate $\boldsymbol{\eta}_{opt}(\boldsymbol{r})$ for representative non-adaptive spatial filters. For the minimum-norm filter, the weight $\boldsymbol{w}(\boldsymbol{r},\boldsymbol{\eta})$ is expressed as

$$\boldsymbol{w}(\boldsymbol{r},\boldsymbol{\eta}) = \boldsymbol{G}^{-1}\boldsymbol{l} = \boldsymbol{G}^{-1}\boldsymbol{L}(\boldsymbol{r})\boldsymbol{\eta}(\boldsymbol{r}). \tag{13.16}$$

Then, $\boldsymbol{\eta}_{opt}(\boldsymbol{r})$ is derived by maximizing the outputs $\boldsymbol{w}^T(\boldsymbol{r},\boldsymbol{\eta})\boldsymbol{R}\boldsymbol{w}(\boldsymbol{r},\boldsymbol{\eta})$, i.e.,

$$\boldsymbol{\eta}_{opt}(\boldsymbol{r}) = \arg\max_{\boldsymbol{\eta}(\boldsymbol{r})} \boldsymbol{\eta}^T(\boldsymbol{r})[\boldsymbol{L}^T(\boldsymbol{r})\boldsymbol{G}^{-1}\boldsymbol{R}\boldsymbol{G}^{-1}\boldsymbol{L}(\boldsymbol{r})]\boldsymbol{\eta}(\boldsymbol{r}). \tag{13.17}$$

As discussed in Section 13.3, the above maximization has a closed-form solution, in which the optimum orientation $\boldsymbol{\eta}_{opt}(\boldsymbol{r})$ is expressed as

$$\boldsymbol{\eta}_{opt}(\boldsymbol{r}) = \boldsymbol{\vartheta}_{max}\{\boldsymbol{L}^T(\boldsymbol{r})\boldsymbol{G}^{-1}\boldsymbol{R}\boldsymbol{G}^{-1}\boldsymbol{L}(\boldsymbol{r})\}, \tag{13.18}$$

where $\boldsymbol{\vartheta}_{max}\{\cdot\}$ indicates the eigenvector corresponding to the maximum eigenvalue of the matrix in the curly braces. For the weight-normalized minimum-norm filter, $\boldsymbol{\eta}_{opt}(\boldsymbol{r})$ is obtained using

$$\boldsymbol{\eta}_{opt}(\boldsymbol{r}) = \arg\max_{\boldsymbol{\eta}(\boldsymbol{r})} \frac{\boldsymbol{\eta}^T(\boldsymbol{r})[\boldsymbol{L}^T(\boldsymbol{r})\boldsymbol{G}^{-1}\boldsymbol{R}\boldsymbol{G}^{-1}\boldsymbol{L}(\boldsymbol{r})]\boldsymbol{\eta}(\boldsymbol{r})}{\boldsymbol{\eta}^T(\boldsymbol{r})[\boldsymbol{L}^T(\boldsymbol{r})\boldsymbol{G}^{-2}\boldsymbol{L}(\boldsymbol{r})]\boldsymbol{\eta}(\boldsymbol{r})}, \tag{13.19}$$

and its closed-form solution is given by

$$\boldsymbol{\eta}_{opt}(\boldsymbol{r}) = \boldsymbol{\vartheta}_{max}\{\boldsymbol{L}^T(\boldsymbol{r})\boldsymbol{G}^{-1}\boldsymbol{R}\boldsymbol{G}^{-1}\boldsymbol{L}(\boldsymbol{r}), \boldsymbol{L}^T(\boldsymbol{r})\boldsymbol{G}^{-2}\boldsymbol{L}(\boldsymbol{r})\}. \tag{13.20}$$

For sLORETA, $\boldsymbol{\eta}_{opt}(\boldsymbol{r})$ is obtained using

$$\boldsymbol{\eta}_{opt}(\boldsymbol{r}) = \arg\max_{\boldsymbol{\eta}(\boldsymbol{r})} \frac{\boldsymbol{\eta}^T(\boldsymbol{r})[\boldsymbol{L}^T(\boldsymbol{r})\boldsymbol{G}^{-1}\boldsymbol{R}\boldsymbol{G}^{-1}\boldsymbol{L}(\boldsymbol{r})]\boldsymbol{\eta}(\boldsymbol{r})}{\boldsymbol{\eta}^T(\boldsymbol{r})[\boldsymbol{L}^T(\boldsymbol{r})\boldsymbol{G}^{-1}\boldsymbol{L}(\boldsymbol{r})]\boldsymbol{\eta}(\boldsymbol{r})}, \tag{13.21}$$

and its closed-form solution is given by

$$\boldsymbol{\eta}_{opt}(\boldsymbol{r}) = \boldsymbol{\vartheta}_{max}\{\boldsymbol{L}^T(\boldsymbol{r})\boldsymbol{G}^{-1}\boldsymbol{R}\boldsymbol{G}^{-1}\boldsymbol{L}(\boldsymbol{r}), \boldsymbol{L}^T(\boldsymbol{r})\boldsymbol{G}^{-1}\boldsymbol{L}(\boldsymbol{r})\}. \tag{13.22}$$

For the spatial matched filter, $\boldsymbol{\eta}_{opt}(\boldsymbol{r})$ is obtained using

$$\boldsymbol{\eta}_{opt}(\boldsymbol{r}) = \arg\max_{\boldsymbol{\eta}(\boldsymbol{r})} \frac{\boldsymbol{\eta}^T(\boldsymbol{r})[\boldsymbol{L}^T(\boldsymbol{r})\boldsymbol{R}\boldsymbol{L}(\boldsymbol{r})]\boldsymbol{\eta}(\boldsymbol{r})}{\boldsymbol{\eta}^T(\boldsymbol{r})[\boldsymbol{L}^T(\boldsymbol{r})\boldsymbol{L}(\boldsymbol{r})]\boldsymbol{\eta}(\boldsymbol{r})}, \tag{13.23}$$

and its closed-form solution is given by

$$\boldsymbol{\eta}_{opt}(\boldsymbol{r}) = \boldsymbol{\vartheta}_{max}\{\boldsymbol{L}^T(\boldsymbol{r})\boldsymbol{R}\boldsymbol{L}(\boldsymbol{r}), \boldsymbol{L}^T(\boldsymbol{r})\boldsymbol{L}(\boldsymbol{r})\}. \tag{13.24}$$

13.2.2 Equivalence between the vector and scalar minimum-norm filters

Generally, the scalar and the vector formulations are not equivalent for most of the spatial filters. However, for the minimum-norm filter, we can show the equivalence between these two formulations in the following manner. Using Eqs. (2.78) and

(3.17), the optimum orientation for the vector formulation, $\bar{\boldsymbol{\eta}}_{opt}(\boldsymbol{r})$, is obtained such that

$$\begin{aligned}\bar{\boldsymbol{\eta}}_{opt}(\boldsymbol{r}) &= \arg\max_{\boldsymbol{\eta}(\boldsymbol{r})} \boldsymbol{\eta}^T(\boldsymbol{r})\langle \widehat{\boldsymbol{s}}(\boldsymbol{r},t)\widehat{\boldsymbol{s}}^T(\boldsymbol{r},t)\rangle \boldsymbol{\eta}(\boldsymbol{r}) \\ &= \arg\max_{\boldsymbol{\eta}(\boldsymbol{r})} \boldsymbol{\eta}^T(\boldsymbol{r})[\boldsymbol{L}^T(\boldsymbol{r})\boldsymbol{G}^{-1}\langle \boldsymbol{b}(t)\boldsymbol{b}^T(t)\rangle \boldsymbol{G}^{-1}\boldsymbol{L}(\boldsymbol{r})]\boldsymbol{\eta}(\boldsymbol{r}) \\ &= \arg\max_{\boldsymbol{\eta}(\boldsymbol{r})} \boldsymbol{\eta}^T(\boldsymbol{r})[\boldsymbol{L}^T(\boldsymbol{r})\boldsymbol{G}^{-1}\boldsymbol{R}\boldsymbol{G}^{-1}\boldsymbol{L}(\boldsymbol{r})]\boldsymbol{\eta}(\boldsymbol{r}). \end{aligned} \tag{13.25}$$

A comparison between the above equation and Eq. (13.17) shows that

$$\bar{\boldsymbol{\eta}}_{opt}(\boldsymbol{r}) = \boldsymbol{\eta}_{opt}(\boldsymbol{r}).$$

Then, using this $\bar{\boldsymbol{\eta}}_{opt}(\boldsymbol{r})$, we have

$$\bar{\boldsymbol{\eta}}_{opt}^T(\boldsymbol{r})\widehat{\boldsymbol{s}}(\boldsymbol{r},t) = \boldsymbol{\eta}_{opt}^T(\boldsymbol{r})\boldsymbol{L}^T(\boldsymbol{r})\boldsymbol{G}^{-1}\boldsymbol{b}(t) = \boldsymbol{l}^T(\boldsymbol{r})\boldsymbol{G}^{-1}\boldsymbol{b}(t) = \widehat{s}(\boldsymbol{r},t). \tag{13.26}$$

This equation shows that $\bar{\boldsymbol{\eta}}_{opt}^T(\boldsymbol{r})\widehat{\boldsymbol{s}}(\boldsymbol{r},t)$ is equal to the scalar spatial filter output $\widehat{s}(\boldsymbol{r},t)$, thus demonstrating the equivalence between the vector and the scalar formulations. Although the equivalence between the scalar and the vector formulations can be shown for the minimum-norm filter, these two formulations are not equivalent for the other non-adaptive filters. For example, in the spatial matched filter, the estimated power matrix can be expressed as

$$\widehat{\boldsymbol{\Sigma}}_s(\boldsymbol{r}) = \boldsymbol{W}^T(\boldsymbol{r})\boldsymbol{R}\boldsymbol{W}(\boldsymbol{r}) = \frac{1}{\|\boldsymbol{L}(\boldsymbol{r})\|^2}\boldsymbol{L}^T(\boldsymbol{r})\boldsymbol{R}\boldsymbol{L}(\boldsymbol{r}). \tag{13.27}$$

Therefore, the optimum orientation from the vector output, $\bar{\boldsymbol{\eta}}_{opt}(\boldsymbol{r})$, is obtained using

$$\bar{\boldsymbol{\eta}}_{opt}(\boldsymbol{r}) = \arg\max_{\boldsymbol{\eta}(\boldsymbol{r})} \left[\boldsymbol{\eta}^T\widehat{\boldsymbol{\Sigma}}_s(\boldsymbol{r})\boldsymbol{\eta}\right] = \boldsymbol{\vartheta}_{max}\{\boldsymbol{L}^T(\boldsymbol{r})\boldsymbol{R}\boldsymbol{L}(\boldsymbol{r})\}. \tag{13.28}$$

It is easy to see that this $\bar{\boldsymbol{\eta}}_{opt}(\boldsymbol{r})$ is different from the $\boldsymbol{\eta}_{opt}(\boldsymbol{r})$ obtained in Eq. (13.24). For other non-adaptive filters, it can be shown in the same manner that the two kinds of formulations are not equivalent.

13.3 Rayleigh-Ritz formula

This section provides a proof of the Rayleigh-Ritz formula, which is according to [37]. We define $\boldsymbol{A}$ and $\boldsymbol{B}$ as positive definite matrices of the same dimensions. We introduce the following notations and use them throughout the book:

- The minimum and maximum eigenvalues of a matrix $\boldsymbol{A}$ are denoted $\lambda_{min}\{\boldsymbol{A}\}$ and $\lambda_{max}\{\boldsymbol{A}\}$.

- The eigenvectors corresponding to the minimum and maximum eigenvalues of a matrix $\boldsymbol{A}$ are denoted $\boldsymbol{\vartheta}_{min}\{\boldsymbol{A}\}$ and $\boldsymbol{\vartheta}_{max}\{\boldsymbol{A}\}$.

- The minimum and maximum generalized eigenvalues of a matrix $\boldsymbol{A}$ with a metric $\boldsymbol{B}$ are denoted $\lambda_{min}\{\boldsymbol{A},\boldsymbol{B}\}$ and $\lambda_{max}\{\boldsymbol{A},\boldsymbol{B}\}$, and the corresponding eigenvectors are denoted $\boldsymbol{\vartheta}_{min}\{\boldsymbol{A},\boldsymbol{B}\}$ and $\boldsymbol{\vartheta}_{max}\{\boldsymbol{A},\boldsymbol{B}\}$.

Here, if the matrix $\boldsymbol{B}$ is nonsingular, the following relationships hold:

$$\begin{aligned} \lambda_{max}\{\boldsymbol{A},\boldsymbol{B}\} &= \lambda_{max}\{\boldsymbol{B}^{-1}\boldsymbol{A}\}, \\ \boldsymbol{\vartheta}_{max}\{\boldsymbol{A},\boldsymbol{B}\} &= \boldsymbol{\vartheta}_{max}\{\boldsymbol{B}^{-1}\boldsymbol{A}\}, \\ \lambda_{min}\{\boldsymbol{A},\boldsymbol{B}\} &= \lambda_{min}\{\boldsymbol{B}^{-1}\boldsymbol{A}\}, \\ \text{and} \quad \boldsymbol{\vartheta}_{min}\{\boldsymbol{A},\boldsymbol{B}\} &= \boldsymbol{\vartheta}_{min}\{\boldsymbol{B}^{-1}\boldsymbol{A}\}. \end{aligned}$$

Then, using $\boldsymbol{x}$ to denote a column vector with its dimension commensurate with the size of the matrices, this appendix shows that

$$\min_{\boldsymbol{x}} \frac{\boldsymbol{x}^T\boldsymbol{A}\boldsymbol{x}}{\boldsymbol{x}^T\boldsymbol{B}\boldsymbol{x}} = \lambda_{min}\{\boldsymbol{A},\boldsymbol{B}\}, \tag{13.29}$$

and

$$\arg\min_{\boldsymbol{x}} \frac{\boldsymbol{x}^T\boldsymbol{A}\boldsymbol{x}}{\boldsymbol{x}^T\boldsymbol{B}\boldsymbol{x}} = \boldsymbol{\vartheta}_{min}\{\boldsymbol{A},\boldsymbol{B}\}. \tag{13.30}$$

Since the value of the ratio $(\boldsymbol{x}^T\boldsymbol{A}\boldsymbol{x})/(\boldsymbol{x}^T\boldsymbol{B}\boldsymbol{x})$ is not affected by the norm of $\boldsymbol{x}$, we set the norm of $\boldsymbol{x}$ so as to satisfy the relationship $\boldsymbol{x}^T\boldsymbol{B}\boldsymbol{x} = 1$. Then, the minimization problem in Eq. (13.29) is rewritten as

$$\min_{\boldsymbol{x}} \boldsymbol{x}^T\boldsymbol{A}\boldsymbol{x} \quad \text{subject to} \quad \boldsymbol{x}^T\boldsymbol{B}\boldsymbol{x} = 1. \tag{13.31}$$

We change this constrained minimization problem to an unconstrained minimization problem by introducing the Lagrange multiplier κ. We define the Lagrangian $\mathbb{L}(\boldsymbol{x},\kappa)$ such that

$$\mathbb{L}(\boldsymbol{x},\kappa) = \boldsymbol{x}^T\boldsymbol{A}\boldsymbol{x} - \kappa(\boldsymbol{x}^T\boldsymbol{B}\boldsymbol{x} - 1). \tag{13.32}$$

The minimization in Eq. (13.31) is equivalent to minimizing $\mathbb{L}(\boldsymbol{x},\kappa)$ with no constraints.

To obtain the minimum of $\mathbb{L}(\boldsymbol{x},\kappa)$, we calculate the derivatives,

$$\frac{\partial \mathbb{L}(\boldsymbol{x},\kappa)}{\partial \boldsymbol{x}} = 2(\boldsymbol{A}\boldsymbol{x} - \kappa\boldsymbol{B}\boldsymbol{x}), \tag{13.33}$$

$$\text{and} \quad \frac{\partial \mathbb{L}(\boldsymbol{x},\kappa)}{\partial \kappa} = -(\boldsymbol{x}^T\boldsymbol{B}\boldsymbol{x} - 1). \tag{13.34}$$

By setting these derivatives to zero, we can derive the relationships, $\boldsymbol{A}\boldsymbol{x} = \kappa\boldsymbol{B}\boldsymbol{x}$ and $\kappa = \boldsymbol{x}^T\boldsymbol{A}\boldsymbol{x}$. Therefore, the minimum value of $\boldsymbol{x}^T\boldsymbol{A}\boldsymbol{x}$ is equal to the minimum eigenvalue of $\boldsymbol{Ax} = \kappa \boldsymbol{Bx}$, and the $\boldsymbol{x}$ that attains this minimum value is equal to the eigenvector corresponding to this minimum eigenvalue. Namely, we have

$$\min_{\boldsymbol{x}} \frac{\boldsymbol{x}^T \boldsymbol{Ax}}{\boldsymbol{x}^T \boldsymbol{Bx}} = \lambda_{min}\{\boldsymbol{A}, \boldsymbol{B}\}$$

and

$$\arg\min_{\boldsymbol{x}} \frac{\boldsymbol{x}^T \boldsymbol{Ax}}{\boldsymbol{x}^T \boldsymbol{Bx}} = \boldsymbol{\vartheta}_{min}\{\boldsymbol{A}, \boldsymbol{B}\}.$$

Using exactly the same derivation, it is easy to show that

$$\max_{\boldsymbol{x}} \frac{\boldsymbol{x}^T \boldsymbol{Ax}}{\boldsymbol{x}^T \boldsymbol{Bx}} = \lambda_{max}\{\boldsymbol{A}, \boldsymbol{B}\} \tag{13.35}$$

and

$$\arg\max_{\boldsymbol{x}} \frac{\boldsymbol{x}^T \boldsymbol{Ax}}{\boldsymbol{x}^T \boldsymbol{Bx}} = \boldsymbol{\vartheta}_{max}\{\boldsymbol{A}, \boldsymbol{B}\}. \tag{13.36}$$

13.4 Supplementary formulae when only one or two sources exist

The definition of the generalized cosine between the two column vectors $\boldsymbol{a}_1$ and $\boldsymbol{a}_2$ with the metric $\boldsymbol{A}$, where $\boldsymbol{A}$ is a positive definite matrix is

$$\cos(\boldsymbol{a}_1, \boldsymbol{a}_2 | \boldsymbol{A}) = \frac{|\boldsymbol{a}_1^T \boldsymbol{A} \boldsymbol{a}_2|}{\sqrt{(\boldsymbol{a}_1^T \boldsymbol{A} \boldsymbol{a}_1)(\boldsymbol{a}_2^T \boldsymbol{A} \boldsymbol{a}_2)}}. \tag{13.37}$$

Here, the inequality $\cos(\boldsymbol{a}_1, \boldsymbol{a}_2 | \boldsymbol{A}) \leq 1$ holds, because the Swartz inequality holds for any positive definite matrix $\boldsymbol{A}$, i.e.,

$$(\boldsymbol{a}_1^T \boldsymbol{A} \boldsymbol{a}_1)(\boldsymbol{a}_2^T \boldsymbol{A} \boldsymbol{a}_2) \geq (\boldsymbol{a}_1^T \boldsymbol{A} \boldsymbol{a}_2)^2 \tag{13.38}$$

When $\boldsymbol{A}$ is equal to the identity matrix $\boldsymbol{I}$, the generalized cosine, $\cos(\boldsymbol{a}_1, \boldsymbol{a}_2 | \boldsymbol{I})$, is simply written as $\cos(\boldsymbol{a}_1, \boldsymbol{a}_2)$, which is equal to

$$\cos(\boldsymbol{a}_1, \boldsymbol{a}_2) = \frac{|\boldsymbol{a}_1^T \boldsymbol{a}_2|}{\sqrt{(\boldsymbol{a}_1^T \boldsymbol{a}_1)(\boldsymbol{a}_2^T \boldsymbol{a}_2)}}. \tag{13.39}$$

The generalized sine is defined as

$$\sin^2(\boldsymbol{a}_1, \boldsymbol{a}_2 | \boldsymbol{A}) = 1 - \cos^2(\boldsymbol{a}_1, \boldsymbol{a}_2 | \boldsymbol{A}) \tag{13.40}$$

and

$$\sin^2(\boldsymbol{a}_1, \boldsymbol{a}_2) = 1 - \cos^2(\boldsymbol{a}_1, \boldsymbol{a}_2). \tag{13.41}$$

Next, we provide several supplementary formulae for when a single source exists. Let us assume that a single source exists at $\boldsymbol{r}_1$ and its power is σ_1^2. The lead-field vector at $\boldsymbol{r}_1$ in the source direction is denoted $\boldsymbol{f}$, namely, $\boldsymbol{f} = \boldsymbol{l}(\boldsymbol{r}_1)$. Then, using Eq. (2.43), the measurement covariance matrix $\boldsymbol{R}$ is expressed as

$$\boldsymbol{R} = \sigma_1^2 \boldsymbol{f}\boldsymbol{f}^T + \sigma_0^2 \boldsymbol{I}. \tag{13.42}$$

Its inverse is given by

$$\boldsymbol{R}^{-1} = \frac{1}{\sigma_0^2}(\boldsymbol{I} - \frac{\alpha}{1+\alpha}\frac{\boldsymbol{f}\boldsymbol{f}^T}{\|\boldsymbol{f}\|^2}), \tag{13.43}$$

where $\alpha = (\sigma_1^2/\sigma_0^2)\|\boldsymbol{f}\|^2$. The square inverse is derived as

$$\boldsymbol{R}^{-2} = \frac{1}{\sigma_0^4}(\boldsymbol{I} - \frac{(2+\alpha)\alpha}{(1+\alpha)^2}\frac{\boldsymbol{f}\boldsymbol{f}^T}{\|\boldsymbol{f}\|^2}). \tag{13.44}$$

Making use of the formula in Eq. (13.43), for an arbitrary vector $\boldsymbol{l}$ whose dimension is commensurate with the size of the covariance matrix, we have,

$$\boldsymbol{f}^T\boldsymbol{R}^{-1}\boldsymbol{f} = \frac{\|\boldsymbol{f}\|^2}{\sigma_0^2}\frac{1}{1+\alpha}, \tag{13.45}$$

$$\boldsymbol{l}^T\boldsymbol{R}^{-1}\boldsymbol{l} = \frac{1}{\sigma_0^2}(\|\boldsymbol{l}\|^2 - \frac{\alpha}{1+\alpha}\frac{(\boldsymbol{l}^T\boldsymbol{f})^2}{\|\boldsymbol{f}\|^2}) = \frac{\|\boldsymbol{l}\|^2}{\sigma_0^2}(1 - \frac{\alpha}{1+\alpha}\cos^2(\boldsymbol{l},\boldsymbol{f})), \tag{13.46}$$

$$\text{and} \quad \boldsymbol{l}^T\boldsymbol{R}^{-1}\boldsymbol{f} = \frac{1}{\sigma_0^2}(\boldsymbol{l}^T\boldsymbol{f} - \frac{\alpha}{1+\alpha}\frac{(\boldsymbol{l}^T\boldsymbol{f})(\boldsymbol{f}^T\boldsymbol{f})}{\|\boldsymbol{f}\|^2}) = \frac{\boldsymbol{l}^T\boldsymbol{f}}{\sigma_0^2(1+\alpha)}. \tag{13.47}$$

Also, using Eq. (13.44) we have

$$\boldsymbol{f}^T\boldsymbol{R}^{-2}\boldsymbol{f} = \frac{\|\boldsymbol{f}\|^2}{\sigma_0^4}\frac{1}{(1+\alpha)^2}, \tag{13.48}$$

$$\text{and} \quad \boldsymbol{l}^T\boldsymbol{R}^{-2}\boldsymbol{l} = \frac{\|\boldsymbol{l}\|^2}{\sigma_0^4(1+\alpha)^2}[1 + (2\alpha+\alpha^2)\sin(\boldsymbol{l},\boldsymbol{f})]. \tag{13.49}$$

We also provide several supplementary formulae when two sources exist at $\boldsymbol{r}_1$ and $\boldsymbol{r}_2$. Their powers are denoted σ_1^2 and σ_2^2. The lead-field vectors at the source locations and orientations are denoted $\boldsymbol{f}$ and $\boldsymbol{g}$, respectively, where $\boldsymbol{f} = \boldsymbol{l}(\boldsymbol{r}_1)$ and $\boldsymbol{g} = \boldsymbol{l}(\boldsymbol{r}_2)$. Then, the measurement covariance matrix is given by:

$$\boldsymbol{R} = \sigma_1^2\boldsymbol{f}\boldsymbol{f}^T + \sigma_2^2\boldsymbol{g}\boldsymbol{g}^T + \sigma_0^2\boldsymbol{I} = \sigma_1^2\boldsymbol{f}\boldsymbol{f}^T + \sigma_0^2\boldsymbol{D}, \tag{13.50}$$

and its inverse is expressed as

$$\boldsymbol{R}^{-1} = \frac{1}{\sigma_0^2}(\boldsymbol{D}^{-1} - \boldsymbol{D}^{-1}\boldsymbol{f}\boldsymbol{f}^T\boldsymbol{D}^{-1}\frac{(\sigma_1^2/\sigma_0^2)}{1+\bar{\alpha}_1}), \tag{13.51}$$

where

$$\begin{aligned}
\boldsymbol{D} &= \boldsymbol{I} + \frac{\sigma_2^2}{\sigma_0^2}\boldsymbol{g}\boldsymbol{g}^T, &(13.52)\\
\boldsymbol{D}^{-1} &= \boldsymbol{I} - \frac{\alpha_2}{(1+\alpha_2)}\frac{\boldsymbol{g}\boldsymbol{g}^T}{\|\boldsymbol{g}\|^2}, &(13.53)\\
\bar{\alpha}_1 &= \frac{\sigma_1^2}{\sigma_0^2}\boldsymbol{f}^T\boldsymbol{D}^{-1}\boldsymbol{f}, &(13.54)\\
\alpha_1 &= \frac{\sigma_1^2}{\sigma_0^2}\|\boldsymbol{f}\|^2, &(13.55)\\
\text{and} \quad \alpha_2 &= \frac{\sigma_2^2}{\sigma_0^2}\|\boldsymbol{g}\|^2. &(13.56)
\end{aligned}$$

Using Eq. (13.51), we can obtain, for an arbitrary vector $\boldsymbol{l}$, the following equations:

$$\boldsymbol{l}^T\boldsymbol{R}^{-1}\boldsymbol{l} = \frac{\|\boldsymbol{l}\|^2}{\Gamma}\left[[1 - \frac{\alpha_2}{1+\alpha_2}\cos^2(\boldsymbol{l},\boldsymbol{g})] + \alpha_1 C(\boldsymbol{l},\boldsymbol{f},\boldsymbol{g})\right], \quad (13.57)$$

where

$$\Gamma = \sigma_0^2(1+\bar{\alpha}_1) = \sigma_0^2 + \sigma_1^2\|\boldsymbol{f}\|^2[1 - \frac{\alpha_2}{1+\alpha_2}\cos^2(\boldsymbol{f},\boldsymbol{g})], \quad (13.58)$$

and

$$\begin{aligned}
C(\boldsymbol{l},\boldsymbol{f},\boldsymbol{g}) = 1 - \cos^2(\boldsymbol{l},\boldsymbol{f}) - \frac{\alpha_2}{1+\alpha_2}\cos^2(\boldsymbol{f},\boldsymbol{g}) - \frac{\alpha_2}{1+\alpha_2}\cos^2(\boldsymbol{l},\boldsymbol{g})\\
+ 2\frac{\alpha_2}{1+\alpha_2}\cos(\boldsymbol{l},\boldsymbol{f})\cos(\boldsymbol{l},\boldsymbol{g})\cos(\boldsymbol{f},\boldsymbol{g}). \quad (13.59)
\end{aligned}$$

Note that $C(\boldsymbol{l},\boldsymbol{f},\boldsymbol{g}) = 0$ when $\boldsymbol{l} = \boldsymbol{f}$ or $\boldsymbol{l} = \boldsymbol{g}$. Also, using Eq. (13.51), we have

$$\begin{aligned}
\boldsymbol{R}^{-2} &= \frac{1}{\sigma_0^4}[\boldsymbol{D}^{-2} - \boldsymbol{D}^{-2}\boldsymbol{f}\boldsymbol{f}^T\boldsymbol{D}^{-1}\frac{(\sigma_1^2/\sigma_0^2)}{1+\bar{\alpha}_1}\\
&- \boldsymbol{D}^{-1}\boldsymbol{f}\boldsymbol{f}^T\boldsymbol{D}^{-2}\frac{(\sigma_1^2/\sigma_0^2)}{1+\bar{\alpha}_1} + \boldsymbol{D}^{-1}\boldsymbol{f}\boldsymbol{f}^T\boldsymbol{D}^{-1}(\boldsymbol{f}^T\boldsymbol{D}^{-2}\boldsymbol{f})\frac{(\sigma_1^2/\sigma_0^2)^2}{(1+\bar{\alpha}_1)^2}]. \quad (13.60)
\end{aligned}$$

Thus, for an arbitrary vector $\boldsymbol{l}$, we can obtain

$$\begin{aligned}
\boldsymbol{l}^T\boldsymbol{R}^{-2}\boldsymbol{l} &= \frac{1}{\Gamma^2}[(1+\bar{\alpha}_1)^2(\boldsymbol{l}^T\boldsymbol{D}^{-2}\boldsymbol{l})\\
&- 2(\frac{\sigma_1^2}{\sigma_0^2})(1+\bar{\alpha}_1)(\boldsymbol{l}^T\boldsymbol{D}^{-2}\boldsymbol{f})(\boldsymbol{l}^T\boldsymbol{D}^{-1}\boldsymbol{f}) + (\frac{\sigma_1^2}{\sigma_0^2})^2(\boldsymbol{l}^T\boldsymbol{D}^{-1}\boldsymbol{f})^2(\boldsymbol{f}^T\boldsymbol{D}^{-2}\boldsymbol{f})], \quad (13.61)
\end{aligned}$$

where

$$\boldsymbol{l}^T\boldsymbol{D}^{-1}\boldsymbol{f}=\|\boldsymbol{l}\|\|\boldsymbol{f}\|\left[\cos(\boldsymbol{f},\boldsymbol{l})-\frac{\alpha_2}{1+\alpha_2}\cos(\boldsymbol{f},\boldsymbol{g})\cos(\boldsymbol{g},\boldsymbol{l})\right], \tag{13.62}$$

$$\boldsymbol{l}^T\boldsymbol{D}^{-2}\boldsymbol{f}=\|\boldsymbol{l}\|\|\boldsymbol{f}\|\left[\cos(\boldsymbol{f},\boldsymbol{l})-\frac{\alpha_2(2+\alpha_2)}{(1+\alpha_2)^2}\cos(\boldsymbol{f},\boldsymbol{g})\cos(\boldsymbol{g},\boldsymbol{l})\right], \tag{13.63}$$

$$\text{and}\quad \boldsymbol{l}^T\boldsymbol{D}^{-2}\boldsymbol{l}=\|\boldsymbol{l}\|^2\left[1-\frac{\alpha_2(2+\alpha_2)}{(1+\alpha_2)^2}\cos^2(\boldsymbol{l},\boldsymbol{g})\right]. \tag{13.64}$$

Using Eq. (13.61), we can also obtain

$$\boldsymbol{f}^T\boldsymbol{R}^{-2}\boldsymbol{f}=\frac{1}{\Gamma^2}(\boldsymbol{f}^T\boldsymbol{D}^{-2}\boldsymbol{f})=\frac{\|\boldsymbol{f}\|^2}{\Gamma^2}\left[1-\frac{\alpha_2(2+\alpha_2)}{(1+\alpha_2)^2}\cos^2(\boldsymbol{f},\boldsymbol{g})\right]. \tag{13.65}$$

13.5 Robustness of the prewhitening signal covariance estimation to the control-only-sources scenario

In Section 8.2, it is assumed that control-state measurements contain only the contributions from the background interference and sensor noise, and the interference-plus-noise covariance matrix $\boldsymbol{R}_{i+n}$ can be obtained from such control-state measurements. This assumption, however, may not always be valid in real-life task-and-control-type measurements. There may be a situation in which there are some sources that appear only in the control state and do not appear in the task state. Such sources are called the control-only sources. This appendix shows that the prewhitening signal covariance estimation is robust to the existence of control-only sources.

We assume that there are P sources that exist only in the control state and do not appear in the task state. We also assume that the signal is still low-rank, i.e., $Q+P<M$. When control-only sources exist, the control state measurements $\boldsymbol{b}_c(t)$ can be expressed as

$$\boldsymbol{b}_c(t)=\boldsymbol{b}_I(t)+\boldsymbol{n}(t)+\boldsymbol{b}_\Delta(t), \tag{13.66}$$

where $\boldsymbol{b}_\Delta(t)$ indicates the signal generated by control-only sources. Assuming that the activity of control-only sources is uncorrelated with the interference and sensor noise, the covariance matrix relationships are then expressed as:

$$\begin{aligned}\text{Control:}\quad \boldsymbol{R}_c&=\boldsymbol{R}_{i+n}+\boldsymbol{R}_\Delta,\\ \text{Task:}\quad \boldsymbol{R}&=\boldsymbol{R}_s+\boldsymbol{R}_{i+n},\end{aligned} \tag{13.67}$$

where

$$\boldsymbol{R}_\Delta=\langle\boldsymbol{b}_\Delta(t)\boldsymbol{b}_\Delta^T(t)\rangle. \tag{13.68}$$

Using Eq. (13.67), we have

$$\boldsymbol{R}=\boldsymbol{R}_s+\boldsymbol{R}_c-\boldsymbol{R}_\Delta, \tag{13.69}$$

and thus

$$\widetilde{\boldsymbol{R}} = \widetilde{\boldsymbol{R}}_s + \boldsymbol{I} - \widetilde{\boldsymbol{R}}_\Delta, \tag{13.70}$$

where $\widetilde{\boldsymbol{R}}_\Delta = \boldsymbol{R}_c^{-1/2}\boldsymbol{R}_\Delta\boldsymbol{R}_c^{-1/2}$. Since $\boldsymbol{R}_\Delta$ is a positive semi-definite matrix with rank P and $\boldsymbol{R}_c^{-1/2}$ is a non-singular matrix, $\widetilde{\boldsymbol{R}}_\Delta$ is a positive semi-definite matrix with rank P. Thus, $\widetilde{\boldsymbol{R}}_\Delta$ is decomposed as

$$\widetilde{\boldsymbol{R}}_\Delta = \sum_{j=1}^{P} \beta_j \boldsymbol{k}_j \boldsymbol{k}_j^T, \tag{13.71}$$

where β_j, $(j = 1, \ldots, P)$ are the P non-zero eigenvalues of $\widetilde{\boldsymbol{R}}_\Delta$, and $\boldsymbol{k}_j$ are the corresponding eigenvectors. Substituting Eqs.(8.22) and (13.71) into (13.70), we have

$$\widetilde{\boldsymbol{R}} = \sum_{j=1}^{Q} \gamma_j \boldsymbol{h}_j \boldsymbol{h}_j^T + \sum_{j=P+1}^{M} \boldsymbol{k}_j \boldsymbol{k}_j^T + \sum_{j=1}^{P} (1-\beta_j)\boldsymbol{k}_j \boldsymbol{k}_j^T. \tag{13.72}$$

When the control-only sources are well separated from the signal sources of interest, the relationship

$$span\{\boldsymbol{k}_1, \ldots, \boldsymbol{k}_P\} \perp span\{\boldsymbol{h}_1, \ldots, \boldsymbol{h}_Q\}, \tag{13.73}$$

approximately holds. Under this assumption, we will show that the set of vectors

$$\{\boldsymbol{h}_1, \ldots, \boldsymbol{h}_Q, \boldsymbol{d}_1, \ldots, \boldsymbol{d}_{M-P-Q}, \boldsymbol{k}_1, \ldots, \boldsymbol{k}_P\} \tag{13.74}$$

are the eigenvectors of $\widetilde{\boldsymbol{R}}$, where $\{\boldsymbol{d}_1, \ldots, \boldsymbol{d}_{M-P-Q}\}$ is the orthonormal basis set of the subspace: $span\{\boldsymbol{h}_{Q+1}, \ldots, \boldsymbol{h}_M\} \cap span\{\boldsymbol{k}_{P+1}, \ldots, \boldsymbol{k}_M\}$.

We first show that the relationship

$$\widetilde{\boldsymbol{R}}\boldsymbol{h}_i = (\gamma_i + 1)\boldsymbol{h}_i \qquad (i = 1, \ldots, Q), \tag{13.75}$$

holds. That is, we show that the vectors $\boldsymbol{h}_i$ (where $i = 1, \ldots, Q$) are the eigenvectors of $\widetilde{\boldsymbol{R}}$ and their corresponding eigenvalues are $\gamma_i + 1$. To show this, we calculate the right multiplication of $\widetilde{\boldsymbol{R}}$ with $\boldsymbol{h}_i$ $(i = 1, \ldots, Q)$, and using Eq. (13.72) this multiplication results in

$$\widetilde{\boldsymbol{R}}\boldsymbol{h}_i = (\sum_{j=1}^{Q} \gamma_j \boldsymbol{h}_j \boldsymbol{h}_j^T)\boldsymbol{h}_i + (\sum_{j=P+1}^{M} \boldsymbol{k}_j \boldsymbol{k}_j^T)\boldsymbol{h}_i + (\sum_{j=1}^{P} (1-\beta_j)\boldsymbol{k}_j \boldsymbol{k}_j^T)\boldsymbol{h}_i. \tag{13.76}$$

The first term on the right-hand side becomes $\gamma_i \boldsymbol{h}_i$. The third term on the right-hand side of Eq. (13.76) vanishes due to the orthogonality assumption in Eq. (13.73).

The second term becomes $\boldsymbol{h}_i$, i.e.,

$$(\sum_{j=P+1}^{M} \boldsymbol{k}_j \boldsymbol{k}_j^T)\boldsymbol{h}_i = \boldsymbol{h}_i. \tag{13.77}$$

To show this, we first point out that the relationship,

$$span\{\boldsymbol{h}_1, \ldots, \boldsymbol{h}_Q\} \cap span\{\boldsymbol{k}_1, \ldots, \boldsymbol{k}_P\} = \emptyset \qquad (13.78)$$

holds, where $\emptyset$ indicates the empty set[1]. Because this relationship holds, we have $\boldsymbol{h}_j\,(j = 1, \ldots, Q) \notin span\{\boldsymbol{k}_1, \ldots, \boldsymbol{k}_P\}$, and $\boldsymbol{h}_j\ (j = 1, \ldots, Q)$ belongs to the complementary subspace of $span\{\boldsymbol{k}_1, \ldots, \boldsymbol{k}_P\}$, which is equal to $span\{\boldsymbol{k}_{Q+1}, \ldots, \boldsymbol{k}_M\}$. Namely, $\boldsymbol{h}_j \in span\{\boldsymbol{k}_{Q+1}, \ldots, \boldsymbol{k}_M\}$. Therefore, because $\sum_{j=P+1}^{M} \boldsymbol{k}_j\boldsymbol{k}_j^T$ is the projector onto $span\{\boldsymbol{k}_{Q+1}, \ldots, \boldsymbol{k}_M\}$, the application of this projector to $\boldsymbol{h}_i$ results in $\boldsymbol{h}_i$ itself. Thus, we show that the vectors $\boldsymbol{h}_i$ (where $i = 1, \ldots, Q$) are the eigenvectors of $\widetilde{\boldsymbol{R}}$, and the corresponding eigenvalues are $\gamma_i + 1$.

Next, we show that the relationship,

$$\widetilde{\boldsymbol{R}}\boldsymbol{d}_i = \boldsymbol{d}_i \qquad (i = 1, \ldots, M - P - Q), \qquad (13.82)$$

holds. That is, we show that the vectors $\boldsymbol{d}_i$ $(i = 1, \ldots, M - P - Q)$ are the eigenvectors of $\widetilde{\boldsymbol{R}}$, and the corresponding eigenvalues are equal to 1. To show this relationship, we calculate the right multiplication of $\widetilde{\boldsymbol{R}}$ with $\boldsymbol{d}_i$, which is equal to

$$\widetilde{\boldsymbol{R}}\boldsymbol{d}_i = (\sum_{j=1}^{Q} \gamma_j \boldsymbol{h}_j\boldsymbol{h}_j^T)\boldsymbol{d}_i + (\sum_{j=P+1}^{M} \boldsymbol{k}_j\boldsymbol{k}_j^T)\boldsymbol{d}_i + (\sum_{j=1}^{P}(1-\beta_j)\boldsymbol{k}_j\boldsymbol{k}_j^T)\boldsymbol{d}_i. \qquad (13.83)$$

Since $\boldsymbol{d}_i$ is orthogonal to both the subspace spanned by $\boldsymbol{h}_j$ $(j = 1, \ldots, Q)$ and that spanned by $\boldsymbol{k}_j$ $(j = 1, \ldots, P)$, the only non-zero term on the right-hand side is the second term, which is equal to $\boldsymbol{d}_i$, because $\boldsymbol{d}_i \in span\{\boldsymbol{k}_{P+1}, \ldots, \boldsymbol{k}_M\}$. Thus, we have proved Eq. (13.82).

Finally, we calculate the right multiplication of $\widetilde{\boldsymbol{R}}$ with $\boldsymbol{k}_i$ $(i = 1, \ldots, P)$, which produces

$$\widetilde{\boldsymbol{R}}\boldsymbol{k}_i = (\sum_{j=1}^{Q} \gamma_j \boldsymbol{h}_j\boldsymbol{h}_j^T)\boldsymbol{k}_i + (\sum_{j=P+1}^{M} \boldsymbol{k}_j\boldsymbol{k}_j^T)\boldsymbol{k}_i + (\sum_{j=1}^{P}(1-\beta_j)\boldsymbol{k}_j\boldsymbol{k}_j^T)\boldsymbol{k}_i. \qquad (13.84)$$

---

[1] To prove this relationship, we use the fact that the vectors

$$\boldsymbol{l}(\boldsymbol{r}_1), \ldots, \boldsymbol{l}(\boldsymbol{r}_Q), \boldsymbol{l}(\boldsymbol{r}_1^c), \ldots, \boldsymbol{l}(\boldsymbol{r}_P^c)$$

are linearly independent, where $\boldsymbol{r}_1, \ldots, \boldsymbol{r}_Q$ are the locations of the signal sources and $\boldsymbol{r}_1^c, \ldots, \boldsymbol{r}_P^c$ are the locations of the control-only sources. Since the matrix $\boldsymbol{R}_c^{-1/2}$ is non-singular, the vectors

$$\boldsymbol{R}_c^{-1/2}\boldsymbol{l}(\boldsymbol{r}_1), \ldots, \boldsymbol{R}_c^{-1/2}\boldsymbol{l}(\boldsymbol{r}_Q), \boldsymbol{R}_c^{-1/2}\boldsymbol{l}(\boldsymbol{r}_1^c), \ldots, \boldsymbol{R}_c^{-1/2}\boldsymbol{l}(\boldsymbol{r}_P^c)$$

are linearly independent, and consequently the following relationship holds,

$$span\{\boldsymbol{R}_c^{-1/2}\boldsymbol{l}(\boldsymbol{r}_1), \ldots, \boldsymbol{R}_c^{-1/2}\boldsymbol{l}(\boldsymbol{r}_Q)\} \cap span\{\boldsymbol{R}_c^{-1/2}\boldsymbol{l}(\boldsymbol{r}_1^c), \ldots, \boldsymbol{R}_c^{-1/2}\boldsymbol{l}(\boldsymbol{r}_P^c)\} = \emptyset. \qquad (13.79)$$

Since we have

$$span\{\boldsymbol{h}_1, \ldots, \boldsymbol{h}_Q\} = span\{\boldsymbol{R}_c^{-1/2}\boldsymbol{l}(\boldsymbol{r}_1), \ldots, \boldsymbol{R}_c^{-1/2}\boldsymbol{l}(\boldsymbol{r}_Q)\}, \qquad (13.80)$$

$$\text{and} \quad span\{\boldsymbol{k}_1, \ldots, \boldsymbol{k}_P\} = span\{\boldsymbol{R}_c^{-1/2}\boldsymbol{l}(\boldsymbol{r}_1^c), \ldots, \boldsymbol{R}_c^{-1/2}\boldsymbol{l}(\boldsymbol{r}_P^c)\}, \qquad (13.81)$$

Eq. (13.79) is equal to the relationship in Eq. (13.78).

On the right-hand side, the first term becomes zero due to the orthogonality assumption in Eq. (13.73), and the second term becomes zero due to the orthogonality relationship between the signal and the noise subspaces. Thus, we have

$$\widetilde{\boldsymbol{R}}\boldsymbol{k}_i = (1-\beta_i)\boldsymbol{k}_i. \tag{13.85}$$

Therefore, $\boldsymbol{k}_i$ $(i = 1, \ldots, P)$ are eigenvectors of $\widetilde{\boldsymbol{R}}$ and the corresponding eigenvalues are $1 - \beta_i$. We can also show that these eigenvalues are positive but less than 1, i.e., $0 < 1 - \beta_j < 1$, although we do not include the proof here. In summary, we have shown that the vectors

$$\{\boldsymbol{h}_1, \ldots, \boldsymbol{h}_Q, \boldsymbol{d}_1, \ldots, \boldsymbol{d}_{M-P-Q}, \boldsymbol{k}_P, \ldots, \boldsymbol{k}_1\}$$

are the eigenvectors of $\widetilde{\boldsymbol{R}}$. The corresponding eigenvalues, in decreasing order, are:

$$\gamma_1 + 1, \ldots, \gamma_Q + 1, \underbrace{1, \ldots, 1}_{M-Q-P}, 1 - \beta_P, \ldots, 1 - \beta_1. \tag{13.86}$$

Here, the largest Q eigenvalues $\gamma_1 + 1, \ldots, \gamma_Q + 1$ are greater than 1, and therefore, Eq. (8.26) is still effective at retrieving $\boldsymbol{R}_s$, even when control-only sources exist.

In general, however, the subspace angle between $span\{\boldsymbol{h}_1, \ldots, \boldsymbol{h}_Q\}$ and $span\{\boldsymbol{k}_1, \ldots, \boldsymbol{k}_P\}$ may not be so large and the assumption that these two subspaces are orthogonal may not be satisfied. In such cases, Eq. (13.75) changes to

$$\widetilde{\boldsymbol{R}}\boldsymbol{h}_i = (\gamma_i + 1)\boldsymbol{h}_i + \sum_{j=1}^{P}(1-\beta_j)(\boldsymbol{k}_j\boldsymbol{k}_j^T)\boldsymbol{h}_i = (\gamma_i + 1)\boldsymbol{h}_i + \sum_{j=1}^{P}\left[(1-\beta_j)\boldsymbol{k}_j^T\boldsymbol{h}_i\right]\boldsymbol{k}_j. \tag{13.87}$$

This equation shows that $\boldsymbol{h}_i$ $(i = 1, \ldots, Q)$ is no longer the eigenvector of $\widetilde{\boldsymbol{R}}$ and the second term on the right-hand side of Eq. (13.87) indicates the error term. Thus, if the relationship,

$$(\gamma_i + 1) \gg |\sum_{j=1}^{P}(1-\beta_j)\boldsymbol{k}_j^T\boldsymbol{h}_i| \tag{13.88}$$

holds, the error term is negligibly smaller than the first term, and the $\boldsymbol{h}_i$ are still approximately the signal-level eigenvectors of $\widetilde{\boldsymbol{R}}$. Conversely, when the error term is not small, the signal-covariance estimate obtained from Eq (8.26) could be erroneous.

13.6 Derivation of GLRT scanning function in Eq. (12.45)

Here, we derive the scanning function in Eq. (12.45). The derivation here is according to [86]. We first define a $\mathcal{D} \times K$ matrix $\boldsymbol{P}$ such that

$$\boldsymbol{P} = (\boldsymbol{\Phi}\boldsymbol{\Phi}^T)^{-1/2}\boldsymbol{\Phi}, \tag{13.89}$$

where $\boldsymbol{\Phi}$ is defined in Eq. (12.39), and K is the number of time point and $\mathcal{D}$ is the number of basis functions, which is naturally much smaller than K. We define a $(K-\mathcal{D}) \times K$ matrix $\boldsymbol{Q}$ whose rows consist of the basis functions orthogonal to $\boldsymbol{\varphi}(t)$. Then, we have the relationships,

$$\boldsymbol{P}\boldsymbol{P}^T = \boldsymbol{I}_{\mathcal{D}}, \tag{13.90}$$

$$\boldsymbol{Q}\boldsymbol{Q}^T = \boldsymbol{I}_{K-\mathcal{D}}, \tag{13.91}$$

$$\text{and} \quad \boldsymbol{P}^T\boldsymbol{P} + \boldsymbol{Q}^T\boldsymbol{Q} = \boldsymbol{I}_K, \tag{13.92}$$

where $\boldsymbol{I}_{\mathcal{D}}$, $\boldsymbol{I}_{K-\mathcal{D}}$, and $\boldsymbol{I}_K$ indicate the $\mathcal{D} \times \mathcal{D}$, $(K-\mathcal{D}) \times (K-\mathcal{D})$, and $K \times K$ identity matrices, respectively. Defining $\boldsymbol{B}_P = \boldsymbol{B}\boldsymbol{P}^T$ and $\boldsymbol{B}_Q = \boldsymbol{B}\boldsymbol{Q}^T$ (where $\boldsymbol{B}$ is the spatio-temporal data set: $\boldsymbol{B} = [\boldsymbol{b}(t_1), \boldsymbol{b}(t_2), \ldots, \boldsymbol{b}(t_K)]$), we thus have

$$\boldsymbol{B} = \boldsymbol{B}(\boldsymbol{P}^T\boldsymbol{P} + \boldsymbol{Q}^T\boldsymbol{Q}) = [\boldsymbol{B}_P, \boldsymbol{B}_Q]\begin{bmatrix} \boldsymbol{P} \\ \boldsymbol{Q} \end{bmatrix} \tag{13.93}$$

and

$$\boldsymbol{B}\boldsymbol{B}^T = [\boldsymbol{B}_P\boldsymbol{B}_Q]\begin{bmatrix} \boldsymbol{P} \\ \boldsymbol{Q} \end{bmatrix}[\boldsymbol{P}^T\boldsymbol{Q}^T]\begin{bmatrix} \boldsymbol{B}_P^T \\ \boldsymbol{B}_Q^T \end{bmatrix} = \boldsymbol{B}_P\boldsymbol{B}_P^T + \boldsymbol{B}_Q\boldsymbol{B}_Q^T. \tag{13.94}$$

We also define $\boldsymbol{E}$ such that $\boldsymbol{E} = \boldsymbol{L}(\boldsymbol{L}^T\boldsymbol{L})^{-1/2}$ where $\boldsymbol{L}$ is the lead-field matrix with the explicit notation of $(\boldsymbol{r})$ omitted for simplicity.

In Eq. (12.40), we derive the covariance matrix estimate for the alternative hypothesis H_1, such that

$$\boldsymbol{\Sigma} = \frac{1}{K}(\boldsymbol{B} - \boldsymbol{L}\boldsymbol{X}\boldsymbol{\Phi})(\boldsymbol{B} - \boldsymbol{L}\boldsymbol{X}\boldsymbol{\Phi})^T.$$

Using $\boldsymbol{E}$ and $\boldsymbol{P}$, we have

$$\boldsymbol{L}\boldsymbol{X}\boldsymbol{\Phi} = \boldsymbol{L}(\boldsymbol{L}^T\boldsymbol{L})^{-1/2}(\boldsymbol{L}^T\boldsymbol{L})^{1/2}\boldsymbol{X}(\boldsymbol{\Phi}\boldsymbol{\Phi}^T)^{1/2}(\boldsymbol{\Phi}\boldsymbol{\Phi}^T)^{-1/2}\boldsymbol{\Phi} = \boldsymbol{E}\widetilde{\boldsymbol{X}}\boldsymbol{P}, \tag{13.95}$$

where

$$\widetilde{\boldsymbol{X}} = (\boldsymbol{L}^T\boldsymbol{L})^{1/2}\boldsymbol{X}(\boldsymbol{\Phi}\boldsymbol{\Phi}^T)^{1/2}.$$

Therefore, $\boldsymbol{\Sigma}$ is expressed as

$$\boldsymbol{\Sigma} = \frac{1}{K}(\boldsymbol{B} - \boldsymbol{E}\widetilde{\boldsymbol{X}}\boldsymbol{P})(\boldsymbol{B} - \boldsymbol{E}\widetilde{\boldsymbol{X}}\boldsymbol{P})^T. \tag{13.96}$$

Next, defining

$$\boldsymbol{S} = \boldsymbol{B}_Q\boldsymbol{B}_Q^T, \tag{13.97}$$

and using the relationship

$$\begin{aligned}
(\boldsymbol{B}-\boldsymbol{E}\widetilde{\boldsymbol{X}}\boldsymbol{P})(\boldsymbol{B}-\boldsymbol{E}\widetilde{\boldsymbol{X}}\boldsymbol{P})^T &= \boldsymbol{B}\boldsymbol{B}^T - \boldsymbol{B}(\boldsymbol{E}\widetilde{\boldsymbol{X}}\boldsymbol{P})^T - (\boldsymbol{E}\widetilde{\boldsymbol{X}}\boldsymbol{P})\boldsymbol{B}^T + (\boldsymbol{E}\widetilde{\boldsymbol{X}}\boldsymbol{P})(\boldsymbol{E}\widetilde{\boldsymbol{X}}\boldsymbol{P})^T \\
&= (\boldsymbol{B}_P\boldsymbol{B}_P^T + \boldsymbol{B}_Q\boldsymbol{B}_Q^T) - \boldsymbol{B}_P(\boldsymbol{E}\widetilde{\boldsymbol{X}})^T - (\boldsymbol{E}\widetilde{\boldsymbol{X}})(\boldsymbol{B}_P^T) + (\boldsymbol{E}\widetilde{\boldsymbol{X}})(\boldsymbol{E}\widetilde{\boldsymbol{X}})^T \\
&= \boldsymbol{S} + (\boldsymbol{B}_P - \boldsymbol{E}\widetilde{\boldsymbol{X}})(\boldsymbol{B}_P - \boldsymbol{E}\widetilde{\boldsymbol{X}})^T \\
&= \boldsymbol{S}^{1/2}[\boldsymbol{I} + \boldsymbol{S}^{-1/2}(\boldsymbol{B}_P - \boldsymbol{E}\widetilde{\boldsymbol{X}})(\boldsymbol{B}_P - \boldsymbol{E}\widetilde{\boldsymbol{X}})^T\boldsymbol{S}^{-1/2}]\boldsymbol{S}^{1/2},
\end{aligned} \tag{13.98}$$

we have

$$
\begin{aligned}
|K\boldsymbol{\Sigma}| &= |\boldsymbol{S}^{1/2}[\boldsymbol{I} + \boldsymbol{S}^{-1/2}(\boldsymbol{B}_P - \boldsymbol{E}\widetilde{\boldsymbol{X}})(\boldsymbol{B}_P - \boldsymbol{E}\widetilde{\boldsymbol{X}})^T\boldsymbol{S}^{-1/2}]\boldsymbol{S}^{1/2}| \\
&= |\boldsymbol{S}||\boldsymbol{I} + \boldsymbol{S}^{-1/2}(\boldsymbol{B}_P - \boldsymbol{E}\widetilde{\boldsymbol{X}})(\boldsymbol{B}_P - \boldsymbol{E}\widetilde{\boldsymbol{X}})^T\boldsymbol{S}^{-1/2}| \\
&= |\boldsymbol{S}||\boldsymbol{I} + (\boldsymbol{B}_P - \boldsymbol{E}\widetilde{\boldsymbol{X}})^T\boldsymbol{S}^{-1}(\boldsymbol{B}_P - \boldsymbol{E}\widetilde{\boldsymbol{X}})|. \quad (13.99)
\end{aligned}
$$

Let us define $\mathfrak{F}(\widetilde{\boldsymbol{X}})$ such that

$$\mathfrak{F}(\widetilde{\boldsymbol{X}}) = \boldsymbol{I} + (\boldsymbol{B}_P - \boldsymbol{E}\widetilde{\boldsymbol{X}})^T\boldsymbol{S}^{-1}(\boldsymbol{B}_P - \boldsymbol{E}\widetilde{\boldsymbol{X}}),$$

and try to find the minimum value of $\mathfrak{F}(\widetilde{\boldsymbol{X}})$ with respect to $\widetilde{\boldsymbol{X}}$. We rewrite this $\mathfrak{F}(\widetilde{\boldsymbol{X}})$ such that

$$\mathfrak{F}(\widetilde{\boldsymbol{X}}) = \boldsymbol{I} + \boldsymbol{B}_P^T\boldsymbol{S}^{-1}\boldsymbol{B}_P + (\widetilde{\boldsymbol{X}} - \widetilde{\boldsymbol{X}}_0)^T(\boldsymbol{E}^T\boldsymbol{S}^{-1}\boldsymbol{E})(\widetilde{\boldsymbol{X}} - \widetilde{\boldsymbol{X}}_0) - \widetilde{\boldsymbol{X}}_0^T(\boldsymbol{E}^T\boldsymbol{S}^{-1}\boldsymbol{E})\widetilde{\boldsymbol{X}}_0, \quad (13.100)$$

where

$$\widetilde{\boldsymbol{X}}_0 = (\boldsymbol{E}^T\boldsymbol{S}^{-1}\boldsymbol{E})^{-1}\boldsymbol{E}^T\boldsymbol{S}^{-1}\boldsymbol{B}_P. \quad (13.101)$$

Therefore, the case of $\widetilde{\boldsymbol{X}} = \widetilde{\boldsymbol{X}}_0$ minimizes $\mathfrak{F}(\widetilde{\boldsymbol{X}})$, and the minimum value $\mathfrak{F}(\widetilde{\boldsymbol{X}}_0)$ is

$$\mathfrak{F}(\widetilde{\boldsymbol{X}}_0) = \boldsymbol{I} + \boldsymbol{B}_P^T\boldsymbol{S}^{-1}\boldsymbol{B}_P - \widetilde{\boldsymbol{X}}_0^T(\boldsymbol{E}^T\boldsymbol{S}^{-1}\boldsymbol{E})\widetilde{\boldsymbol{X}}_0. \quad (13.102)$$

The scanning function in Eq. (12.44) can then be expressed as

$$\mathcal{S}(\boldsymbol{r}) = \frac{|\boldsymbol{B}\boldsymbol{B}^T|}{\min_{\boldsymbol{X}}|K\boldsymbol{\Sigma}|} = \frac{|\boldsymbol{B}_P\boldsymbol{B}_P^T + \boldsymbol{S}|}{|\boldsymbol{S}||\boldsymbol{I} + \boldsymbol{B}_P^T\boldsymbol{S}^{-1}\boldsymbol{B}_P - \widetilde{\boldsymbol{X}}_0^T(\boldsymbol{E}^T\boldsymbol{S}^{-1}\boldsymbol{E})\widetilde{\boldsymbol{X}}_0|}. \quad (13.103)$$

Substituting $\widetilde{\boldsymbol{X}}_0$ in Eq. (13.101) into the right-hand side of the above equation, after some manipulation, we finally get

$$
\begin{aligned}
\mathcal{S}(\boldsymbol{r}) &= \frac{|\boldsymbol{I} + \boldsymbol{B}_P^T\boldsymbol{S}^{-1}\boldsymbol{B}_P|}{|\boldsymbol{I} + \boldsymbol{B}_P^T\boldsymbol{S}^{-1}\boldsymbol{B}_P - \boldsymbol{B}_P^T\boldsymbol{S}^{-1}\boldsymbol{E}(\boldsymbol{E}^T\boldsymbol{S}^{-1}\boldsymbol{E})^{-1}\boldsymbol{E}^T\boldsymbol{S}^{-1}\boldsymbol{B}_P|} \\
&= \frac{|\boldsymbol{I} + \boldsymbol{B}_P^T\boldsymbol{S}^{-1}\boldsymbol{B}_P|}{|\boldsymbol{I} + \boldsymbol{B}_P^T\boldsymbol{S}^{-1}\boldsymbol{B}_P - \boldsymbol{B}_P^T\boldsymbol{S}^{-1}\boldsymbol{L}(\boldsymbol{L}^T\boldsymbol{S}^{-1}\boldsymbol{L})^{-1}\boldsymbol{L}^T\boldsymbol{S}^{-1}\boldsymbol{B}_P|}, \quad (13.104)
\end{aligned}
$$

where we use the relationship[2]

$$\boldsymbol{E}(\boldsymbol{E}^T\boldsymbol{S}^{-1}\boldsymbol{E})^{-1}\boldsymbol{E}^T = \boldsymbol{L}(\boldsymbol{L}^T\boldsymbol{S}^{-1}\boldsymbol{L})^{-1}\boldsymbol{L}^T.$$

[2] To show this relationship, we first derive,

$$\boldsymbol{E}^T\boldsymbol{S}^{-1}\boldsymbol{E} = [\boldsymbol{L}(\boldsymbol{L}^T\boldsymbol{L})^{-1/2}]^T\boldsymbol{S}^{-1}\boldsymbol{L}(\boldsymbol{L}^T\boldsymbol{L})^{-1/2} = (\boldsymbol{L}^T\boldsymbol{L})^{-1/2}\boldsymbol{L}^T\boldsymbol{S}^{-1}\boldsymbol{L}(\boldsymbol{L}^T\boldsymbol{L})^{-1/2}.$$

Thus, we have,

$$
\begin{aligned}
\boldsymbol{E}(\boldsymbol{E}^T\boldsymbol{S}^{-1}\boldsymbol{E})^{-1}\boldsymbol{E}^T &= \boldsymbol{L}(\boldsymbol{L}^T\boldsymbol{L})^{-1/2}\left[(\boldsymbol{L}^T\boldsymbol{L})^{-1/2}\boldsymbol{L}^T\boldsymbol{S}^{-1}\boldsymbol{L}(\boldsymbol{L}^T\boldsymbol{L})^{-1/2}\right]^{-1}(\boldsymbol{L}^T\boldsymbol{L})^{-1/2}\boldsymbol{L}^T \\
&= \boldsymbol{L}(\boldsymbol{L}^T\boldsymbol{S}^{-1}\boldsymbol{L})^{-1}\boldsymbol{L}^T.
\end{aligned}
$$

Because the matrices $\boldsymbol{I}+\boldsymbol{B}_P^T\boldsymbol{S}^{-1}\boldsymbol{B}_P$ and $\boldsymbol{L}^T\boldsymbol{S}^{-1}\boldsymbol{L}$ are non-singular matrices, we can use the identity concerning the matrix determinant, i.e.,

$$|\boldsymbol{I}+\boldsymbol{B}_P^T\boldsymbol{S}^{-1}\boldsymbol{B}_P||\boldsymbol{L}^T\boldsymbol{S}^{-1}\boldsymbol{L}-\boldsymbol{L}^T\boldsymbol{S}^{-1}\boldsymbol{B}_P(\boldsymbol{I}+\boldsymbol{B}_P^T\boldsymbol{S}^{-1}\boldsymbol{B}_P)^{-1}\boldsymbol{B}_P^T\boldsymbol{S}^{-1}\boldsymbol{L}|$$
$$=|\boldsymbol{L}^T\boldsymbol{S}^{-1}\boldsymbol{L}||\boldsymbol{I}+\boldsymbol{B}_P^T\boldsymbol{S}^{-1}\boldsymbol{B}_P-\boldsymbol{B}_P^T\boldsymbol{S}^{-1}\boldsymbol{L}(\boldsymbol{L}^T\boldsymbol{S}^{-1}\boldsymbol{L})^{-1}\boldsymbol{L}^T\boldsymbol{S}^{-1}\boldsymbol{B}_P|.$$

Therefore, the scanning function in Eq. (13.104) can be rewritten as

$$\mathcal{S}(\boldsymbol{r})=\frac{|\boldsymbol{L}^T\boldsymbol{S}^{-1}\boldsymbol{L}|}{|\boldsymbol{L}^T\boldsymbol{S}^{-1}\boldsymbol{L}-\boldsymbol{L}^T\boldsymbol{S}^{-1}\boldsymbol{B}_P(\boldsymbol{I}+\boldsymbol{B}_P^T\boldsymbol{S}^{-1}\boldsymbol{B}_P)^{-1}\boldsymbol{B}_P^T\boldsymbol{S}^{-1}\boldsymbol{L}|}$$
$$=\frac{|\boldsymbol{L}^T\boldsymbol{S}^{-1}\boldsymbol{L}|}{|\boldsymbol{L}^T\left[\boldsymbol{S}^{-1}-\boldsymbol{S}^{-1}\boldsymbol{B}_P(\boldsymbol{I}+\boldsymbol{B}_P^T\boldsymbol{S}^{-1}\boldsymbol{B}_P)^{-1}\boldsymbol{B}_P^T\boldsymbol{S}^{-1}\right]\boldsymbol{L}|}. \qquad (13.105)$$

Using the matrix inversion formula,

$$\boldsymbol{S}^{-1}-\boldsymbol{S}^{-1}\boldsymbol{B}_P(\boldsymbol{I}+\boldsymbol{B}_P^T\boldsymbol{S}^{-1}\boldsymbol{B}_P)^{-1}\boldsymbol{B}_P^T\boldsymbol{S}^{-1}=(\boldsymbol{S}+\boldsymbol{B}_P\boldsymbol{B}_P^T)^{-1},$$

we finally obtain

$$\mathcal{S}(\boldsymbol{r})=\frac{|\boldsymbol{L}^T\boldsymbol{S}^{-1}\boldsymbol{L}|}{|\boldsymbol{L}^T(\boldsymbol{S}+\boldsymbol{B}_P\boldsymbol{B}_P^T)^{-1}\boldsymbol{L}|}=\frac{|\boldsymbol{L}^T\boldsymbol{S}^{-1}\boldsymbol{L}|}{|\boldsymbol{L}^T(\boldsymbol{B}\boldsymbol{B}^T)^{-1}\boldsymbol{L}|}. \qquad (13.106)$$

The equation above is the same as the scanning function in Eq. (12.45) because the following relationship holds:

$$\frac{1}{K}\boldsymbol{S}=\frac{1}{K}\boldsymbol{B}_Q\boldsymbol{B}_Q^T=\frac{1}{K}(\boldsymbol{B}\boldsymbol{B}^T-\boldsymbol{B}_P\boldsymbol{B}_P^T)=\frac{1}{K}(\boldsymbol{B}\boldsymbol{B}^T-\boldsymbol{B}\boldsymbol{\Phi}^T(\boldsymbol{\Phi}\boldsymbol{\Phi}^T)^{-1}\boldsymbol{\Phi}\boldsymbol{B}^T)$$
$$=\widehat{\boldsymbol{R}}-\boldsymbol{R}_{\Phi B}^T\boldsymbol{R}_{\Phi\Phi}^{-1}\boldsymbol{R}_{\Phi B}=\boldsymbol{A}.$$

13.7 Bioelectromagnetic forward modeling

Estimation of the sensor lead-field is called the forward problem[92], which is the problem of computing the sensor outputs from a point source located at a known location. The forward problem stands in sharp contrast to the inverse problem, which is the problem of estimating source spatial distributions from the sensor data. It is obvious that to solve the inverse problem we need to have a reasonably accurate estimate of the sensor lead field. In this chapter, we describe how we can obtain an estimate of the lead field for an MEG sensor array. Although this section strays somewhat from the scope of the book, we include it for readers' convenience. The first five subsections follow the derivation reported in Sarvas[93], and we derive the well-known Sarvas formula for calculating the sensor lead field for the spherically-symmetric homogeneous conductor. We briefly mention how to compute the lead field from a realistic head-shape model. An extension to derive the EEG sensor array is also described.

13.7.1 Quasi-static Maxwell's equations

When neural activities exist, they generate an external electro-motive force $\boldsymbol{F}_{ext}$. Then, the total force $\boldsymbol{F}$ received by a particle with charge q located in an electric field $\boldsymbol{E}$ is

$$\boldsymbol{F} = q\boldsymbol{E} + \boldsymbol{F}_{ext} = q(\boldsymbol{E} + \boldsymbol{E}_{ext}), \tag{13.107}$$

where $\boldsymbol{E}_{ext}$ represents the external electro-motive force expressed as an electric field. The electric current density generally is proportional to $\boldsymbol{F}$. Thus, denoting the proportionality constant as ρ/q, the relationship

$$\boldsymbol{J} = \frac{\rho}{q}\boldsymbol{F} = \rho(\boldsymbol{E} + \boldsymbol{E}_{ext}) = \rho\boldsymbol{E} + \boldsymbol{J}_e \tag{13.108}$$

holds where the proportionality constant, ρ, is called the conductivity. In Eq. (13.108), $\boldsymbol{J}$ is the total current density and $\boldsymbol{J}_e$ represents the external electro-motive force expressed in units of current density. This $\boldsymbol{J}_e$ is called the primary current or the impressed current, and the electric current represented by $\rho\boldsymbol{E}$, which does not directly represent the neural activities, is called the return current or the volume current. This $\boldsymbol{J}_e$ is the quantity referred to as the source and is denoted $\boldsymbol{s}(\boldsymbol{r})$ throughout this book. In this section, however, we use the notation $\boldsymbol{J}_e$ to explicitly express the physical nature of the source. Here, we will derive the relationship between the magnetic field $\boldsymbol{B}$ and $\boldsymbol{J}_e$, starting with the quasi-static Maxwell's equations

$$\boldsymbol{E} = -\nabla V, \tag{13.109}$$

$$\text{and} \quad \nabla \times \boldsymbol{B} = \mu_0 \boldsymbol{J}, \tag{13.110}$$

where μ_0 is the magnetic permeability of free space, ∇ is the gradient operator, and V is the electric potential. The relationship between the total current $\boldsymbol{J}$ and the resultant magnetic field $\boldsymbol{B}$ is given by the well-known Biot-Savart law, which is derived from Eq. (13.110) and expressed as

$$\boldsymbol{B}(\boldsymbol{r}') = \frac{\mu_0}{4\pi}\int_\Omega \boldsymbol{J}(\boldsymbol{r}) \times \frac{\boldsymbol{r}' - \boldsymbol{r}}{|\boldsymbol{r}' - \boldsymbol{r}|^3}\, d^3r = \frac{\mu_0}{4\pi}\int_\Omega \boldsymbol{J}(\boldsymbol{r}) \times \boldsymbol{G}(\boldsymbol{r}', \boldsymbol{r})\, d^3r, \tag{13.111}$$

where the kernel $\boldsymbol{G}(\boldsymbol{r}', \boldsymbol{r})$ is expressed as

$$\boldsymbol{G}(\boldsymbol{r}', \boldsymbol{r}) = \frac{\boldsymbol{r}' - \boldsymbol{r}}{|\boldsymbol{r}' - \boldsymbol{r}|^3} = \nabla|\boldsymbol{r}' - \boldsymbol{r}|^{-1}. \tag{13.112}$$

In Eq. (13.111), Ω indicates the whole brain region, and the volume integral is performed in the region Ω.

13.7.2 Magnetic field in an infinite homogeneous conductor

We first consider the case where a whole space is filled with a conductor with a constant conductivity ρ, and derive the relationship between the primary source current $\boldsymbol{J}_e$ and the magnetic field $\boldsymbol{B}$. Let us consider the identity

$$\int_\Omega \nabla \times (\boldsymbol{J}(\boldsymbol{r})|\boldsymbol{r}'-\boldsymbol{r}|^{-1})\, d^3r$$
$$= \int_\Omega |\boldsymbol{r}'-\boldsymbol{r}|^{-1}\nabla \times \boldsymbol{J}(\boldsymbol{r})\, d^3r + \int_\Omega (\nabla|\boldsymbol{r}'-\boldsymbol{r}|^{-1}) \times \boldsymbol{J}(\boldsymbol{r})\, d^3r, \quad (13.113)$$

where the gradient operator is applied to $\boldsymbol{r}$, and the volume integral is applied to a region Ω. The volume integral on the left-hand side of the equation above is changed to a surface integral such that

$$\int_\Omega \nabla \times (\boldsymbol{J}(\boldsymbol{r})|\boldsymbol{r}'-\boldsymbol{r}|^{-1})\, d^3r = \int_{\partial\Omega} d\boldsymbol{S} \times (\boldsymbol{J}(\boldsymbol{r})|\boldsymbol{r}'-\boldsymbol{r}|^{-1}), \quad (13.114)$$

where $\partial\Omega$ indicates the surface of Ω, and the surface integral on the right-hand side is applied to $\partial\Omega$. Here, $d\boldsymbol{S}$ is the surface-element vector on $\partial\Omega$. When we assume that the region Ω is an infinitely extended volume, $\boldsymbol{J}(\boldsymbol{r})$ on the surface becomes zero. Namely, the right-hand side of Eq. (13.114) becomes zero, and we have

$$\int_\Omega \nabla \times (\boldsymbol{J}(\boldsymbol{r})|\boldsymbol{r}'-\boldsymbol{r}|^{-1})\, d^3r = 0. \quad (13.115)$$

Therefore, using Eq. (13.113), the relationship,

$$\int_\Omega \frac{\nabla \times \boldsymbol{J}(\boldsymbol{r})}{|\boldsymbol{r}'-\boldsymbol{r}|}\, d^3r = \int_\Omega \boldsymbol{J}(\boldsymbol{r})\times(\nabla|\boldsymbol{r}'-\boldsymbol{r}|^{-1})\, d^3r = \int_\Omega \boldsymbol{J}(\boldsymbol{r})\times\boldsymbol{G}(\boldsymbol{r}',\boldsymbol{r})\, d^3r, \quad (13.116)$$

holds.

Using this equation, the Biot-Savart law in Eq. (13.111) is rewritten as

$$\boldsymbol{B}(\boldsymbol{r}') = \frac{\mu_0}{4\pi}\int_\Omega \boldsymbol{J}(\boldsymbol{r}) \times \boldsymbol{G}(\boldsymbol{r}',\boldsymbol{r})\, d^3r = \frac{\mu_0}{4\pi}\int_\Omega \frac{\nabla \times \boldsymbol{J}(\boldsymbol{r})}{|\boldsymbol{r}'-\boldsymbol{r}|}\, d^3r. \quad (13.117)$$

Substituting $\boldsymbol{J} = \boldsymbol{J}_e - \rho\nabla V$ into the above equation, and considering the relationship, $\nabla \times \nabla V = 0$, we finally obtain

$$\boldsymbol{B}(\boldsymbol{r}') = \frac{\mu_0}{4\pi}\int_\Omega \frac{\nabla \times \boldsymbol{J}_e(\boldsymbol{r})}{|\boldsymbol{r}'-\boldsymbol{r}|}\, d^3r = \frac{\mu_0}{4\pi}\int_\Omega \boldsymbol{J}_e(\boldsymbol{r}) \times \boldsymbol{G}(\boldsymbol{r}',\boldsymbol{r})\, d^3r, \quad (13.118)$$

where we again use the relationship in Eq. (13.117) with $\boldsymbol{J}_e$ instead of $\boldsymbol{J}$. The equation above indicates that the exact same formula for the Biot-Savart law holds for $\boldsymbol{J}_e$ in the case of an infinite homogeneous conductor. Let us consider the situation in which the source distribution is highly localized and concentrated around $\boldsymbol{r}_0$, so it can be expressed as $\boldsymbol{J}_e(\boldsymbol{r}) = \boldsymbol{Q}\delta(\boldsymbol{r}-\boldsymbol{r}_0)$ where $\boldsymbol{Q}$ is the moment of the source located at $\boldsymbol{r}_0$. Then, Eq. (13.118) changes to

$$\boldsymbol{B}(\boldsymbol{r}') = \frac{\mu_0}{4\pi}\int_\Omega \boldsymbol{Q}\delta(\boldsymbol{r}-\boldsymbol{r}_0) \times \boldsymbol{G}(\boldsymbol{r}',\boldsymbol{r})\, d^3r = \frac{\mu_0}{4\pi}\boldsymbol{Q} \times \frac{\boldsymbol{r}'-\boldsymbol{r}_0}{|\boldsymbol{r}'-\boldsymbol{r}_0|^3}. \quad (13.119)$$

13.7.3 Electric potential in an infinite homogeneous conductor

We again consider the case where a whole space is filled with a conductor with a constant conductivity, ρ, and next derive the relationship between the primary source current, $\boldsymbol{J}_e$, and the electric potential, V. To derive a formula for the electric potential, we start from the Poisson equation,

$$\nabla^2 V = \nabla \cdot \boldsymbol{J}_e(\boldsymbol{r})/\rho. \tag{13.120}$$

The equation above has a well-known solution expressed as

$$V(\boldsymbol{r}') = -\frac{1}{4\pi\rho}\int_\Omega \frac{\nabla \cdot \boldsymbol{J}_e(\boldsymbol{r})}{|\boldsymbol{r}' - \boldsymbol{r}|}\, d^3r. \tag{13.121}$$

Let us consider the identity

$$\int_\Omega \nabla \cdot (\boldsymbol{J}_e(\boldsymbol{r})|\boldsymbol{r}' - \boldsymbol{r}|^{-1})\, d^3r$$
$$= \int_\Omega |\boldsymbol{r}' - \boldsymbol{r}|^{-1}\nabla \cdot \boldsymbol{J}_e(\boldsymbol{r})\, d^3r + \int_\Omega (\nabla|\boldsymbol{r}' - \boldsymbol{r}|^{-1}) \cdot \boldsymbol{J}_e(\boldsymbol{r})\, d^3r. \tag{13.122}$$

The volume integral on the left-hand side of the equation above can be converted to a surface integral by using the Gauss theorem whereby

$$\int_\Omega \nabla \cdot (\boldsymbol{J}_e(\boldsymbol{r})|\boldsymbol{r}' - \boldsymbol{r}|^{-1})\, d^3r = \int_{\partial\Omega} d\boldsymbol{S} \cdot (\boldsymbol{J}_e(\boldsymbol{r})|\boldsymbol{r}' - \boldsymbol{r}|^{-1}). \tag{13.123}$$

We again assume that the region Ω is an infinitely extended volume. Then, $\boldsymbol{J}_e(\boldsymbol{r})$ on the surface becomes zero, and the left-hand side of Eq. (13.122) is equal to zero. Consequently, we have

$$V(\boldsymbol{r}') = \frac{1}{4\pi\rho}\int_\Omega (\nabla|\boldsymbol{r}' - \boldsymbol{r}|^{-1}) \cdot \boldsymbol{J}_e(\boldsymbol{r})\, d^3r = \frac{1}{4\pi\rho}\int_\Omega \boldsymbol{J}_e(\boldsymbol{r}) \cdot \boldsymbol{G}(\boldsymbol{r}', \boldsymbol{r})\, d^3r. \tag{13.124}$$

Considering the case in which the source distribution can be expressed as $\boldsymbol{J}_e(\boldsymbol{r}) = \boldsymbol{Q}\delta(\boldsymbol{r} - \boldsymbol{r}_0)$, Eq. (13.124) changes to

$$V(\boldsymbol{r}') = \frac{1}{4\pi\rho}\int_\Omega \boldsymbol{Q}\delta(\boldsymbol{r} - \boldsymbol{r}_0) \cdot \boldsymbol{G}(\boldsymbol{r}', \boldsymbol{r})\, d^3r = \frac{1}{4\pi\rho}\boldsymbol{Q} \cdot \frac{\boldsymbol{r}' - \boldsymbol{r}_0}{|\boldsymbol{r}' - \boldsymbol{r}_0|^3}. \tag{13.125}$$

13.7.4 Formulae in a bounded conductor with piecewise-constant conductivity

We next consider the magnetic field generated by an inhomogeneous conductor. We assume that the region Ω can be divided into subregions Ω_j, $j = 1, \dots, \mathcal{O}$, and the region Ω_j has conductivity ρ_j. We also assume that the conductivity, $\rho(\boldsymbol{r})$, is zero outside Ω. In this case, substituting $\boldsymbol{J}(\boldsymbol{r}) = \boldsymbol{J}_e(\boldsymbol{r}) - \rho(\boldsymbol{r})\nabla V(\boldsymbol{r})$ into the Biot-Savart law in Eq. (13.111), we obtain

$$\begin{aligned} \boldsymbol{B}(\boldsymbol{r}') &= \frac{\mu_0}{4\pi}\int_\Omega [\boldsymbol{J}_e(\boldsymbol{r}) - \rho(\boldsymbol{r})\nabla V(\boldsymbol{r})] \times \boldsymbol{G}(\boldsymbol{r}',\boldsymbol{r})\, d^3r \\ &= \frac{\mu_0}{4\pi}\int_\Omega \boldsymbol{J}_e(\boldsymbol{r}) \times \boldsymbol{G}(\boldsymbol{r}',\boldsymbol{r})\, d^3r - \frac{\mu_0}{4\pi}\sum_{j=1}^{\mathcal{O}} \rho_j \int_{\Omega_j} \nabla V(\boldsymbol{r}) \times \boldsymbol{G}(\boldsymbol{r}',\boldsymbol{r})\, d^3r. \end{aligned} \tag{13.126}$$

Using

$$\nabla \times [V(\boldsymbol{r})\boldsymbol{G}(\boldsymbol{r}',\boldsymbol{r})] = \nabla V(\boldsymbol{r}) \times \boldsymbol{G}(\boldsymbol{r}',\boldsymbol{r}), \tag{13.127}$$

and using the same identity for deriving Eq. (13.114), the second term on the right-hand side of Eq. (13.126) can be rewritten as

$$\begin{aligned} \int_{\Omega_j} \nabla V(\boldsymbol{r}) \times \boldsymbol{G}(\boldsymbol{r}',\boldsymbol{r})\, d^3r &= \int_{\Omega_j} \nabla \times V(\boldsymbol{r})\boldsymbol{G}(\boldsymbol{r}',\boldsymbol{r})\, d^3r \\ &= \int_{\partial\Omega_j} d\boldsymbol{S} \times V(\boldsymbol{r})\boldsymbol{G}(\boldsymbol{r}',\boldsymbol{r}) = \int_{\partial\Omega_j} V(\boldsymbol{r})\boldsymbol{m}(\boldsymbol{r}) \times \boldsymbol{G}(\boldsymbol{r}',\boldsymbol{r}) dS, \end{aligned} \tag{13.128}$$

where $\partial\Omega_j$ indicates the surface of Ω_j, and $\boldsymbol{m}(\boldsymbol{r})$ is the normal vector of a surface element on $\partial\Omega_j$. Then, substituting this equation into Eq. (13.126), we can derive the following Geselowitz formula[94]:

$$\boldsymbol{B}(\boldsymbol{r}') = \boldsymbol{B}_0(\boldsymbol{r}') - \frac{\mu_0}{4\pi}\sum_{j=1}^{\mathcal{O}} (\rho_j - \rho_j') \int_{\partial\Omega_j} V(\boldsymbol{r})\boldsymbol{m}(\boldsymbol{r}) \times \boldsymbol{G}(\boldsymbol{r}',\boldsymbol{r}) dS, \tag{13.129}$$

where

$$\boldsymbol{B}_0(\boldsymbol{r}') = \frac{\mu_0}{4\pi}\int_\Omega \boldsymbol{J}_e(\boldsymbol{r}) \times \boldsymbol{G}(\boldsymbol{r}',\boldsymbol{r})\, d^3r$$

and ρ_j' is the conductivity just outside of Ω_j. $\boldsymbol{B}_0(\boldsymbol{r}')$ above is the magnetic field for the infinite homogeneous conductor in Eq. (13.118). We can also derive a formula similar to Eq. (13.129) for the case of electric potential V:

$$\begin{aligned} \rho(\boldsymbol{r})V(\boldsymbol{r}') &= \frac{1}{4\pi}\int_\Omega \boldsymbol{J}_e(\boldsymbol{r}) \cdot \boldsymbol{G}(\boldsymbol{r}',\boldsymbol{r})\, d^3r \\ &\quad - \frac{1}{4\pi}\sum_{j=1}^{\mathcal{O}} (\rho_j - \rho_j') \int_{\partial\Omega_j} V(\boldsymbol{r})\boldsymbol{m}(\boldsymbol{r}) \cdot \boldsymbol{G}(\boldsymbol{r}',\boldsymbol{r}) dS. \end{aligned} \tag{13.130}$$

13.7.5 Magnetic field from a homogeneous spherical conductor

Here, we assume that Ω_j is spherically symmetric, and we set the coordinate origin at the center of Ω_j. We have

$$\boldsymbol{B}(\boldsymbol{r}')\cdot\boldsymbol{e}_r = \boldsymbol{B}_0(\boldsymbol{r}')\cdot\boldsymbol{e}_r - \frac{\mu_0}{4\pi}\sum_{j=1}^{\mathcal{O}} (\rho_j - \rho_j') \int_{\partial\Omega_j} V(\boldsymbol{r})\boldsymbol{m}(\boldsymbol{r}) \times \boldsymbol{G}(\boldsymbol{r}',\boldsymbol{r})\cdot\boldsymbol{e}_r dS, \tag{13.131}$$

where $\boldsymbol{e}_r$ is the unit vector in the radial direction, which is defined as $\boldsymbol{e}_r = \boldsymbol{r}/|\boldsymbol{r}|$. Since $\boldsymbol{m}(\boldsymbol{r}) = \boldsymbol{e}_r$ in this case, the second term is equal to zero, and we have the relationship

$$\boldsymbol{B} \cdot \boldsymbol{e}_r = \boldsymbol{B}_0 \cdot \boldsymbol{e}_r. \tag{13.132}$$

This equation indicates that the radial component of the magnetic field is not affected by the volume current and that the radial component is determined solely by the primary current. We then proceed to deriving a closed-form formula for the magnetic field outside a spherically-symmetric homogeneous conductor. The relationship $\nabla' \times \boldsymbol{B}(\boldsymbol{r}') = 0$ (where ∇' applies $\boldsymbol{r}'$) holds outside the volume conductor, because there is no electric current. Thus, $\boldsymbol{B}(\boldsymbol{r}')$ can be expressed in terms of the magnetic scalar potential $U(\boldsymbol{r}')$, as

$$\boldsymbol{B}(\boldsymbol{r}') = -\mu_0 \nabla' U(\boldsymbol{r}'). \tag{13.133}$$

This potential function is derived from

$$U(\boldsymbol{r}') = \frac{1}{\mu_0} \int_0^\infty \boldsymbol{B}(\boldsymbol{r}' + \tau \boldsymbol{e}_r) \cdot \boldsymbol{e}_r d\tau = \frac{1}{\mu_0} \int_0^\infty \boldsymbol{B}_0(\boldsymbol{r} + \tau \boldsymbol{e}_r) \cdot \boldsymbol{e}_r d\tau, \tag{13.134}$$

where we use the relationship in Eq. (13.132). Assuming that a highly localized source exists at $\boldsymbol{r}_0$, by substituting Eq. (13.119) into Eq. (13.134) and integrating it, we finally obtain

$$\begin{aligned} U(\boldsymbol{r}') &= \frac{1}{\mu_0} \int_0^\infty \boldsymbol{B}_0(\boldsymbol{r} + \tau \boldsymbol{e}_r) \cdot \boldsymbol{e}_r d\tau \\ &= \frac{1}{4\pi} \boldsymbol{Q} \times (\boldsymbol{r}' - \boldsymbol{r}_0) \cdot \boldsymbol{e}_r \int_0^\infty \frac{d\tau}{|\boldsymbol{r}' + \tau \boldsymbol{e}_r - \boldsymbol{r}_0|^3} = -\frac{1}{4\pi} \frac{(\boldsymbol{Q} \times \boldsymbol{r}_0) \cdot \boldsymbol{r}'}{A}, \end{aligned} \tag{13.135}$$

where

$$A = |\boldsymbol{r}' - \boldsymbol{r}_0|(|\boldsymbol{r}' - \boldsymbol{r}_0||\boldsymbol{r}'| + |\boldsymbol{r}'|^2 - \boldsymbol{r}_0 \cdot \boldsymbol{r}'). \tag{13.136}$$

The well-known Sarvas formula for $\boldsymbol{B}(\boldsymbol{r}')$ is then obtained by substituting Eq. (13.135) into Eq. (13.133) and performing the gradient operation. The results are expressed as

$$\begin{aligned} \boldsymbol{B}(\boldsymbol{r}') = \frac{\mu_0}{4\pi} \nabla' \frac{(\boldsymbol{Q} \times \boldsymbol{r}_0) \cdot \boldsymbol{r}'}{A} &= \frac{\mu_0}{4\pi} \left[\frac{\boldsymbol{Q} \times \boldsymbol{r}_0}{A} - \frac{1}{A^2} (\boldsymbol{Q} \times \boldsymbol{r}_0) \cdot \boldsymbol{r}' \nabla' A \right] \\ &= \frac{\mu_0}{4\pi A^2} \left[A \boldsymbol{Q} \times \boldsymbol{r}_0 - [(\boldsymbol{Q} \times \boldsymbol{r}_0) \cdot \boldsymbol{r}'] \nabla' A \right], \end{aligned} \tag{13.137}$$

where

$$\begin{aligned} \nabla' A = &\left[\frac{|\boldsymbol{r}' - \boldsymbol{r}_0|^2}{|\boldsymbol{r}'|} + \frac{(\boldsymbol{r}' - \boldsymbol{r}_0) \cdot \boldsymbol{r}'}{|\boldsymbol{r}' - \boldsymbol{r}_0|} + 2|\boldsymbol{r}' - \boldsymbol{r}_0| + 2|\boldsymbol{r}'| \right] \boldsymbol{r}' \\ &- \left[|\boldsymbol{r}' - \boldsymbol{r}_0| + 2|\boldsymbol{r}'| + \frac{(\boldsymbol{r}' - \boldsymbol{r}_0) \cdot \boldsymbol{r}'}{|\boldsymbol{r}' - \boldsymbol{r}_0|} \right] \boldsymbol{r}_0. \end{aligned} \tag{13.138}$$

We can see that when $\boldsymbol{r}_0$ approaches the center of the sphere, $\boldsymbol{B}(\boldsymbol{r}')$ becomes zero, and no magnetic field is generated outside the conductor from a source at the origin. This is an important property of the spherically-symmetric homogeneous conductor model. The other important property of the magnetic field obtained using the spherically-symmetric homogeneous conductor model is that if the source vector $\boldsymbol{Q}$ and the location vector $\boldsymbol{r}_0$ are parallel, i.e., if the primary current source is oriented in the radial direction, no magnetic fields are generated outside the spherical conductor from such a radial source. This is because both of the terms on the left-hand side of Eq. (13.137) contain the vector product $\boldsymbol{Q} \times \boldsymbol{r}_0$, which is equal to zero when $\boldsymbol{Q}$ and $\boldsymbol{r}_0$ are parallel. Therefore, when using the spherically-homogeneous conductor model, instead of the x, y, z directions, we usually use the three orthogonal directions $(\boldsymbol{e}_r, \boldsymbol{e}_\phi, \boldsymbol{e}_\theta)$ to express the source vector. These directions are illustrated in Fig. 13.1. Because the $\boldsymbol{e}_r$ component of a source never creates a measurable magnetic field outside the spherical conductor, we can disregard this component and only deal with the $\boldsymbol{e}_\phi$ and $\boldsymbol{e}_\theta$ components of the source vector.

To obtain the ϕ component of the lead field, $l_m^\phi(\boldsymbol{r})$, for example, we first calculate $\boldsymbol{B}(\boldsymbol{r}_m)$ (where $\boldsymbol{r}_m$ is the mth sensor location) by using Eq. (13.137) and by putting the unit-magnitude source at $\boldsymbol{r}$ directed in the ϕ direction. Namely, letting $\boldsymbol{r}' = \boldsymbol{r}_m$, $\boldsymbol{Q} = \boldsymbol{e}_\phi$, and $\boldsymbol{r}_0 = \boldsymbol{r}$, we calculate $\boldsymbol{B}(\boldsymbol{r}_m)$ using Eq. (13.137). When the sensor is a magnetometer, (which measures only the magnetic field component normal to the sensor), $l_m^\phi(\boldsymbol{r})$ is calculated from

$$l_m^\phi(\boldsymbol{r}) = \boldsymbol{B}(\boldsymbol{r}_m) \cdot \boldsymbol{e}_m^{\mathrm{sen}}, \tag{13.139}$$

where $\boldsymbol{e}_m^{\mathrm{sen}}$ is a unit vector expressing the normal direction of the mth sensor coil. When the sensor is a first-order axial gradiometer with a baseline of D, $l_m^\phi(\boldsymbol{r})$ is calculated from

$$l_m^\phi(\boldsymbol{r}) = \boldsymbol{B}(\boldsymbol{r}_m) \cdot \boldsymbol{e}_m^{\mathrm{sen}} - \boldsymbol{B}(\boldsymbol{r}_m + D\boldsymbol{e}_m^{\mathrm{sen}}) \cdot \boldsymbol{e}_m^{\mathrm{sen}}. \tag{13.140}$$

This $l_m^\phi(\boldsymbol{r})$ represents the sensitivity of the mth sensor to the primary current density located at $\boldsymbol{r}$ and directed in the $\boldsymbol{e}_\phi$ direction. In exactly the same manner, the θ component of the lead field $l_m^\theta(\boldsymbol{r})$ can be obtained.

13.7.6 Magnetic field from a realistically-shaped conductor

Numerical method

The spherically-symmetric homogeneous conductor is generally satisfactory in explaining the measured magnetic field when only superficial sources exist, i.e., when all sources are located relatively close to the sensor array. This is because the curvature of the upper half of the brain is well approximated by a sphere. However, for sources located in lower regions of the brain, the model becomes inaccurate because the curvature of the lower brain regions significantly differs from a sphere. The errors caused by misfits of the model may be reduced by using realistically-shaped volume-conductor models. Such conductor models can be constructed by

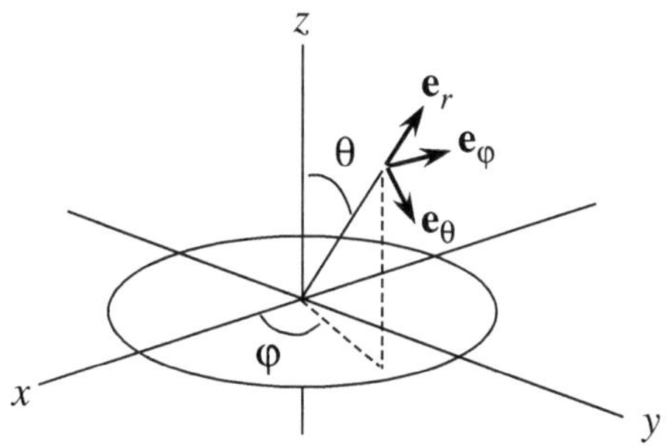

Figure 13.1: The three orthogonal directions $(\boldsymbol{e}_r, \boldsymbol{e}_\phi, \boldsymbol{e}_\theta)$ used to express the source vector when the spherically-symmetric conductor model is used for the forward calculation.

first extracting the brain boundary surface from the subject's 3D MRI. We denote this surface $\partial\Omega$, and the region surrounded by $\partial\Omega$ is denoted by Ω. We assume that the conductivity in Ω is uniform and denoted by ρ. We then use the Geselowitz formula in Eq. (13.129) to calculate the magnetic fields outside the volume conductor:

$$\boldsymbol{B}(\boldsymbol{r}') = \boldsymbol{B}_0(\boldsymbol{r}') - \frac{\mu_0}{4\pi}\rho \int_{\partial\Omega} V(\boldsymbol{r})\boldsymbol{m}(\boldsymbol{r}) \times \boldsymbol{G}(\boldsymbol{r}', \boldsymbol{r}) dS. \tag{13.141}$$

To calculate the second term on the right-hand side of Eq. (13.141), which represents the influence of the volume current, we need to know $V(\boldsymbol{r})$ on $\partial\Omega$, which is obtained by solving Eq. (13.130) for $\boldsymbol{r}'$ on $\partial\Omega$. For $\boldsymbol{r}'$ on the boundary of the surface, the integral equation in Eq. (13.130) is changed to

$$\frac{\rho}{2}V(\boldsymbol{r}') = \frac{1}{4\pi}\int_{\Omega} \boldsymbol{J}_e(\boldsymbol{r}) \cdot \boldsymbol{G}(\boldsymbol{r}', \boldsymbol{r})\, d^3r - \frac{\rho}{4\pi}\int_{\partial\Omega} V(\boldsymbol{r})\boldsymbol{m}(\boldsymbol{r}) \cdot \boldsymbol{G}(\boldsymbol{r}', \boldsymbol{r}) dS. \tag{13.142}$$

In this calculation, we first estimate the electric potential, $V(\boldsymbol{r})$, on the brain boundary surface by iteratively solving Eq. (13.142). We then calculate the magnetic fields outside the brain using Eq. (13.141). The details of these numerical calculations are beyond the scope of this book, and can be found in [95][96]. The numerical method mentioned so far assumes uniform conductivity within the brain boundary, and is called the single-compartment boundary element method (BEM). The single-compartment BEM is usually used in estimating MEG sensor lead fields [97]. It can be extended to the multiple-compartment BEM and such models are usually used for estimating the EEG sensor lead fields. The BEM-based realistically-shaped volume conductor models generally provide significant improvements in the accuracy of the forward calculation particularly for deep sources [98][97], although they are computationally expensive. Improvements in the computational efficiency of the BEM have been reported [99][100].

Perturbation method

One interesting approach, which does not require heavy numerical computations, has been proposed by Nolte *et al.*[101] to compute the lead field for MEG sensors. This approach computes the lead field for a realistic volume conductor as a sum of the lead field for the spherical homogeneous conductor plus a (typically small) correction factor. To explain this method, first we formulate the relationship between the primary current density, $\boldsymbol{J}_e(\boldsymbol{r})$, and the magnetic field, $\boldsymbol{B}(\boldsymbol{r}')$, such that

$$\boldsymbol{B}(\boldsymbol{r}') \cdot \boldsymbol{m} = \int \boldsymbol{J}_e(\boldsymbol{r}) \cdot \boldsymbol{d}(\boldsymbol{r}', \boldsymbol{r}, \boldsymbol{m})\, d^3r, \tag{13.143}$$

where $\cdot$ represents the vector inner product and $\boldsymbol{B}(\boldsymbol{r}')\cdot\boldsymbol{m}$ is the field magnitude in the direction represented by the unit vector $\boldsymbol{m}$. In the equation above, $\boldsymbol{d}(\boldsymbol{r}', \boldsymbol{r}, \boldsymbol{m})$ is a vector quantity called the lead field in their paper[101]. Denoting $\boldsymbol{d}(\boldsymbol{r}', \boldsymbol{r}, \boldsymbol{m}) = [d_x(\boldsymbol{r}', \boldsymbol{r}, \boldsymbol{m}), d_y(\boldsymbol{r}', \boldsymbol{r}, \boldsymbol{m}), d_z(\boldsymbol{r}', \boldsymbol{r}, \boldsymbol{m})]$, the relationship between our definition of lead field in Section 2.1.2 and this $\boldsymbol{d}(\boldsymbol{r}', \boldsymbol{r}, \boldsymbol{m})$ is expressed as

$$l_m^x(\boldsymbol{r}) = d_x(\boldsymbol{r}_m, \boldsymbol{r}, \boldsymbol{e}_m^{\mathrm{sen}}), \tag{13.144}$$

$$l_m^y(\boldsymbol{r}) = d_y(\boldsymbol{r}_m, \boldsymbol{r}, \boldsymbol{e}_m^{\mathrm{sen}}), \tag{13.145}$$

$$\text{and}\quad l_m^z(\boldsymbol{r}) = d_z(\boldsymbol{r}_m, \boldsymbol{r}, \boldsymbol{e}_m^{\mathrm{sen}}), \tag{13.146}$$

where $\boldsymbol{r}_m$ and $\boldsymbol{e}_m^{\mathrm{sen}}$ are the location and orientation of the mth sensor. With the above relationship in mind, we call this vector quantity, $\boldsymbol{d}(\boldsymbol{r}', \boldsymbol{r}, \boldsymbol{m})$, the lead field in this section. For an infinite homogeneous conductor, using Eq. (13.118), we have

$$\boldsymbol{B}(\boldsymbol{r}') \cdot \boldsymbol{m} = \frac{\mu_0}{4\pi} \int_\Omega \boldsymbol{J}_e(\boldsymbol{r}) \times \boldsymbol{G}(\boldsymbol{r}', \boldsymbol{r}) \cdot \boldsymbol{m}\, d^3r = \frac{\mu_0}{4\pi} \int_\Omega \boldsymbol{J}_e(\boldsymbol{r}) \cdot \boldsymbol{G}(\boldsymbol{r}', \boldsymbol{r}) \times \boldsymbol{m}\, d^3r. \tag{13.147}$$

Comparing the above equation and Eq. (13.143) leads to

$$\boldsymbol{d}(\boldsymbol{r}', \boldsymbol{r}, \boldsymbol{m}) = \frac{\mu_0}{4\pi} \boldsymbol{G}(\boldsymbol{r}', \boldsymbol{r}) \times \boldsymbol{m}. \tag{13.148}$$

The above lead field for an infinite homogeneous conductor is denoted $\boldsymbol{d}_0(\boldsymbol{r}', \boldsymbol{r}, \boldsymbol{m})$ for later use.

Let us get back to Eq. (13.143) and rewrite it, using the Biot-Savart law in Eq. (13.111), as

$$\int (\boldsymbol{J}_e \cdot \boldsymbol{d})\, d^3r = \int (\boldsymbol{J} \times \boldsymbol{G}) \cdot \boldsymbol{m}\, d^3r, \tag{13.149}$$

where we omit the arguments in the vector quantities for simplicity. We decompose the lead field, $\boldsymbol{d}$, into a divergence-free part, $\boldsymbol{d}_{df}$, and a curl-free part, $\boldsymbol{d}_{cf}$. Since the curl-free part, $\boldsymbol{d}_{cf}$, can be expressed using a scalar function U such as $\boldsymbol{d}_{cf} = -\nabla U$, the lead field is expressed as

$$\boldsymbol{d} = \boldsymbol{d}_{df} - \nabla U. \tag{13.150}$$

Substituting the above equation and $\boldsymbol{J} = \boldsymbol{J}_e - \rho\nabla V$ into Eq. (13.149), we can finally derive the relationships

$$\boldsymbol{d} = \boldsymbol{d}_0 - \nabla U \tag{13.151}$$

$$\text{and} \quad \nabla \cdot \rho\boldsymbol{d} = 0. \tag{13.152}$$

Combining these two equations and considering the fact that $\nabla \cdot \boldsymbol{d}_0 = 0$, we obtain

$$\nabla^2 U = 0. \tag{13.153}$$

This equation shows that the scalar function U should be a harmonic function, which is a constraint on U. The key fact here is that Eq. (13.151) can be extended to

$$\boldsymbol{d} = \boldsymbol{d}_X - \nabla U, \tag{13.154}$$

where $\boldsymbol{d}_X$ is the lead field in an arbitrary volume conductor and U is a harmonic function in this conductor. Actually, the equation above is a restatement of the well-known fact that an arbitrary magnetic field is decomposed into a divergence-free part and a curl-free part. Therefore, the lead field in a realistic volume conductor can be obtained as the lead field for the spherical conductor plus a perturbed component. On the other hand, Eq. (13.152) can be used to derive the boundary condition. Since $\nabla \cdot \rho\boldsymbol{d} = 0$ holds, the normal component of $\rho\boldsymbol{d}$ is continuous across the border of the volume conductor. Since $\rho\boldsymbol{d}$ is equal to zero outside the volume conductor, the normal component of $\rho\boldsymbol{d}$ is equal to zero at the boundary of the volume conductor, i.e.,

$$\boldsymbol{m}(\boldsymbol{r}) \cdot \rho(\boldsymbol{r})\boldsymbol{d}(\boldsymbol{r}) = 0, \tag{13.155}$$

holds for all surface points $\boldsymbol{r}$, where $\boldsymbol{m}(\boldsymbol{r})$ is the surface normal vector at $\boldsymbol{r}$.

To derive the lead field for the realistically-shaped volume conductor, we start from the relationship

$$\boldsymbol{d}(\boldsymbol{r}) = \boldsymbol{d}_{\mathrm{sph}}(\boldsymbol{r}) - \nabla U(\boldsymbol{r}), \tag{13.156}$$

where $\boldsymbol{d}_{sph}(\boldsymbol{r})$ is the lead field for a spherical volume conductor that can be easily obtained from Eq. (13.137). The scalar function $U(\boldsymbol{r})$ is a harmonic function chosen so that the total lead field, $\boldsymbol{d}(\boldsymbol{r})$, fulfills the boundary condition in Eq. (13.155). We expand $U(\boldsymbol{r})$, using the basis-function expansion,

$$U(\boldsymbol{r}) = \sum_j a_j U_j(\boldsymbol{r}), \tag{13.157}$$

where the natural choice for the basis functions are the spherical harmonics, although the choice of the basis functions is arbitrary as long as they are harmonics. Then, the problem of computing the lead field is reduced to the problem of determining the appropriate values for the expansion coefficient a_j. The expansion coefficient, a_j, can be determined from the boundary condition. That is, defining the locations on the surface of the volume conductor as $\boldsymbol{r}_n$, where $k = 1, \ldots, N$, the coefficient a_j is determined so as to minimize the error in the boundary condition; the error is expressed as

$$\sum_n [\boldsymbol{m}(\boldsymbol{r}_n) \cdot \boldsymbol{d}_{\text{sph}}(\boldsymbol{r}_n) - \sum_j a_j \boldsymbol{m}(\boldsymbol{r}_n) \cdot \nabla U_j(\boldsymbol{r}_n)]^2. \tag{13.158}$$

Once the coefficient a_j is derived, the magnetic field at the sensor locations due to a source at $\boldsymbol{r}$ is computed using Eqs. (13.156), (13.157), and (13.143). The principles of the approach described here can be extended to compute the EEG lead field[102].

Multiple local-sphere model

The other approach that does not require heavy numerical computations, in contrast to BEM, is the method of using multiple spheres. Although this approach has been empirically known for years in the biomagnetic community, its rationale was first reported by Huang *et al.* [103]. To explain this method, we start from the Geselowitz formula for the single-shell model(in Eq. (13.141)), which is expressed as

$$\boldsymbol{B}(\boldsymbol{r}') = \boldsymbol{B}_0(\boldsymbol{r}') + \frac{\mu_0}{4\pi} \int_{\partial\Omega_{is}} J_V(\boldsymbol{r}) \boldsymbol{m}(\boldsymbol{r}) \times \boldsymbol{G}(\boldsymbol{r}', \boldsymbol{r}) dS, \tag{13.159}$$

where $J_V(\boldsymbol{r}) = -\rho V(\boldsymbol{r})$ is the magnitude of the volume current on the inner surface of the skull. Here, ρ is the conductivity inside the surface, and the conductivity outside the surface is assumed to be zero. In the equation above, the second term of the right-hand side represents the magnetic field caused by this volume current. Here, the surface integral is computed over the inner skull surface, which is denoted $\partial\Omega_{is}$ and $\boldsymbol{r}$ is the location coordinate of a point on the surface.

The method of using multiple local spheres seeks multiple overlapping spheres that best fit the volume current term in Eq. (13.159). Let us assume that the mth sensor is located at $\boldsymbol{r}_m$ and its orientation is represented by $\boldsymbol{e}_m^{\text{sen}}$. We seek the spherical conductor that best approximates the volume current contributions. That is, the following relationship should hold:

$$\int_{\partial\Omega_{is}} \boldsymbol{e}_m^{\text{sen}} \cdot \boldsymbol{m}(\boldsymbol{r}) \times \boldsymbol{G}(\boldsymbol{r}_m, \boldsymbol{r}) dS = \int_{\partial\Omega_{sp}} \boldsymbol{e}_m^{\text{sen}} \cdot \boldsymbol{m}_{sp}(\boldsymbol{r}_{sp}) \times \boldsymbol{G}(\boldsymbol{r}_m, \boldsymbol{r}_{sp}) dS. \tag{13.160}$$

Here, the integral on the right-hand side is computed on the surface of the sphere, which is denoted $\partial\Omega_{sp}$, and the normal vector $\boldsymbol{m}(\boldsymbol{r})$ and the surface coordinate should be converted into a new coordinate having an origin at the center of the sphere. The subscript *sp* indicates values in this new coordinate system. In deriving Eq. (13.160), we assume that the true volume current is similar to the volume current for the spherical conductor, i.e., $J_V(\boldsymbol{r}) = J_V^{sp}(\boldsymbol{r})$.

To solve Eq. (13.160) numerically, we introduce a discrete surface mesh, which represents the inner skull surface of a subject's MRI. The least-squares cost function is expressed as

$$\mathcal{F} = \sum_n [\boldsymbol{e}_m^{\text{sen}} \cdot \boldsymbol{m}(n) \times \frac{\boldsymbol{r}_m - \boldsymbol{r}(n)}{|\boldsymbol{r}_m - \boldsymbol{r}(n)|^3} - \boldsymbol{e}_m^{\text{sen}} \cdot \boldsymbol{m}_{sp}(n) \times \frac{\boldsymbol{r}_m - \boldsymbol{r}_{sp}(n)}{|\boldsymbol{r}_m - \boldsymbol{r}_{sp}(n)|^3}]^2, \quad (13.161)$$

where $\boldsymbol{r}(n)$ and $\boldsymbol{m}(n)$ are the location and the orientation of the nth mesh point, respectively. The variables $\boldsymbol{r}_{sp}(n)$ and $\boldsymbol{m}_{sp}(n)$ are the location and the orientation of the nth mesh point expressed in the new coordinate system, where the origin is located at the center of the sphere. Denoting the center and the radius of the sphere as $\boldsymbol{c}_m$ and R_m, respectively, the conversion between the original and the new coordinates is carried out by using

$$\boldsymbol{m}_{sp}(n) = \frac{\boldsymbol{r}(n) - \boldsymbol{c}_m}{|\boldsymbol{r}(n) - \boldsymbol{c}_m|} \quad (13.162)$$

and

$$\boldsymbol{r}_{sp}(n) = R_m \boldsymbol{m}_{sp}(n) + \boldsymbol{c}_m. \quad (13.163)$$

Optimum estimates of $\boldsymbol{c}_m$ and R_m can be obtained by minimizing the cost function in Eq. (13.161). Because these optimum estimates are obtained for each sensor, each of these parameters has an index m (the sensor numbering), and this optimization is repeated for all MEG sensors to obtain a set of overlapping spheres. Huang *et al.* compared this multiple sphere model with the three-compartment BEM, and they found that the difference is generally very small, in fact less than 0.5%[103].

13.7.7 Electric potential for a multiple-shell conductor

The lead field for EEG sensors can be calculated in the following manner. The most commonly used head model for the EEG forward calculation is a multiple shell model consisting of a set of concentric spheres, each with homogeneous and isotropic conductivity. For a given multiple-shell head model, the potential on the outer surface at $\boldsymbol{r}'$ due to a source located at $\boldsymbol{r}_0$ with its moment $\boldsymbol{Q}$ is denoted $V(\boldsymbol{r}', \boldsymbol{r}_0, \boldsymbol{Q})$, and the potential on the outer surface of the single-shell conductor due to a source located at $\boldsymbol{r}_0$ with its moment $\boldsymbol{Q}$ is denoted $V^S(\boldsymbol{r}', \boldsymbol{r}_0, \boldsymbol{Q})$. Then, the potential on the outer surface of the multiple shell is expressed as

$$V(\boldsymbol{r}', \boldsymbol{r}_0, \boldsymbol{Q}) \approx V^S(\boldsymbol{r}', \mu_1 \boldsymbol{r}_0, \lambda_1 \boldsymbol{Q}) + V^S(\boldsymbol{r}', \mu_2 \boldsymbol{r}_0, \lambda_2 \boldsymbol{Q}) + V^S(\boldsymbol{r}', \mu_3 \boldsymbol{r}_0, \lambda_3 \boldsymbol{Q}). \quad (13.164)$$

The equation above indicates that $V(\boldsymbol{r}', \boldsymbol{r}_0, \boldsymbol{Q})$ is obtained by summation of the potentials of the three single-shell cases in which the source locations and the source moments are equal to $\mu_j \boldsymbol{r}_0$ and $\lambda_j \boldsymbol{Q}$ where ($j = 1, 2$, and 3). These μ_j and λ_j are scalar constants called the Berg parameters. The single-shell potential $V^S(\boldsymbol{r}', \boldsymbol{r}_0, \boldsymbol{Q})$ is given by

$$V^S(\boldsymbol{r}', \boldsymbol{r}_0, \boldsymbol{Q}) = V_r^S(\boldsymbol{r}', \boldsymbol{r}_0, \boldsymbol{Q}) + V_t^S(\boldsymbol{r}', \boldsymbol{r}_0, \boldsymbol{Q}), \quad (13.165)$$

where

$$V_r^S(\boldsymbol{r}',\boldsymbol{r}_0,\boldsymbol{Q}) = \frac{1}{4\pi\rho|\boldsymbol{r}_0|^2}\left[2\frac{(\boldsymbol{r}'-\boldsymbol{r}_0)\cdot\boldsymbol{r}_0}{|\boldsymbol{r}'-\boldsymbol{r}_0|^3} + \frac{1}{|\boldsymbol{r}'-\boldsymbol{r}_0|} - \frac{1}{|\boldsymbol{r}'|}\right](\boldsymbol{r}_0\cdot\boldsymbol{Q}), \quad (13.166)$$

and

$$V_t^S(\boldsymbol{r}',\boldsymbol{r}_0,\boldsymbol{Q}) = \frac{1}{4\pi\rho|\boldsymbol{r}_0|^2}\left[\frac{2}{|\boldsymbol{r}'-\boldsymbol{r}_0|^3} + \frac{|\boldsymbol{r}'-\boldsymbol{r}_0|+|\boldsymbol{r}'|}{|\boldsymbol{r}'|A}\right] \left[|\boldsymbol{r}_0|^2(\boldsymbol{r}'\cdot\boldsymbol{Q}) - (\boldsymbol{r}'\cdot\boldsymbol{r}_0)(\boldsymbol{r}_0\cdot\boldsymbol{Q})\right], \quad (13.167)$$

where A is defined in Eq. (13.136), and ρ is the conductivity of the inner-most shell. The derivation of Eqs. (13.164)–(13.167) is beyond the scope of this book. Readers should refer to [92][104]. Regarding the details of the Berg parameters, readers should also refer to [105].

Bibliography

[1] M. Hämäläinen, R. Hari, R. J. Ilmoniemi, J. Knuutila, and O. V. Lounasmaa, "Magnetoencephalography-theory, instrumentation, and applications to noninvasive studies of the working human brain," *Rev. Mod. Phys.*, vol. 65, pp. 413–497, 1993.

[2] T. P. L. Roberts, D. Poeppel, and H. A. Rowley, "Magnetoencephalography and magnetic source imaging," *Neuropsychiatry, Neuropsychology, and Behavioral Neurology*, vol. 11, pp. 49–64, 1998.

[3] J. S. Ebersole and T. A. Pedley, *Current Practice of Clinical Electroencephalography.* Philadelphia, PA: Lippincott Williams & Wilkins, 2002.

[4] M. D. Rugg and M. G. H. Coles, *Electrophysiology of Mind: Event-Related Brain Potentials and Cognition.* Oxford, UK: Oxford University Press, 1996.

[5] S. Baillet, J. C. Mosher, and R. M. Leahy, "Electromagnetic brain mapping," *IEEE Signal Processing Magazine*, vol. 18, pp. 14–30, 2001.

[6] J. Capon, "High-resolution frequency wavenumber spectrum analysis," *Proc. IEEE*, vol. 57, pp. 1408–1419, 1969.

[7] B. D. Van Veen and K. M. Buckley, "Beamforming: A versatile approach to spatial filtering," *IEEE ASSP Magazine*, vol. 5, pp. 4–24, April 1988.

[8] S. E. Robinson and D. F. Rose, "Current source image estimation by spatially filtered MEG," in *Biomagnetism Clinical Aspects* (M. Hoke *et al.*, eds.), pp. 761–765, Elsevier Science Publishers, 1992.

[9] B. D. van Veen, W. van Drongelen, M. Yuchtman, and A. Suzuki, "Localization of brain electrical activity via linearly constrained minimum variance spatial filtering," *IEEE Trans. Biomed. Eng.*, vol. 44, pp. 867–880, 1997.

[10] M. E. Spencer, R. M. Leahy, J. C. Mosher, and P. S. Lewis, "Adaptive filters for monitoring localized brain activity from surface potential time series," in *Conference Record for 26th Annual Asilomer Conference on Signals, Systems, and Computers*, pp. 156–161, November 1992.

[11] H. Cox, "Resolving power and sensitivity to mismatch of optimum array processors," *J. Acoust. Soc. Am.*, vol. 54, pp. 771–785, 1973.

[12] A. Paulraj, B. Ottersten, R. Roy, A. Swindlehurst, G. Xu, and T. Kailath, "Subspace methods for directions-of-arrival estimation," in *Handbook of Statistics* (N. K. Bose and C. R. Rao, eds.), pp. 693–739, Netherlands: Elsevier Science Publishers, 1993.

[13] C. D. Meyer, *Matrix Analysis and Applied Linear Algebra.* Philadelphia, PA: Society for Industrial and Applied Mathematics, 2000.

[14] R. O. Schmidt, "Multiple emitter location and signal parameter estimation," *IEEE Trans. Antenn. Propagat.*, vol. 34, pp. 276–280, 1986.

[15] R. O. Schmidt, *A signal subspace approach to multiple emitter location and spectral estimation.* PhD thesis, Stanford University, Stanford, CA, 1981.

[16] J. C. Mosher, P. S. Lewis, and R. M. Leahy, "Multiple dipole modeling and localization from spatio-temporal MEG data," *IEEE Trans. Biomed. Eng.*, vol. 39, pp. 541–557, 1992.

[17] G. Backus and F. Gilbert, "The resolving power of gross earth data," *Geophys. J. R. Astron. Soc.*, vol. 16, pp. 169–205, 1968.

[18] R. G. de Peralta Menendez, O. Hauk, S. G. Andino, H. Vogt, and C. Michel, "Linear inverse solutions with optimal resolution kernels applied to electromagnetic tomography," *Human Brain Mapping*, vol. 5, pp. 454–467, 1997.

[19] M. S. Hämäläinen and R. J. Ilmoniemi, "Interpreting measured magnetic fields of the brain: Estimates of current distributions," Tech. Rep. TKK-F-A559, Helsinki University of Technology, 1984.

[20] G. T. Herman, *Image Reconstruction from projections.* New York, USA: Academic Press, 1980.

[21] A. C. Kak and M. Slaney, *Principles of Computerized Tomographic Imaging.* New York, USA: IEEE Press, 1988.

[22] A. N. Tikhonov, "Solution of incorrectly formulated problems and the regularization method," *Soviet Mathematics Doklady*, vol. 4, pp. 1035–38, 1963.

[23] A. N. Tikhonov and V. Y. Arsenin, *Solutions of Ill-Posed Problems.* New York, NY: John Wiley & Sons Ltd., 1977.

[24] B. Jeffs, R. Leahy, and M. Singh, "An evaluation of methods for neuromagnetic image reconstruction," *IEEE Trans. Biomed. Eng.*, vol. 34, pp. 713–723, 1987.

[25] A. M. Dale, A. K. Liu, B. R. Fischl, R. L. Buckner, J. W. Belliveau, J. D. Lewine, and E. Halgren, "Dynamic statistical parametric mapping: Combining fMRI and MEG for high-resolution imaging of cortical activity," *Neuron*, vol. 26, pp. 55–67, 2000.

[26] R. D. Pascual-Marqui, "Standardized low resolution brain electromagnetic tomography (sloreta): technical details," *Methods and Findings in Experimental and Clinical Pharmacology*, vol. 24, pp. 5–12, 2002.

[27] D. G. Manolakis, V. K. Ingle, and S. M. Kogon, *Statistical and Adaptive Signal Processing.* Boston: McGraw-Hill Companies Inc., 2000.

[28] D. H. Johnson and D. E. Dudgeon, *Array Signal Processing: Concepts and Techniques.* Englewood Cliffs, New Jersey: Prentice-Hall, 1993.

[29] W. Menke, *Geophisical data analysis: discrete inverse theory, Revised edition.* San Diego, California: Academic Press, 1989.

[30] R. E. Greenblatt, A. Ossadtchi, and M. E. Pflieger, "Local linear estimators for the bioelectromagnetic inverse problem," *IEEE Trans. Signal Process.*, vol. 53, pp. 3403–3412, 2005.

[31] H. L. van Trees, *Optimum Array Processing.* New York: John Wiley & Sons, Inc., 2002.

[32] G. Borgiotti and L. J. Kaplan, "Superresolution of uncorrelated interference sources by using adaptive array technique," *IEEE Trans. Antenn. and Propagat.*, vol. 27, pp. 842–845, 1979.

[33] M. D. Zoltowski, "On the performance analysis of the MVDR beamformer in the presence of correlated interference," *IEEE Trans. Signal Process.*, vol. 36, pp. 945–947, 1988.

[34] S. E. Robinson and J. Vrba, "Functional neuroimaging by synthetic aperture magnetometry (SAM)," in *Recent Advances in Biomagnetism* (T. Yoshimoto *et al.*, eds.), (Sendai), pp. 302–305, Tohoku University Press, 1999.

[35] K. Sekihara and B. Scholz, "Generalized Wiener estimation of three-dimensional current distribution from biomagnetic measurements," in *Biomag 96: Proceedings of the Tenth International Conference on Biomagnetism* (C. J. Aine *et al.*, eds.), (New York), pp. 338–341, Springer-Verlag, 1996.

[36] K. Sekihara, S. S. Nagarajan, D. Poeppel, and A. Marantz, "Asymptotic SNR of scalar and vector minimum-variance beamformers for neuromagnetic source reconstruction," *IEEE Trans. Biomed. Eng.*, vol. 51, pp. 1726–1734, 2004.

[37] F. R. Gantmacher, *The Theory of Matrices.* New York, NY: Chelsea Publishing Company, 1960.

[38] O. T. Frost, "An algorithm for linearly constrained adaptive array processing," *Proc. IEEE*, vol. 60, pp. 926–935, 1972.

[39] W. van Drongelen, M. Yuchtman, B. D. van Veen, and A. C. van Huffelen, "A spatial filtering technique to detect and localize multiple sources in the brain," *Brain Topography*, vol. 9, pp. 39–49, 1996.

[40] K. Sekihara, S. S. Nagarajan, D. Poeppel, A. Marantz, and Y. Miyashita, "Reconstructing spatio-temporal activities of neural sources using an MEG vector beamformer technique," *IEEE Trans. Biomed. Eng.*, vol. 48, pp. 760–771, 2001.

[41] J. Vrba and S. Robinson, "Differences between synthetic aperture magnetometry (SAM) and linear beamformer," in *Proceedings of 12th International Conference on Biomagnetism* (R. Hari *et al.*, eds.), pp. 681–684, Helsinki University of Technology, 2001.

[42] H.-C. Lin, "Spatial correlations in adaptive arrays," *IEEE Trans. Antenn. Propagat.*, vol. 30, pp. 212–223, 1982.

[43] H. Lütkepohl, *Handbook of Matrices.* New York, NY: John Wiley & Sons Inc., 1996.

[44] B. D. Carlson, "Covariance matrix estimation errors and diagonal loading in adaptive arrays," *IEEE Trans. Aerospace and Electronic Systems*, vol. 24, pp. 397–401, 1988.

[45] H. Cox, R. M. Zeskind, and M. M. Owen, "Robust adaptive beamforming," *IEEE Trans. Signal Process.*, vol. 35, pp. 1365–1376, 1987.

[46] L. Chang and C. C. Yeh, "Effect of pointing errors on the performance of the projection beamformer," *IEEE Trans. Antenn. Propagat.*, vol. 41, pp. 1045–1056, 1993.

[47] K. Sekihara, S. S. Nagarajan, D. Poeppel, A. Marantz, and Y. Miyashita, "Application of an MEG eigenspace beamformer to reconstructing spatio-temporal activities of neural sources," *Human Brain Mapping*, vol. 15, pp. 199–215, 2002.

[48] L. Chang and C. C. Yeh, "Performance of DMI and eigenspace-based beamformers," *IEEE Trans. Antenn. Propagat.*, vol. 40, pp. 1336–1347, 1992.

[49] J. L. Yu and C. C. Yeh, "Generalized eigenspace-based beamformers," *IEEE Trans. Signal Process.*, vol. 43, pp. 2453–2461, 1995.

[50] K. Sekihara, S. S. Nagarajan, D. Poeppel, and A. Marantz, "Performance of an MEG adaptive-beamformer source reconstruction technique in the presence of additive low-rank interference," *IEEE Trans. Biomed. Eng.*, vol. 51, pp. 90–99, 2004.

[51] J. C. de Munck, P. C. M. Vijn, and F. H. L. da Silva, "A random dipole model for spontaneous brain activity," *IEEE Trans. Biomed. Eng.*, vol. 39, pp. 791–804, 1992.

[52] B. Lütkenhöner, "Magnetic field arising from current dipoles randomly distributed in a homogeneous spherical volume conductor," *J. Appl. Phys.*, vol. 75, pp. 7204–7210, 1994.

[53] K. Sekihara, K. E. Hild II, and S. S. Nagarajan, "A novel adaptive beamformer for MEG source reconstruction effective when large background brain activities exist," *IEEE Trans. Biomed. Eng.*, vol. 53, pp. 1755–64, 2006.

[54] G. Pfurtscheller and F. H. Lopes da Silva, "Event-related EEG/MEG synchronization and desynchronization: basic principles," *Clin. Neurophysiol.*, vol. 110, pp. 1842–1857, 1999.

[55] K. Sekihara, K. E. Hild II, S. S. Dalal, and S. S. Nagarajan, "Performance of prewhitening beamforming in MEG dual experimental conditions," in *Conference Proceedings of Third International IEEE EMBS Conference on Neural Engineering*, (Kohala Coast, Hawaii), May 2007.

[56] K. Sekihara, S. S. Nagarajan, D. Poeppel, and A. Marantz, "Performance of an MEG adaptive-beamformer technique in the presence of correlated neural activities: Effects on signal intensity and time-course estimates," *IEEE Trans. Biomed. Eng.*, vol. 49, pp. 1534–1546, 2002.

[57] B. Widrow, K. M. Duval, R. P. Gooch, and W. C. Newman, "Signal cancellation phenomena in adaptive antennas: Causes and cures," *IEEE Trans. Antenn. and Propagat.*, vol. 30, pp. 469–478, 1982.

[58] A. Paulraj, V. U. Reddy, and T. Kailath, "Analysis of signal cancellation due to multipath in optimum beamformers for moving arrays," *IEEE Journal of Oceanic Engineering*, vol. 12, pp. 163–172, 1987.

[59] S. S. Dalal, K. Sekihara, and S. S. Nagarajan, "Modified beamformers for coherent source region suppression," *IEEE Trans. Biomed. Eng.*, vol. 53, pp. 1357–63, 2006.

[60] J. Gross, J. Kujara, M. Hämäläinen, L. Timmermann, A. Schnitzler, and R. Salmelin, "Dynamic imaging of coherent sources: Studying neural interactions in the human brain," *Proceedings of National Academy of Science*, vol. 98, pp. 694–699, 2001.

[61] K. Jerbi, J.-P. Lachaux, K. N'Diaye, D. Pantazis, R. M. Leahy, L. Garnero, and S. Baillet, "Coherent neural representation of hand speed in humans revealed by meg imaging," *Proceedings of the National Academy of Sciences*, vol. 104, pp. 7676–7681, 2007.

[62] A. G. Guggisberg, S. M. Honma, A. M. Findlay, S. S. Dalal, H. E. Kirsch, M. S. Berger, and S. S. Nagarajan, "Mapping functional connectivity in patients with brain lesions," *Annals of Neurology*, accepted for publication.

[63] J. Kujara, J. Gross, and R. Salmelin, "Localization of correlated network activity at the cortical level with MEG," *NeuroImage*. accepted for publication.

[64] P. Belardinelli, L. Ciancetta, V. Pizzella, C. D. Gratta, and G. L. Romani, "Localizing complex neural circuits with MEG data," *Cognitive processing*, vol. 7, pp. 53–9, 2006.

[65] P. Belardinelli, L. Ciancetta, M. Staudt, V. Pizzella, A. Londei, N. B. G. L. Romani, and C. Braun, "Cerebro-muscular and cerebro-cerebral coherence in patients with pre- and perinatally acquired unilateral brain lesions," *NeuroImage*, vol. 37, pp. 1301–14, 2007.

[66] G. Nolte, O. B. L. Wheaton, Z. Mari, S. Vorbach, and M. Hallett, "Identifying true brain intereaction from EEG data using the imaginary part of coherency," *Clinical Neurophysiology*, vol. 115, pp. 2292–2307, 2004.

[67] A. C. Rencher, *Methods of multivariate analysis*. New York, NY: John Wiley & Sons Inc., 1995.

[68] K. V. Mardia, J. T. Kent, and J. M. Bibby, *Multivariate analysis*. London, UK: Academic Press, 2003.

[69] J. Capon and N. R. Goodman, "Probability distribution for estimators of the frequency-wavenumber spectrum," *Proc. IEEE*, vol. 58, pp. 1785–86, 1970.

[70] T. Limpiti, B. D. V. Veen, and R. T. Wakai, "Cortical patch basis model for spatially extended neural activity," *IEEE Trans. Biomed. Eng.*, vol. 53, pp. 1740–1754, 2006.

[71] G. R. Barnes and A. Hillebrand, "Statistical flattening of MEG beamformer images," *Human Brain Mapping*, vol. 18, pp. 1–12, 2003.

[72] C. R. Genovese, N. A. Lazar, and T. E. Nichols, "Thresholding of statistical maps in functional neuroimaging using the false discovery rate," *NeuroImage*, vol. 15, pp. 870–878, 2002.

[73] Y. Benjamini and Y. Hochberg, "Controlling the false discovery rate: a practical and powerful approach to multiple testing," *J. R. Stat. Soc. B*, vol. 57, pp. 289–300, 1995.

[74] D. Pantazis, T. E. Nichols, S. Baillet, and R. M. Leahy, "Spatiotemporal localization of significant activation in MEG using permutation tests," in *Proceedings of the 18th Conference on Information Processing in Medical Imaging* (C. J. Taylor and J. A. Noble, eds.), (Ambleside, UK), pp. 512–523, Springer Series in Medical Imaging, July 2003.

[75] B. C. Blair and W. Karniski, "Distribution-free statistical analysis of surface and volumetric maps," in *Functional Neuroimaging* (R. W. T. *et al.*, ed.), Academic Press, 1994.

[76] T. E. Nichols and S. Hayasaka, "Controlling the familywise error rate in functional neuroimaging: a comparative review," *Statistical Methods in Medical Research*, vol. 12, pp. 419–446, 2003.

[77] H. Schimmel, "The (±) reference:Accuracy of estimated mean components in average response studies," *Science*, vol. 157, pp. 92–94, 1967.

[78] R. E. Greenblatt and M. E. Pflieger, "Randomization-based hypothesis testing from event-related data," *Brain Topography*, vol. 16, pp. 225–232, 2004.

[79] K. D. Singh, G. R. Barnes, and A. Hillebrand, "Group imaging of task-related changes in cortical synchronisation using nonparametric permutation testing," *NeuroImage*, vol. 19, pp. 1589–1601, 2003.

[80] K. Sekihara, M. Sahani, and S. S. Nagarajan, "A simple non-parametric statistical thresholding for MEG spatial-filter source reconstruction images," *NeuroImage*, vol. 27, pp. 368–376, 2005.

[81] R. N. McDonough and A. D. Whalen, *Detection of signals in noise.* London, UK: Academic Press Inc., 1995.

[82] S. Haykin, *Adaptive filter theory, Third Edition.* Upper Saddle River, New Jersey: Prentice-Hall, Inc., 1996.

[83] M. Wax, *Detection and estimation of superimposed signals.* PhD thesis, Stanford University, Stanford, CA, 1985.

[84] K. Sekihara and B. Scholz, "Generalized Wiener estimation of three-dimensional current distribution from biomagnetic measurements," *IEEE Trans. Biomed. Eng.*, vol. 43, pp. 281–291, 1996.

[85] G. H. Golub and V. Pereyra, "The differentiation of pseudo-inverses and nonlinear least squares problems whose variables separate," *SIAM J. Numer. Anal.*, vol. 10, pp. 413–432, 1973.

[86] E. J. Kelly and K. M. Forsythe, "Adaptive detection and parameter estimation for multidimensional signal models," Tech. Rep. Technical Report 848, Massachusetts Institute of Technology, Lincoln Laboratory, 1989.

[87] A. Dogandzic and A. Nehorai, "Estimating evoked dipole responses in unknown spatially correlated noise with MEG/EEG arrays," *IEEE Trans. Signal Process.*, vol. 48, pp. 13–25, 2000.

[88] D. J. Thompson, "Spectrum estimation and harmonic analysis," *Proc. IEEE*, vol. 70, pp. 1055–1096, 1982.

[89] D. B. Percival and A. T. Walden, *Spectral analysis for physical applications – multitaper and conventional univariate analysis.* Cambridge, Great Britain: Cambridge University Press, 1993.

[90] B. V. Baryshnikov, B. D. V. Veen, and R. T. Wakai, "Maximum-likelihood estimation of low-rank signals for multiepoch EEG/MEG analysis," *IEEE Trans. Biomed. Eng.*, vol. 51, pp. 1981–1993, 2004.

[91] L. L. Scharf, *Statistical Signal Processing: detection, estimation, and time series analysis.* New York: Addison-Wesley Publishing Company, 1991.

[92] J. C. Mosher, R. M. Leahy, and P. S. Lewis, "EEG and MEG: Forward solutions for inverse methods," *IEEE Trans. Biomed. Eng.*, vol. 46, pp. 245–259, 1999.

[93] J. Sarvas, "Basic mathematical and electromagnetic concepts of the biomagnetic inverse problem," *Phys. Med. Biol.*, vol. 32, pp. 11–22, 1987.

[94] D. B. Geselowitz, "On the magnetic field generated outside an inhomogeneous volume conductor by internal current sources," *IEEE Trans. Biomed. Eng.*, vol. 2, pp. 346–347, 1970.

[95] A. Barnard, I. Duck, M. Lynn, and W. Timlake, "The application of electromagnetic theory to electrocardiography II. Numerical solution of the integral equations," *Biophys. J.*, vol. 7, pp. 433–462, 1967.

[96] M. Hämäläinen and J. Sarvas, "Realistic conductivity geometry model of the human head for interpretation of neuromagnetic data," *IEEE Trans. Biomed. Eng.*, vol. 36, pp. 165–171, 1989.

[97] M. Fuchs, R. Drenckhahn, H.-A. Wischmann, and M. Wagner, "An improved boundary element method for realistic volume-conductor modeling," *IEEE Trans. Biomed. Eng.*, vol. 45, pp. 980–997, 1998.

[98] B. N. Cuffin, "EEG localization accuracy improvements using realistically shaped head models," *IEEE Trans. Biomed. Eng.*, vol. 43, pp. 299–303, 1996.

[99] C. P. Bradley, G. M. Harris, and A. J. Pillan, "The computational performance of a high-order coupled FEM/BEM procedure in electropotential problems," *IEEE Trans. Biomed. Eng.*, vol. 48, pp. 1238–1250, 2001.

[100] D. van't Ent, J. C. de Munck, and A. L. Kaas, "A fast method to derive realistic BEM models for E/MEG source reconstruction," *IEEE Trans. Biomed. Eng.*, vol. 48, pp. 1434–1443, 2001.

[101] G. Nolte, "The magnetic lead field theorem in the quasi-static approximation and its use for magnetoencephalography forward calculation in realistic volume conductors," *Phys. Med. Biol.*, vol. 48, pp. 3637–3652, 2003.

[102] G. Nolte and G. Dassios, "Analytic expansion of the EEG lead field for realistic volume conductors," *Phys. Med. Biol.*, vol. 50, pp. 3807–3823, 2005.

[103] M. X. Huang, J. C. Mosher, and R. M. Leahy, "A sensor-weighted overlapping-sphere head model and exhaustive head model comparison for meg," *Phys. Med. Biol.*, vol. 44, pp. 423–440, 1999.

[104] Z. Zhang, "A fast method to compute surface potentials generated by dipoles within multilayer anisotropic spheres," *Phys. Med. Biol.*, vol. 40, pp. 335–349, 1995.

[105] P. Berg and M. Scherg, "A fast method for forward computation of multiple-shell spherical head models," *Electroenceph. Clin. Neurophysiol.*, vol. 90, pp. 58–64, 1994.

Index